Methods in Molecular Biology

Series Editor
John M. Walker
School of Life and Medical Sciences
University of Hertfordshire
Hatfield, Hertfordshire AL10 9AB, UK

For further volumes:
http://www.springer.com/series/7651

Plant Metabolomics

Methods and Protocols

Edited by

Carla António

Plant Metabolomics Laboratory, Instituto de Tecnologia Química e Biológica António Xavier, Universidade Nova de Lisboa (ITQB NOVA), Oeiras, Portugal

Humana Press

Editor
Carla António
Plant Metabolomics Laboratory
Instituto de Tecnologia Química e Biológica António Xavier
Universidade Nova de Lisboa (ITQB NOVA)
Oeiras, Portugal

ISSN 1064-3745 ISSN 1940-6029 (electronic)
Methods in Molecular Biology
ISBN 978-1-4939-7818-2 ISBN 978-1-4939-7819-9 (eBook)
https://doi.org/10.1007/978-1-4939-7819-9

Library of Congress Control Number: 2018940501

Printed on acid-free paper

This Humana Press imprint is published by the registered company Springer Science+Business Media, LLC part of Springer Nature.
The registered company address is: 233 Spring Street, New York, NY 10013, U.S.A.

Dedication

In Memory of the Late Professor Tsutomu Masujima.

Carla António

Preface

Studies of the plant metabolome include the analysis of a wide range of chemical species with very diverse physico-chemical properties, and therefore, powerful analytical methodologies are required for the separation, characterization, and quantification of this vast compound diversity present in plant biological samples.

Due to its high sensitivity, mass spectrometry (MS) is one main analytical technology driving plant metabolomics studies, typically coupled to liquid chromatography (LC-MS) and gas chromatography (GC-MS). However, currently, there is a growing interest in narrowing metabolomics analyses down to the tissue or single-cell level. These can be achieved using mass spectrometry imaging (MSI) approaches and live single-cell mass spectrometry (LSC-MS), respectively. MSI is a developing technique designed to achieve spatial analysis of the metabolites present in a plant tissue (or specific section) that is commonly performed *in situ* with ionization techniques, such as matrix-assisted laser desorption ionization (MALDI) and laser ablation electrospray ionization (LAESI). Indeed, major achievements in LSC-MS, pioneered by the group of the late Professor Tsutomu Masujima, have paved the way for sensitive, high-resolution single-cell plant analysis that will increasingly challenge our current view of the cell mechanisms on a single-cell level. Nonetheless, the need for standardizing key steps in plant metabolomics experiments is still considered a critical issue.

This book unites diverse MS-based protocols, from sample preparation and extraction of plant primary and secondary metabolites (e.g., a range of sugars, lipids, phytohormones, and volatile organic compounds) to metabolite analysis, using exclusively highly sensitive MS-based methodologies, followed by frontline bioinformatics and/or mathematical modeling approaches to produce reliable biological data interpretation.

I hope that both new and well-established researchers in the plant metabolomics field find in this book helpful guidelines and useful laboratory-based protocols for setting up their routine plant metabolomics experiments.

Last but not least, I would like to express my heartfelt thanks to all contributors who made this book possible and Professor John Walker for the opportunity to take on the role of editor for what I hope will be an ultimate source of updated MS-based protocols in the subject of Plant Metabolomics.

Oeiras, Portugal *Carla António*

Contents

Contributors

YASMINE ABOULEILA • *Laboratory for Single-Cell Mass Spectrometry, Quantitative Biology Center, RIKEN, Osaka, Japan*

LUIS GUILHERME FURLAN DE ABREU • *Brazilian Bioethanol Science and Technology Laboratory (CTBE), Brazilian Center for Research in Energy and Materials (CNPEM), Campinas, SP, Brazil*

FRANCISCO DE ABREU E LIMA • *Max Planck Institute of Molecular Plant Physiology, Potsdam-Golm, Germany*

SÉBASTIEN ACKET • *Sorbonne Universités, Université de Technologie de Compiègne, UMR CNRS 7025, Génie Enzymatique et Cellulaire, Compiègne Cedex, France*

ASAPH AHARONI • *Department of Plant and Environmental Sciences, Faculty of Biochemistry, Weizmann Institute of Science, Rehovot, Israel*

AHMED ALI • *Laboratory for Single-Cell Mass Spectrometry, Quantitative Biology Center, RIKEN, Osaka, Japan*

SALEH ALSEEKH • *Max Planck Institute of Molecular Plant Physiology, Potsdam-Golm, Germany; Center of Plant System Biology and Biotechnology, Plovdiv, Bulgaria*

CARLA ANTÓNIO • *Plant Metabolomics Laboratory, Instituto de Tecnologia Química e Biológica António Xavier, Universidade Nova de Lisboa (ITQB NOVA), Oeiras, Portugal*

CAMILLE BÉNARD • *Service de Pharmacologie et Immunoanalyse (SPI), MetaboHUB-Paris, CEA, INRA, Université Paris Saclay, Gif-sur-Yvette Cedex, France; UMR 1332 Biologie du Fruit et Pathologie, Centre INRA de Nouvelle Aquitaine-Bordeaux, INRA, Univ. Bordeaux, Villenave d'Ornon, France; Plateforme Métabolome Bordeaux, PHENOME, MetaboHUB, Centre INRA de Nouvelle Aquitaine-Bordeaux, INRA, CNRS, Univ. Bordeaux, Villenave d'Ornon, France*

THIERRY BERTON • *UMR 1332 Biologie du Fruit et Pathologie, Centre INRA de Nouvelle Aquitaine-Bordeaux, INRA, Univ. Bordeaux, Villenave d'Ornon, France; Plateforme Métabolome Bordeaux, PHENOME, MetaboHUB, Centre INRA de Nouvelle Aquitaine-Bordeaux, INRA, CNRS, Univ. Bordeaux, Villenave d'Ornon, France*

BERIN A. BOUGHTON • *Metabolomics Australia, School of BioSciences, The University of Melbourne, Parkville, VIC, Australia*

YARIV BROTMAN • *Max Planck Institute of Molecular Plant Physiology, Potsdam-Golm, Germany; Department of Life Sciences, Ben-Gurion University of the Negev, Beersheba, Israel*

CÉCILE CABASSON • *Plateforme Métabolome Bordeaux, PHENOME, MetaboHUB, Centre INRA de Nouvelle Aquitaine-Bordeaux, INRA, CNRS, Univ. Bordeaux, Villenave d'Ornon, France; UMR 1332 Biologie du Fruit et Pathologie, Centre INRA de Nouvelle-Aquitaine Bordeaux, INRA, Univ. Bordeaux, Villenave d'Ornon, France*

CAMILA CALDANA • *Brazilian Bioethanol Science and Technology Laboratory (CTBE), Brazilian Center for Research in Energy and Materials (CNPEM), Campinas, SP, Brazil; Max-Planck Partner Group, Brazilian Bioethanol Science and Technology Laboratory (CTBE/CNPEM), Campinas, Brazil; Max Planck Institute of Molecular Plant Physiology, Potsdam-Golm, Germany*

CARMEN DÍEZ-SIMÓN • *Laboratory of Plant Physiology, Wageningen University and Research, Wageningen, The Netherlands*

DANILO M. DALOSO • *Departamento de Bioquímica e Biologia Molecular, Universidade Federal do Ceará, Fortaleza, CE, Brazil*

DANIEL ANTHONY DIAS • *Metabolomics Australia, School of BioSciences, The University of Melbourne, Parkville, VIC, Australia; School of Health and Biomedical Sciences, Discipline of Laboratory Medicine, RMIT University, Bundoora, VIC, Australia*

ALEXANDER ERBAN • *Max Planck Institute of Molecular Plant Physiology, Potsdam-Golm, Germany*

DESALEGN W. ETALO • *Netherlands Institute of Ecology, Wageningen, The Netherlands*

LISA FÜRTAUER • *Department of Ecogenomics and Systems Biology, Faculty of Life Sciences, University of Vienna, Vienna, Austria*

INES FEHRLE • *Max Planck Institute of Molecular Plant Physiology, Potsdam-Golm, Germany*

REGINA FEIL • *Max Planck Institute of Molecular Plant Physiology, Potsdam-Golm, Germany*

OLIVIER FERNANDEZ • *UMR 1332 Biologie du Fruit et Pathologie, Centre INRA de Nouvelle Aquitaine-Bordeaux, INRA, Univ. Bordeaux, Villenave d'Ornon, France*

ALISDAIR R. FERNIE • *Max Planck Institute of Molecular Plant Physiology, Potsdam-Golm, Germany*

KRISTÝNA FLOKOVÁ • *Laboratory of Growth Regulators, Centre of the Region Haná for Biotechnological and Agricultural Research, Institute of Experimental Botany of the Czech Academy of Sciences, Olomouc, Czech Republic; Faculty of Science, Palacký University, Olomouc, Czech Republic; Plant Hormone Biology, Swammerdam Institute for Life Sciences, University of Amsterdam, Amsterdam, The Netherlands*

PATRICK GIAVALISCO • *Max Planck Institute of Molecular Plant Physiology, Potsdam-Golm, Germany*

YVES GIBON • *UMR 1332 Biologie du Fruit et Pathologie, Centre INRA de Nouvelle Aquitaine-Bordeaux, INRA, Univ. Bordeaux, Villenave d'Ornon, France; Plateforme Métabolome Bordeaux, PHENOME, MetaboHUB, Centre INRA de Nouvelle Aquitaine-Bordeaux, INRA, CNRS, Univ. Bordeaux, Villenave d'Ornon, France*

ROBERT D. HALL • *Laboratory of Plant Physiology, Wageningen University and Research, Wageningen, The Netherlands; Business Unit Bioscience, Wageningen University and Research, Wageningen, The Netherlands; Netherlands Metabolomics Centre, Leiden, The Netherlands*

CARSTEN JAEGER • *Molecular Cancer Research Center (MKFZ), Charité-Universitätsmedizin Berlin, Berlin, Germany; Berlin Institute of Health (BIH), Berlin, Germany*

NIRUPAMA SAMANMALIE JAYASINGHE • *Metabolomics Australia, School of BioSciences, The University of Melbourne, Parkville, VIC, Australia; Racing Analytical Services Ltd., Flemington, VIC, Australia*

TIAGO F. JORGE • *Plant Metabolomics Laboratory, Instituto de Tecnologia Química e Biológica António Xavier, Universidade Nova de Lisboa (ITQB NOVA), Oeiras, Portugal*

JOACHIM KOPKA • *Max Planck Institute of Molecular Plant Physiology, Potsdam-Golm, Germany*

LYDIA LEIFELS • *Max Planck Institute of Molecular Plant Physiology, Potsdam-Golm, Germany*

VALÉRIA F. LIMA • *Departamento de Bioquímica e Biologia Molecular, Universidade Federal do Ceará, Fortaleza, CE, Brazil*

JAN LISEC • *Molecular Cancer Research Center (MKFZ), Charité-Universitätsmedizin Berlin, Berlin, Germany; Division 1.7 Analytical Chemistry, Federal Institute for Materials Research and Testing (BAM), Berlin, Germany*

JOHN EDWARD LUNN • *Max Planck Institute of Molecular Plant Physiology, Potsdam-Golm, Germany*

MARINA C. M. MARTINS • *Brazilian Bioethanol Science and Technology Laboratory (CTBE), Brazilian Center for Research in Energy and Materials (CNPEM), Campinas, SP, Brazil*

KEIKO MASUDA • *Laboratory for Single-Cell Mass Spectrometry, Quantitative Biology Center, RIKEN, Osaka, Japan*

TSUTOMU MASUJIMA • *Laboratory for Single-Cell Mass Spectrometry, Quantitative Biology Center, RIKEN, Osaka, Japan*

HIMASHA MENDIS • *Metabolomics Australia, School of BioSciences, The University of Melbourne, Parkville, VIC, Australia*

THOMAS NÄGELE • *Department of Ecogenomics and Systems Biology, Faculty of Life Sciences, University of Vienna, Vienna, Austria; Vienna Metabolomics Center, University of Vienna, Vienna, Austria; Department Biology I, Ludwig-Maximilians-Universität München, Planegg-Martinsried, Austria*

ZORAN NIKOLOSKI • *Max Planck Institute of Molecular Plant Physiology, Potsdam-Golm, Germany; Bioinformatics Group, Institute of Biochemistry and Biology, University of Potsdam, Potsdam, Germany*

ONDŘEJ NOVÁK • *Laboratory of Growth Regulators, Centre of the Region Haná for Biotechnological and Agricultural Research, Institute of Experimental Botany of the Czech Academy of Sciences, Olomouc, Czech Republic; Faculty of Science, Palacký University, Olomouc, Czech Republic*

YOZO OKAZAKI • *Metabolomics Research Group, RIKEN Center for Sustainable Resource Science, Yokohama, Japan; Graduate School of Bioresources, Mie University, Tsu, Mie, Japan*

ISABEL ORF • *Ben-Gurion University of the Negev, Jacob Blaustein Institutes for Desert Research, French Associates Institute for Agriculture & Biotechnology of Drylands, Midreshet Ben-Gurion, Israel; Bactevo Ltd, The Merrifield Centre, Rosemary Lane, Cambridge, UK*

SONIA OSORIO • *Departamento de Biología Molecular y Bioquímica, Instituto de Hortofruticultura Subtropical y Mediterranea (IHSM), Universidad de Málaga-Consejo Superior de Investigaciones Científicas, Málaga, Spain*

ANA MARGARIDA RODRIGUES • *Plant Metabolomics Laboratory, Instituto de Tecnologia Química e Biológica António Xavier, Universidade Nova de Lisboa (ITQB NOVA), Oeiras, Portugal*

UTE ROESSNER • *Metabolomics Australia, School of BioSciences, The University of Melbourne, Parkville, VIC, Australia; School of BioSciences, The University of Melbourne, Parkville, VIC, Australia*

ILANA ROGACHEV • *Department of Plant and Environmental Sciences, Faculty of Biochemistry, Weizmann Institute of Science, Rehovot, Israel*

YANNICK ROSSEZ • *Sorbonne Universités, Université de Technologie de Compiègne, UMR CNRS 7025, Génie Enzymatique et Cellulaire, Compiègne Cedex, France*

THUSITHA W. T. RUPASINGHE • *Metabolomics Australia, School of BioSciences, The University of Melbourne, Parkville, VIC, Australia*

KAZUKI SAITO • *Metabolomics Research Group, RIKEN Center for Sustainable Resource Science, Yokohama, Japan; Graduate School of Pharmaceutical Sciences, Chiba University, Chiba, Japan*

MOHAMED A. SALEM • *Max Planck Institute of Molecular Plant Physiology, Potsdam-Golm, Germany; Department of Pharmacognosy, Faculty of Pharmacy, Cairo University, Cairo, Egypt*

SAMUEL M. D. SEAVER • *Mathematics and Computer Science Division, Argonne National Laboratory, Argonne, IL, USA; Computation Institute, The University of Chicago, Chicago, IL, USA*

NIR SHAHAF • *Department of Plant and Environmental Sciences, Faculty of Biochemistry, Weizmann Institute of Science, Rehovot, Israel*

TAKAFUMI SHIMIZU • *Graduate School of Biological Sciences, Nara Institute of Science and Technology, Ikoma, Nara, Japan*

LEONARDO PEREZ DE SOUZA • *Max Planck Institute of Molecular Plant Physiology, Potsdam-Golm, Germany*

NADINE TÖPFER • *Faculty of Biochemistry, Department of Plant and Environmental Sciences, Weizmann Institute of Science, Rehovot, Israel*

DINAIZ THINAGARAN • *Metabolomics Australia, School of BioSciences, The University of Melbourne, Parkville, VIC, Australia*

TAKAYUKI TOHGE • *Max Planck Institute of Molecular Plant Physiology, Potsdam-Golm, Germany; Graduate School of Biological Sciences, Nara Institute of Science and Technology, Ikoma, Nara, Japan*

JOSÉ G. VALLARINO • *Max Planck Institute of Molecular Plant Physiology, Potsdam-Golm, Germany*

RIC C. H. DE VOS • *Business Unit Bioscience, Wageningen University and Research, Wageningen, The Netherlands; Netherlands Metabolomics Centre, Leiden, The Netherlands*

MUTSUMI WATANABE • *Graduate School of Biological Sciences, Nara Institute of Science and Technology, Ikoma, Nara, Japan*

WOLFRAM WECKWERTH • *Department of Ecogenomics and Systems Biology, Faculty of Life Sciences, University of Vienna, Vienna, Austria; Vienna Metabolomics Center, University of Vienna, Vienna, Austria*

JAKOB WEISZMANN • *Department of Ecogenomics and Systems Biology, Faculty of Life Sciences, University of Vienna, Vienna, Austria; Vienna Metabolomics Center, University of Vienna, Vienna, Austria*

THOMAS C. R. WILLIAMS • *Departamento de Botânica, Universidade de Brasília, Brasília, DF, Brazil*

LUCIA DANIELA WOLF • *Brazilian Bioethanol Science and Technology Laboratory (CTBE), Brazilian Center for Research in Energy and Materials (CNPEM), Campinas, SP, Brazil*

SI WU • *Max Planck Institute of Molecular Plant Physiology, Potsdam-Golm, Germany; Department of Life Sciences, Ben-Gurion University of the Negev, Beersheba, Israel*

TOSHIO YANAGIDA • *Laboratory for Single-Cell Mass Spectrometry, Quantitative Biology Center, RIKEN, Osaka, Japan*

Part I

Review Chapters

Chapter 1

The Importance of Experimental Design, Quality Assurance, and Control in Plant Metabolomics Experiments

Marina C. M. Martins, Camila Caldana, Lucia Daniela Wolf, and Luis Guilherme Furlan de Abreu

Abstract

The output of metabolomics relies to a great extent upon the methods and instrumentation to identify, quantify, and access spatial information on as many metabolites as possible. However, the most modern machines and sophisticated tools for data analysis cannot compensate for inappropriate harvesting and/or sample preparation procedures that modify metabolic composition and can lead to erroneous interpretation of results. In addition, plant metabolism has a remarkable degree of complexity, and the number of identified compounds easily surpasses the number of samples in metabolomics analyses, increasing false discovery risk. These aspects pose a large challenge when carrying out plant metabolomics experiments. In this chapter, we address the importance of a proper experimental design taking into consideration preventable complications and unavoidable factors to achieve success in metabolomics analysis. We also focus on quality control and standardized procedures during the metabolomics workflow.

Key words Experimental design, Quality control, Noise factors, Variability, Biological replication, Metabolomics workflow

1 Introduction

Plants possess an extraordinarily complex metabolism and coordinate various biochemical processes among different specialized tissues, cells, and subcellular compartments to synthesize metabolites that will ensure their survival against stresses and communication in the surrounding environment. This substantial level of functional organization is required to enable the production of a wide spectrum of chemical compounds [1, 2] with specialized roles, which are present in a broad dynamic range, placing plants as excellent materials to perform metabolomics experiments. Undoubtedly, plants are the source of the most complex individual mixtures [2], and although substantial information on the metabolic universe characterized so far has been obtained using model species, considerable efforts to increase the utility of metabolomics

Carla António (ed.), *Plant Metabolomics: Methods and Protocols*, Methods in Molecular Biology, vol. 1778, https://doi.org/10.1007/978-1-4939-7819-9_1, © Springer Science+Business Media, LLC, part of Springer Nature 2018

in endemic and crop species has been made [3–5]. If on the one hand there is a vast possibility of using metabolomics to discover new gene functions and natural compounds with application for several industries, in practice, it is a tremendous challenge to obtain a great coverage of such a diverse metabolome. Despite of the considerable advance in analytical technologies to increase the fraction of metabolites measured, it is estimated that the most comprehensive methods are able to detect only 15–30% of the small molecules presented in one species [6–9]. Furthermore, the vast number of compounds that differ in their physicochemical (e.g., size, polarity, and stability) and biological properties (e.g., cellular abundance and tissue composition) prevents the use of a single extraction method and analytical platform to uncover the metabolome. As a cornerstone of systems biology [10, 11], metabolomics is usually integrated with other "omics" tools, such as proteomics and transcriptomics, adding an extra layer of difficulty in setting up the appropriate experimental design. In this chapter, we briefly point out several relevant aspects for suitable experimental design and effective controls along all steps of metabolomics experiments.

2 Formulation of the Biological Question

Various important elements have to be considered when planning plant metabolomics experiments. The first fundamental step is to start with a good hypothesis in order to set up the right experimental design and effective controls. It is crucial to explore the biological context of the metabolomics study to know which kind of sample will be analyzed and compared to, and how this will be performed. The easy access to omics technologies occasionally results in their indiscriminate use without carefully considering the biological question. It is very common that experiments are initially designed for other purposes and eventually their sample leftovers are utilized for metabolomics. Depending on the goals, the storage of samples for long periods of time can affect the stability of compounds, seriously jeopardizing the success of the metabolomics analysis. Furthermore, due to the fact that metabolomics has a wide range of applications [12–21], it is emerging as a popular tool in plant science. However, it is relevant to bear in mind that in some cases the isolated use of metabolomics will not completely solve a certain biological question or even might not be the best approach to validate a hypothesis. One example is the use of metabolomics to determine gene function. If the target is a transcription factor, although this approach could provide some hints about which metabolic pathways are regulated, a chromatin immunoprecipitation (ChIP) assay would be a more straightforward tool for identifying direct targets of this gene—first, because it is possible that none of the putative targets are encoding for metabolically related enzymes, and second, even considering that few targets do have a regulatory role in metabolism, the outcome of a ChIP assay would enable a more oriented approach for extracting the metabolic fraction of interest and choose a suitable analytical platform to validate the function of this transcription factor.

Without a clear biological question, there is a high chance of obtaining different possible interpretations for the observed changes that would not reveal important information as regards the biological system.

3 Experimental Design

The aim of the experimental design is to make analysis and interpretation as straightforward and clear as possible ensuring that the collected information will accurately represent the studied biological condition. Therefore, one key aspect in planning metabolomics studies is to inspect all possible sources of variation that can exert a great impact on the hypothesis. One can easily point out variability coming from biological (e.g., genotype, developmental stages, and environmental conditions), technical (e.g., analytical platform and batch/block effect), and stochastic sources (unpredictable factors such as sample mislabeling or power outage). A systematic manner of detecting such sources is to follow a typical workflow for a metabolomics experiment (Fig. 1). The precise identification of those factors would make it possible to reduce putative sources of noise or at least take them into account during the data analysis procedure.

3.1 Experimental Conditions

The processes of growing plants (in the field or in controlled environmental chambers), applying treatments, and determining which tissues will be harvested and how are of outstanding

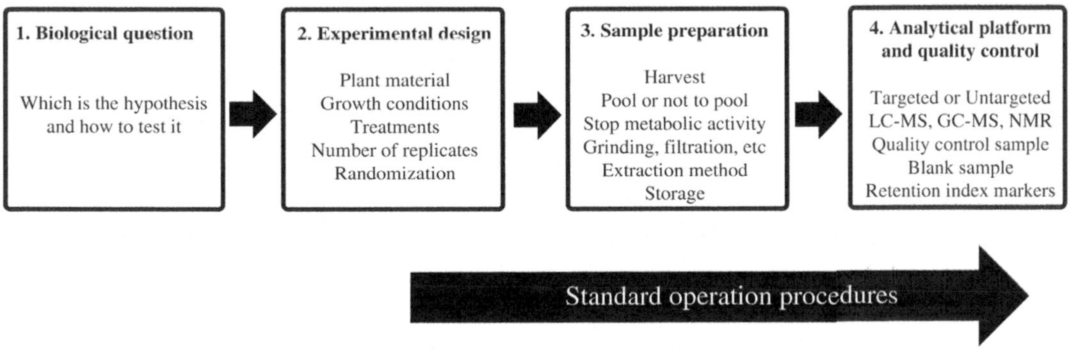

Fig. 1 Workflow of metabolomics studies. The biological question is essential for defining the experimental design, sample preparation, and analytical platform. Controls should be included during all the steps of the experiment. If possible, procedures should be documented in standard operating procedures (SOPs) to allow careful execution of each step always in a same manner

importance for the successful identification of metabolites and translation of the data into biologically relevant information. As sessile organisms, plants are constantly facing changes in light intensity/quality (cloudy days), temperature, and availability of resources along the diel cycle and seasons. Therefore, plant metabolism has evolved a high degree of plasticity to provide a buffer against sudden fluctuations in the environment, supporting rapid adaptation. It is mandatory to consider the environmental conditions and its influence on the metabolome when planning an experiment. The degree of complexity of those responses is further subjected to extra layers of difficulty when analyzing the effect of various abiotic (e.g., drought, temperature, and salinity) or biotic (e.g., symbiotic or pathogenic relationships) stresses. In respect to biotic stress, for example, it is common that a notable set of metabolites is shared between the plant and the interacting partner, hindering the identification of changes targeted to each organism. Furthermore, interactions involving such metabolites are often restricted to few plant cells, and the harvested samples can be a mix of noninfected and infected cells, which differ in their metabolism. Due to the heterogeneity of possible responses, it is important to set up appropriate controls, which might include, for example, infected tissues, tissues with a mock treatment, and isolated pathogenic cells [22]. In such a scenario, it is infeasible to have a perfect experimental design involving all variables and technical limitations, which should be given due thought during the data analysis workflow.

Plant metabolism is highly sophisticated [23], specialized to different organs, tissues, and cell types, which possess an exceptional degree of compartmentation [24, 25]. Leaves are source organs in which photosynthesis takes place, increasing photoassimilates and exporting sugars to sink tissues influenced by the time of the day. In contrast, roots are sink organs capable of incorporating metabolites also from the surrounding environment. Thus, it is useful to evaluate if model systems such as cell and seedling cultures would reflect the real physiological context, enabling to extrapolate the findings to the entire organism.

Special attention is required to decide whether an instant snapshot (steady-state) would be sufficient to effectively address the hypothesis or a high temporal resolution (time-courses) is needed to capture the dynamic nature of the metabolic response. For example, there are many studies based on analyses of specific time points to characterize gene function on metabolism using transgenics/mutants [26–28] and compare natural variation [29–32]. Time series experiments have been useful to follow the response of plants to pathogens [33–35] and diel or circadian changes in metabolism [36–38].

According to the number of variables analyzed in a metabolomics study, it might be advantageous to perform a pilot experiment to delineate the feasibility, time, and adverse events.

3.2 Determining The Size of The Experiment

The number of variables analyzed such as genotypes, treatments, single harvest, or kinetics will drive the size of the experiment. As biological variation is inherent to all organisms, it is challenging to be acutely aware of the variation degree within and across experiments. In plants, biological variation arises from the distinct developmental stages, growth conditions, genotypes, and tissues used for analysis. Moreover, quantitative and qualitative variations in metabolite composition occur at intraspecies and interspecies levels [4, 39, 40].

The multiple measurements obtained from metabolomics analyses do not make the experiments more powerful in a way that less replication is required. In fact, as the number of metabolites identified normally exceeds the number of samples, there is a high false discovery risk. Therefore, it is necessary to evaluate how many replicates does it take to achieve an equilibrium among sample number, cost, and time-consuming data analysis. It is advised that biological replication be preferred to technical replication as it improves the efficiency of statistical analysis [41]. For plant metabolomics studies, a minimum ideal number of biological replicates should be 6 [42]. Pooling samples is a common procedure employed to compensate for biological variation and limited amount of samples, provided that all individuals were carefully grown under equal conditions, thereby avoiding that single individuals are evaluated as in mutagenized or inbred populations [32, 43]. However, pooling can affect data quality and inference (e.g., if the pool contains any unusual sample), and sometimes might not be useful depending on the material analyzed (e.g., plant suspension cells).

3.3 Randomization

Randomization is critical for decreasing experimental error and biological variability, and for maintaining sample homogeneity in both small and large-scale experiments. Even in controlled environmental chambers, light intensity is not completely homogeneous, and it is advised that pots/trays are rotated during the experiment to assure that all plants will be subjected to the same variations in climate. In greenhouse or field experiments, plants are exposed to greater variation in environmental conditions, which are unavoidable but can readily be recognized. Randomization guarantees that treatment and control samples are comparable in terms of observed and unobserved features, decreasing for instance differences in soil (e.g., fertility and drainage) and environment heterogeneity.

A common strategy is to arrange the experimental units into blocks in which each treatment/genotype is arbitrarily allocated in every block at the same number of times, ensuring the intrablock and interblock comparison of treatments on units (plots). In this design, the effect of different levels of the factor of interest can be assessed, and variations in block factors are considered in the analysis. Block factors are those that do exert some effect on the response and are of no interest to the experimenter, and therefore, the variability they transmit to the response needs to be minimized or explained within blocks such as operator and time. In field trials, plots can be arranged in a way that both rows and columns can configure systematic sources of variation, and the format of the plots is usually determined according to the access or ease of sampling. Usually, the number of blocks is regarded as biological replicates, and replication of the experiment can be occasionally related to a blocking factor. Furthermore, it is suggested to randomly harvest the samples in each block to minimize block effect.

4 Sample Preparation

Diverse procedures of harvesting and handling samples are generally significant sources of bias in life sciences, which can impact the outcome of the analysis. The main source of technical variation in metabolomics is originated from sample preparation and analytical instrument. The preparation of samples for metabolomics experiments typically consists of sampling a representative fraction of the studied organism/tissue/cell (keeping it intact and stable), preprocessing, and extracting the sample. Modern analytical methods cannot compensate for poor or improper sampling and preparative procedures that alter metabolome composition, preventing comprehensive investigation and true interpretation of the plant metabolism.

4.1 Harvesting and Storage of The Material

Due to the rapid metabolic turnover rate of metabolites, especially photosynthetic intermediates that change in a time frame of seconds [44, 45], it is crucial that sampling is performed as fast as possible. If the experiment comprises a large dataset, it will demand a big team of operators to reduce the time for harvesting, and comprehensive training about what part and how it will be harvested (e.g., the first expanded leaf, internode number 3) should be given in advance. The time of sampling is equally important because many metabolites, including primary and secondary compounds, fluctuate during the diel cycle [36, 46, 47]. Sudden changes in the environmental conditions, like shading the plants shortly during sampling, can also alter metabolic composition.

Several strategies can be applied to maintain the integrity of the sample and preserve its metabolic content at the moment of sampling. The most popular procedure is the use of cryogenics, submerging the samples directly into liquid nitrogen. However, depending on the dimension of the sample and where it is placed (e.g., plastic tubes or aluminium bags), it is possible that complete freezing is not immediate. To overcome this problem, freeze-clamping [48] is an effective alternative, but is not applicable to a large number of samples. Freeze-drying is also employed to stop metabolic activity because the absence of water avoids enzymatic reactions that result in metabolite breakdown [49]. Because plant tissues are very hygroscopic, dried tissues must be stored in desiccators until extraction. However, this process can lead either to the loose of certain compounds or to their irreversible cell wall binding [50]. Some quenching treatments with alcohols at very low temperatures are described in the literature, most of them being suitable for algae [51–53].

According to the nature of the sample, grinding, filtration, fractionation, or preconcentration might be necessary [54]. Tissue grinding allows sample homogenization and increase of the contact surface, improving the extraction efficiency. There is a wide range of grinding instruments available in the market that facilitate handling with a large number of samples (e.g., ball mills and automated cryogenic grinders), but a simpler possible choice is to use a pre-cooled mortar and pestle with liquid nitrogen. All types of containers employed for grinding must be properly cleaned prior to use to avoid sample contamination.

When fresh material is used, it must be ensured that the tissue remains frozen during all steps of harvesting and sample preparation. Additionally, it is important to monitor the sample storage conditions because the stability of metabolites is generally not estimated, and long-time storage can alter metabolic composition.

It is recommended to use good quality plastic tubes that will endure extreme temperatures, such as liquid nitrogen. Screw cryogenic microtubes might guarantee the undesirable loose of material due to explosion caused by expansion during thaw. According to the planned analyses, samples can be stored fresh-frozen at −80 °C or, if freeze-dried, at −20 °C or room temperature. In this sense, the use of fresh material is mandatory for studies involving the quantification of semivolatile and volatile compounds [55].

All samples should be correctly identified during all steps of sample harvesting and preparation with appropriate pens or tags that can tolerate extreme temperatures and solvents. With the purpose of ensuring that the concentration of metabolites is comparable in all samples, aliquots have to be prepared for extraction. The precise amount of biological material must be recorded to allow normalization and direct comparison among treatments. Most metabolomics experiments use ground fresh weight material as a means of normalization. In cases in which the water content of samples might vary according to the condition (e.g., high salinity or drought stress) or genotypes, it is recommended to lyophilize the ground tissue prior to the extraction and use the recorded dry weight to normalize the variation among the samples. In studies with cell cultures, often standardization is based on the number of cells or estimation of protein/chlorophyll content [56].

4.2 Metabolite Extraction

It is expected that a good extraction procedure will ensure the analysis of the maximum number of small molecules. However, the diversity in abundance, structure, and chemical behavior of metabolites hampers the adoption of a single extraction method. To efficiently extract the metabolites from cells or tissues of interest, several parameters must be considered throughout the extraction procedure. Usually, the investigation of metabolites follows two possible approaches, targeted or untargeted. In targeted metabolomics, a defined group of known and biochemically annotated features is identified, whereas untargeted analyses are unbiased and aim to provide an overview of all the measurable analytes in a sample, including unknown compounds. Targeted strategies rely on the accuracy of absolute quantifications or on the precision of semiquantitative measurements. Cajka and Fiehn [57] provide relevant information about how absolute quantification can be usually achieved and the advances in mass spectrometry that have significant increased selectivity and sensitivity to perform untargeted approaches. In targeted analyses, the choice of the extraction method will enrich the fraction of metabolites with similar chemical properties. In contrast, despite several efforts to ameliorate the metabolite coverage required in untargeted analyses, the selection of an extraction protocol will favor a specific class of compounds, which introduces some kind of bias [58, 59]. Therefore, a combination of extraction methods and analytical technologies is suggested to maximize the number of detectable metabolic features.

The physicochemical properties (polarity, inertia, selectivity, etc.), the solvent–sample ratio, and the interaction between metabolites and biological matrix should also be taken into account for the selection of the most suitable solvent [49, 60, 61]. Depending on the interactions between metabolites and matrix, the dissolution rate in the extraction solvent may be slow. The increase in extraction time and temperature may favor metabolite extraction, although there is a risk of compound degradation [51, 62]. Due to the increased efficiency of the extraction process provided by hydration and cell wall breaking, the ultrasonic method is widely used. It is also crucial to store the extracts appropriately according to the stability of the targeted metabolites [23, 63]. Since contaminants such as plasticizers or surfactants can impair the analysis, the purity of the solvent used is a factor that must be considered. To avoid contamination and evaporation of the solvents as well as changes in the concentration and pH of the solution, the extraction buffer must be prepared and used in a short period of time [64]. Finally, the extract must be compatible with the analytical technique employed. In liquid chromatography coupled to mass spectrometry (LC-MS) analyses, for example, the effect of ion suppression due to extract composition may occur and affect the ionization efficiency for some compounds. In summary, an effective extraction method would permit the recovery of a great number of compounds and avoid the removal of metabolites due to their physical or chemical properties.

Regardless of the extraction method, the addition of one or more isotopically stable internal standards in the extraction buffer is an important control to monitor extraction reproducibility. Such standard should not be present in the metabolome of the biological sample, and will act both to detect pipetting errors and provide an idea of the extraction efficiency (at least for its compound class). These internal standards are also often used in data analysis for normalization [64].

5 Selection of the Analytical Platform and Quality Assurance

The optimal selection of the analytical technology will largely depend on the goal of the study and is often a compromise between selectivity and speed [65]. Mass spectrometry (MS) has been widely used in plant metabolomics experiments in combination with gas or liquid chromatography (GC and LC, respectively). The specificity of the detection method as well as the resolution of the chromatographic system will dictate the amount of metabolites measured. GC-MS offers robust quantification of several metabolite classes in plants requiring a derivatization step for compounds with poor volatility and thermal stability. The electron ionization (EI) method generally employed in GC-MS analyses generates complex but specific mass spectra that are particular to each molecule. Additionally, the high reproducibility of retention times made it possible the generation of libraries that can be used for compound identification [66, 67]. On the other hand, LC-MS can be extremely versatile to analyze a far-reaching range of compounds (e.g., polar and apolar, low and high molecular weight) without the need of derivatization. However, the electrospray ionization (ESI) method used in most LC-MS analyses is susceptible to ion suppression due to coeluting interfering compounds that compromise the ionization and the transfer of analytes from liquid to the gas phase, thereby decreasing the evaporation efficiency. Due to the huge variation in the conditions used in LC-MS methods (e.g., different LC columns, mobile phase eluents), the fragmentation patterns of the molecules can vary a lot, limiting the use of databases for compound identification [65].

Nuclear magnetic resonance (NMR) has a fundamental quantitative nature reflecting the real molar levels of metabolites without the need of calibration curves [55, 68]. It is a convenient method for structure elucidation but has low sensitivity in comparison to MS methods.

Once the analytical platform is chosen, the application of standardized procedures, both prior to and during metabolomics analyses, will contribute to the accuracy of the analytical measurements and guarantee high quality of the generated data. Nevertheless, in MS-based techniques coupled to chromatography, variations in the performance of the analytical platform (e.g., column efficiency, eluent batch, room temperature) often occur, especially for large sample batches, and thus, instrument performance (e.g., repeatability of retention times and peak areas) should be assessed regularly. In addition, quality control (QC) samples are routinely used to verify the performance of the selected analytical platform and method. QC samples are often incorporated as a supplementary experimental setup, and applied at different times during the injection of an analytical sequence (e.g., at the beginning of the sequence, repeated after the run of ten biological samples, and at the end of the experiment). QC samples help to evaluate whether the analysis was performed reliably and reproducibly, and ideally match the expected level of detection. There are different strategies for the type of QCs and how they are used in data analysis: they can be mixtures of authentic standards at known concentrations representing a collection of different chemical features, or the gold standard representative pool of all samples from the analyzed sequence [23, 69–71]. The latter has the main advantage of avoiding possible destabilization of the chromatography column due to the same biological matrix of remaining injections [72] and is useful to monitor system stability along different running days in large experiments [69]. It is worth to mention that in untargeted approaches involving a massive number of unidentified compounds, in a high range of mass-to-charge ratio (m/z) and unknown concentrations, the incorporation of QCs is not applied [69].

To identify possible contaminants and minimize sample carry-over, the injection of a blank (sample without metabolic extract) at different times during the running sequence is a basic strategy. Those samples could be included in the sequence before the QCs.

A common challenge in both GC- and LC-MS platforms is the unavoidable shifts in the retention time of metabolites, which makes data processing difficult. The employment of retention index markers such as fatty acid methyl esters (FAMEs) or chemical standards has been shown to be extremely helpful to adjust variations in the retention time in GC- and LC-MS studies, respectively, while enabling the creation of standard libraries for metabolite identification.

In contrast to other omics technologies, reliability of metabolite measurements to ensure comparability between laboratories is a bottleneck due to the vast variability of the chemical matrices in different plant species and the large number of analytic protocols and machines that are needed to achieve adequate coverage of metabolites [73]. It is of paramount importance to guarantee that metabolomics data are captured, stored and disseminated on open databases, following community information standards. The

MetaboLights, a database for metabolomics experiments from the European Bioinformatics Institute, contains information about individual compounds (references for chemical, analytical, and biological properties) and data repository from metabolomics studies that can be used for publication reference [74–76]. Such an initiative is crucial and demonstrates the efforts of the metabolomics community to permit reanalysis and comparisons between experiments and laboratories, but more importantly to constrain data quality. Added to that, tools to track samples during the metabolomics workflow are being proposed to assist the work of institutional facilities and multilaboratory consortia [77].

6 Standard Operating Procedures (SOPs)

Basic procedures like controlling environmental conditions for plant cultivation (only possible for experiments in controlled chambers or greenhouses), sample processing workflows (e.g., tube labeling, pipetting, balance usage, adequate preparation of solvents, and extraction method), QC for assessing instrument performance, and procedures for data analysis can bring about potential sources of variability, and therefore, should be described in standard operating procedures (SOPs). SOPs are documents containing precise information about the operations related to the quality of the experiment, which ensure that each step will be meticulously executed and always in the same manner.

Metabolite profiles are powerful tools capable of discriminating among, for example, environmental conditions [78–80] and developmental stages [81–83]. The use of SOPs can avoid misinterpretation of the results and facilitate comparisons between datasets.

An excellent example of the importance of standardization for metabolomics experiments is the work in which ten independent laboratories have grown three of the most commonly studied *Arabidopsis thaliana* ecotypes aiming to distinguish those genotypes based on leaf phenotype, metabolome, and transcriptome [84]. A reference experiment was useful to carefully define the conditions for plant cultivation that were shared in a detailed protocol (e.g., identical seed batches and same substrate for plant germination and cultivation). A core group of four laboratories obtained similar results related to phenotypic measurements that were in agreement with the reference experiment, and allowed the discrimination of the *A. thaliana* ecotypes. In contrast, metabolite profiling of 89 compounds was unable to distinguish the genotypes across laboratories and showed that the environment exerted influence on 55 metabolites. There was a strong interaction between genotype and environment demonstrating that metabolic activity is rapidly modulated due to variability in growth conditions. In respect to transcripts, variance in gene expression was higher for plant-to-plant differences than for laboratory effects and averaging or pooling was necessary to reduce sample variation. This work showed that standardization could assure that interlaboratory variation was not larger than intralaboratory variation. In summary, standardization is necessary and possible if following detailed protocols.

7 Final Considerations

Metabolomics possesses a wide range of applications and has been increasingly employed in plant experiments aiming to understand biochemical or regulatory mechanisms. To successfully obtain relevant information with metabolomics analyses, some elements are crucial: (a) pertinent biological question, (b) appropriate experimental design to guarantee representative samples and preservation of their original metabolic composition (from sampling to the metabolomics analysis), and (c) effective QCs to assure reliable measurements. Efforts from the scientific community to allow comparability between laboratories and standardization of information are progressively being made, with journals increasingly requesting the submission of metabolomics-related data to open source databases and repositories to ensure the generation of high-quality data as well as realistic biological interpretation.

Acknowledgments

C. Caldana was funded by the São Paulo Research Foundation (FAPESP) grant no. 2012/19561-0 and the Max Planck Society.

References

1. Dixon RA, Strack D (2003) Phytochemistry meets genome analysis, and beyond. Phytochemistry 62:815–816
2. Hall R, Beale M, Fiehn O et al (2002) Plant metabolomics: the missing link in functional genomics strategies. Plant Cell 14:1437–1440
3. Li D, Heiling S, Baldwin IT, Gaquerel E (2016) Illuminating a plant's tissue-specific metabolic diversity using computational metabolomics and information theory. Proc Natl Acad Sci U S A 113:E7610–E7618
4. Heiling S, Khanal S, Barsch A et al (2016) Using the knowns to discover the unknowns: MS-based dereplication uncovers structural diversity in 17-hydroxygeranyllinalool diterpene glycoside production in the Solanaceae. Plant J 85:561–577
5. Wen W, Li D, Li X et al (2014) Metabolome-based genome-wide association study of maize kernel leads to novel biochemical insights. Nat Commun 5:3438
6. Fernie AR, Aharoni A, Willmitzer L et al (2011) Recommendations for reporting metabolite data. Plant Cell 23:2477–2482
7. Giavalisco P, Kohl K, Hummel J et al (2009) 13C isotope-labeled metabolomes allowing for improved compound annotation and relative quantification in liquid chromatography-mass spectrometry-based metabolomic research. Anal Chem 81:6546–6551
8. Aharoni A, Ric de Vos CH, Verhoeven HA et al (2002) Nontargeted metabolome analysis by use of fourier transform ion cyclotron mass spectrometry. OMICS 6:217–234

9. Iijima Y, Nakamura Y, Ogata Y et al (2008) Metabolite annotations based on the integration of mass spectral information. Plant J 54:949–962
10. Hong J, Yang L, Zhang D, Shi J (2016) Plant metabolomics: an indispensable system biology tool for plant science. Int J Mol Sci 17(6). https://doi.org/10.3390/ijms17060767
11. Zampieri M, Sekar K, Zamboni N, Sauer U (2017) Frontiers of high-throughput metabolomics. Curr Opin Chem Biol 36:15–23
12. Sampaio BL, Edrada-Ebel R, Da Costa FB (2016) Effect of the environment on the secondary metabolic profile of Tithonia diversifolia: a model for environmental metabolomics of plants. Sci Rep 6:29265
13. Glaubitz U, Erban A, Kopka J et al (2015) Metabolite profiling reveals sensitivity-dependent metabolic shifts in rice (Oryza Sativa L.) cultivars under high night temperature stress. Procedia Environ Sci 29:72
14. Liu X, Vrieling K, Klinkhamer PGL (2017) Interactions between plant metabolites affect herbivores: a study with pyrrolizidine alkaloids and chlorogenic acid. Front Plant Sci 8:903
15. Sade D, Shriki O, Cuadros-Inostroza A et al (2015) Comparative metabolomics and transcriptomics of plant response to Tomato yellow leaf curl virus infection in resistant and susceptible tomato cultivars. Metabolomics 11:81–97
16. Kogovsek P, Pompe-Novak M, Petek M et al (2016) Primary metabolism, phenylpropanoids and antioxidant pathways are regulated in potato as a response to potato virus Y infection. PLoS One 11:e0146135
17. Obata T, Witt S, Lisec J et al (2015) Metabolite profiles of maize leaves in drought, heat, and combined stress field trials reveal the relationship between metabolism and grain yield. Plant Physiol 169:2665–2683
18. Chen W, Wang W, Peng M et al (2016) Comparative and parallel genome-wide association studies for metabolic and agronomic traits in cereals. Nat Commun 7:12767
19. Wen W, Liu H, Zhou Y et al (2016) Combining quantitative genetics approaches with regulatory network analysis to dissect the complex metabolism of the maize kernel. Plant Physiol 170:136–146
20. Meyer RC, Steinfath M, Lisec J et al (2007) The metabolic signature related to high plant growth rate in Arabidopsis thaliana. Proc Natl Acad Sci U S A 104:4759–4764
21. Hill CB, Roessner U (2015) Advances in high-throughput untargeted LC-MS analysis for plant metabolomics. Adv LC-MS Appl Metabolomics 38:58–71
22. Allwood JW, Heald J, Lloyd AJ et al (2012) Separating the inseparable: the metabolomic analysis of plant-pathogen interactions. Methods Mol Biol 860:31–49
23. Stitt M, Sulpice R, Keurentjes J (2010) Metabolic networks: how to identify key components in the regulation of metabolism and growth. Plant Physiol 152:428–444
24. Ap Rees T, Hill SA (1994) Metabolic control analysis of plant metabolism. Plant Cell Environ 17:587–599
25. Lunn JE (2007) Compartmentation in plant metabolism. J Exp Bot 58:35–47
26. Vigani G, Bashir K, Ishimaru Y et al (2016) Knocking down mitochondrial iron transporter (MIT) reprograms primary and secondary metabolism in rice plants. J Exp Bot 67:1357–1368
27. Fukushima A, Kusano M, Mejia RF et al (2014) Metabolomic characterization of knockout mutants in Arabidopsis: development of a metabolite profiling database for knockout mutants in Arabidopsis. Plant Physiol 165:948–961
28. Pham PA, Wahl V, Tohge T et al (2015) Analysis of knockout mutants reveals non-redundant functions of poly(ADP-ribose)polymerase isoforms in Arabidopsis. Plant Mol Biol 89:319–338
29. Schwahn K, de Souza LP, Fernie AR, Tohge T (2014) Metabolomics-assisted refinement of the pathways of steroidal glycoalkaloid biosynthesis in the tomato clade. J Integr Plant Biol 56:864–875
30. Strauch RC, Svedin E, Dilkes B et al (2015) Discovery of a novel amino acid racemase through exploration of natural variation in Arabidopsis thaliana. Proc Natl Acad Sci U S A 112:11726–11731
31. Ghaffari MR, Shahinnia F, Usadel B et al (2016) The metabolic signature of biomass formation in barley. Plant Cell Physiol 57:1943–1960
32. Riedelsheimer C, Czedik-Eysenberg A, Grieder C et al (2012) Genomic and metabolic prediction of complex heterotic traits in hybrid maize. Nat Genet 44:217–220
33. Tzin V, Fernandez-Pozo N, Richter A et al (2015) Dynamic maize responses to aphid feeding are revealed by a time series of transcriptomic and metabolomic assays. Plant Physiol 169:1727–1743
34. Rudd JJ, Kanyuka K, Hassani-Pak K et al (2015) Transcriptome and metabolite profiling of the infection bycle of Zymoseptoria tritici on wheat reveals a biphasic interaction with plant immunity involving differential pathogen

chromosomal contributions and a variation on the hemibiotrophic lifestyle def. Plant Physiol 167:1158–1185
35. Perez-Bueno ML, Pineda M, Diaz-Casado E, Baron M (2015) Spatial and temporal dynamics of primary and secondary metabolism in Phaseolus vulgaris challenged by Pseudomonas syringae. Physiol Plant 153:161–174
36. Bénard C, Bernillon S, Biais B et al (2015) Metabolomic profiling in tomato reveals diel compositional changes in fruit affected by source–sink relationships. J Exp Bot 66:3391–3404
37. Figueroa CM, Feil R, Ishihara H et al (2016) Trehalose 6-phosphate coordinates organic and amino acid metabolism with carbon availability. Plant J 85:410–423
38. Espinoza C, Degenkolbe T, Caldana C et al (2010) Interaction with diurnal and circadian regulation results in dynamic metabolic and transcriptional changes during cold acclimation in arabidopsis. PLoS One 5:1–19
39. Windsor AJ, Reichelt M, Figuth A et al (2005) Geographic and evolutionary diversification of glucosinolates among near relatives of Arabidopsis thaliana (Brassicaceae). Phytochemistry 66:1321–1333
40. Keurentjes JJB, Fu J, de Vos CHR et al (2006) The genetics of plant metabolism. Nat Genet 38:842–849
41. Blainey P, Krzywinski M, Altman N (2014) Points of significance: replication. Nat Meth 11:879–880
42. Lisec J, Schauer N, Kopka J et al (2006) Gas chromatography mass spectrometry–based metabolite profiling in plants. Nat Protoc 1:387–396
43. Riedelsheimer C, Lisec J, Czedik-Eysenberg A et al (2012) Genome-wide association mapping of leaf metabolic profiles for dissecting complex traits in maize. Proc Natl Acad Sci U S A 109:8872–8877
44. Arrivault S, Obata T, Szecowka M et al (2017) Metabolite pools and carbon flow during C4 photosynthesis in maize: 13CO2 labeling kinetics and cell type fractionation. J Exp Bot 68:283–298
45. Heise R, Arrivault S, Szecowka M et al (2014) Flux profiling of photosynthetic carbon metabolism in intact plants. Nat Protoc 9:1803–1824
46. Gibon Y, Usadel B, Blaesing OE et al (2006) Integration of metabolite with transcript and enzyme activity profiling during diurnal cycles in Arabidopsis rosettes. Genome Biol 7:R76
47. Huseby S, Koprivova A, Lee B-R et al (2013) Diurnal and light regulation of sulphur assimilation and glucosinolate biosynthesis in Arabidopsis. J Exp Bot 64:1039–1048
48. Ap Rees T, Fuller WA, Wright BW (1977) Measurements of glycolytic intermediates during the onset of thermogenesis in the spadix of Arum maculatum. Biochim Biophys Acta Bioenerg 461:274–282
49. Glauser G, Boccard J, Wolfender JL, Rudaz S (2013) Metabolomics: application in plant sciences. In: Lämmerhofer M, Weckwerth W (eds) Metabolomics in practice: successful strategies to generate and analyze metabolic data. Wiley-VCH Verlag GmbH & Co. KGaA, Weinheim, pp 313–343
50. T'Kindt R, Morreel K, Deforce D et al (2009) Joint GC-MS and LC-MS platforms for comprehensive plant metabolomics: Repeatability and sample pre-treatment. J Chromatogr B Anal Technol Biomed Life Sci 877:3572–3580
51. Tohge T, Mettler T, Arrivault S et al (2011) From models to crop species : caveats and solutions for translational metabolomics. Front Plant Sci 2:1–15
52. Yang D, Song D, Kind T et al (2015) Lipidomic analysis of chlamydomonas reinhardtii under nitrogen and sulfur deprivation. PLoS One 10:e0137948
53. Quinn RA, Vermeij MJA, Hartmann AC et al (2016) Metabolomics of reef benthic interactions reveals a bioactive lipid involved in coral defence. Proc R Soc B 283:20160469
54. Veyel D, Erban A, Fehrle I et al (2014) Rationales and approaches for studying metabolism in eukaryotic microalgae. Meta 4:184–217
55. Kim HK, Choi YH, Verpoorte R (2010) NMR-based metabolomic analysis of plants. Nat Protoc 5:536–549
56. Huege J, Krall L, Steinhauser M-C et al (2011) Sample amount alternatives for data adjustment in comparative cyanobacterial metabolomics. Anal Bioanal Chem 399:3503–3517
57. Cajka T, Fiehn O (2016) Toward merging untargeted and targeted methods in mass spectrometry-based metabolomics and lipidomics. Anal Chem 88:524–545
58. Beltran A, Suarez M, Rodriguez MA et al (2012) Assessment of compatibility between extraction methods for NMR- and LC/MS-based metabolomics. Anal Chem 84:5838–5844
59. Salem MA, Juppner J, Bajdzienko K, Giavalisco P (2016) Protocol: a fast, comprehensive and reproducible one-step extraction method for the rapid preparation of polar and semi-polar metabolites, lipids, proteins, starch and cell wall polymers from a single sample. Plant Methods 12:45
60. Mushtaq MY, Choi YH, Verpoorte R, Wilson EG (2014) Extraction for metabolomics:

access to the metabolome. Phytochem Anal 25:291–306

61. Kim HK, Verpoorte R (2010) Sample preparation for plant metabolomics. Phytochem Anal 21:4–13
62. Maltese F, van der Kooy F, Verpoorte R (2009) Solvent derived artifacts in natural products chemistry. Nat Prod Commun 4:447–454
63. Bais P, Moon SM, He K et al (2010) PlantMetabolomics.org: a web portal for plant metabolomics experiments. Plant Physiol 152:1807–1816
64. Roberts LD, Souza AL, Gerszten RE, Clish CB (2012) Targeted metabolomics. Curr Protoc Mol Biol. 98:30.2:30.2.1–30.2.24
65. Jorge TF, Rodrigues JA, Caldana C et al (2016) Mass spectrometry-based plant metabolomics: metabolite responses to abiotic stress. Mass Spectrom Rev 35:620–649
66. Kopka J, Schauer N, Krueger S et al (2005) GMD@CSB.DB: the Golm metabolome database. Bioinformatics 21:1635–1638
67. Kind T, Wohlgemuth G, Lee DY et al (2009) FiehnLib: mass spectral and retention index libraries for metabolomics based on quadrupole and time-of-flight gas chromatography/mass spectrometry. Anal Chem 81:10038–10048
68. Simmler C, Napolitano JG, McAlpine JB et al (2014) Universal quantitative NMR analysis of complex natural samples. Curr Opin Biotechnol 25:51–59
69. Alonso-herranz JGV, Barbas C, Grace E (2015) Controlling the quality of metabolomics data: new strategies to get the best out of the QC sample. Metabolomics:518–528
70. Dunn WB, Broadhurst DI, Edison A et al (2017) Quality assurance and quality control processes : summary of a metabolomics community questionnaire. Metabolomics 13:1–6
71. Sangster T, Major H, Plumb R et al (2006) A pragmatic and readily implemented quality control strategy for HPLC-MS and GC-MS-based metabonomic analysis. Analyst 131:1075
72. Gibon Y, Rolin D (2012) Aspects of experimental design for plant metabolomics experiments and guidelines for growth of plant material. Methods Mol Biol 860:13–30
73. Fernie AR, Stitt M (2012) On the discordance of metabolomics with proteomics and transcriptomics: coping with increasing complexity in logic, chemistry, and network interactions. Plant Physiol 158:1139–1145
74. Steinbeck C, Conesa P, Haug K et al (2012) MetaboLights: towards a new COSMOS of metabolomics data management. Metabolomics 8:757–760
75. Kale NS, Haug K, Conesa P et al (2016) MetaboLights: an open-access database repository for metabolomics data. Curr Protoc Bioinform 53:14.13.1–14.13.18
76. Haug K, Salek RM, Conesa P et al (2013) MetaboLights: an open-access general-purpose repository for metabolomics studies and associated meta-data. Nucleic Acids Res 41: D781–D786
77. Hunter A, Dayalan S, De Souza D et al (2017) MASTR-MS: a web-based collaborative laboratory information management system (LIMS) for metabolomics. Metabolomics 13:14
78. Hoermiller II, Naegele T, Augustin H et al (2017) Subcellular reprogramming of metabolism during cold acclimation in Arabidopsis thaliana. Plant Cell Environ 40:602–610
79. Jia X, Sun C, Zuo Y et al (2016) Integrating transcriptomics and metabolomics to characterise the response of Astragalus membranaceus Bge. var. mongolicus (Bge.) to progressive drought stress. BMC Genomics 17:188
80. Todaka D, Zhao Y, Yoshida T et al (2017) Temporal and spatial changes in gene expression, metabolite accumulation and phytohormone content in rice seedlings grown under drought stress conditions. Plant J 90:61–78
81. Cuadros-Inostroza A, Ruiz-Lara S, Gonzalez E et al (2016) GC-MS metabolic profiling of Cabernet Sauvignon and Merlot cultivars during grapevine berry development and network analysis reveals a stage- and cultivar-dependent connectivity of primary metabolites. Metabolomics 12:39
82. Monti LL, Bustamante CA, Osorio S et al (2016) Metabolic profiling of a range of peach fruit varieties reveals high metabolic diversity and commonalities and differences during ripening. Food Chem 190:879–888
83. Wiggins NL, Forrister DL, Endara M-J et al (2016) Quantitative and qualitative shifts in defensive metabolites define chemical defense investment during leaf development in Inga, a genus of tropical trees. Ecol Evol 6:478–492
84. Massonnet C, Vile D, Fabre J et al (2010) Probing the reproducibility of leaf growth and molecular phenotypes : a comparison of three Arabidopsis accessions cultivated in ten laboratories. Plant Physiol 152:2142–2157

Chapter 2

Standard Key Steps in Mass Spectrometry-Based Plant Metabolomics Experiments: Instrument Performance and Analytical Method Validation

Ana Margarida Rodrigues and Carla António

Abstract

Studies of the plant metabolome include the analysis of a wide range of chemical species with very diverse physicochemical properties requiring powerful analytical tools for the separation, characterization, and quantification of this vast compound diversity present in plant matrices. In quantitative metabolomics studies, major efforts are put into optimizing sample extraction and separation as well as instrument conditions to measure specific plant metabolites. Here, challenges in the use of mass spectrometry (MS) as a quantitative tool in plant metabolomics experiments are discussed, and an overview of the most critical steps in the development and validation of MS-based analytical methods is presented.

Key words Mass spectrometry, GC-MS, LC-MS, Method validation, Matrix effects, Analytical recoveries, Stable-isotope-labeled standards

1 Introduction

Quantitative plant metabolomics provides an in-depth understanding of plant biochemistry and metabolism by delivering the accurate measurement of concentrations of a wide spectrum of known metabolites that can occur in different levels in plant samples [1]. The two most commonly used analytical technologies driving quantitative plant metabolomics studies are mass spectrometry (MS) and nuclear magnetic resonance (NMR) spectroscopy. Due to its high sensitivity relative to NMR, MS-based analytical technologies are widely used in plant metabolomics studies, and when coupled to powerful chromatographic techniques, such as gas chromatography–mass spectrometry (GC-MS), liquid chromatography–mass spectrometry (LC-MS), capillary electrophoresis–mass spectrometry (CE-MS), Fourier transform-ion cyclotron resonance mass spectrometry (FT-ICR-MS), and more recently, mass spectrometry imaging (MSI), allow the separation and characterization

Carla António (ed.), *Plant Metabolomics: Methods and Protocols*, Methods in Molecular Biology, vol. 1778,
https://doi.org/10.1007/978-1-4939-7819-9_2,

of the extremely high compound diversity present in the plant metabolome [2].

Despite the analytical technology chosen to quantify specific plant metabolites, the need for standardizing key steps in plant metabolomics experiments, from experimental design and sample extraction to separation and instrument performance, is yet considered a critical issue [3]. In quantitative metabolomics, sample extraction optimization, method validation, chromatographic separation as well as optimization of the instrumental conditions are crucial steps to demonstrate that a particular analytical method produces accurate, precise and reproducible quantitative data of an analyte(s) in a given biological matrix [3–6].

When comparing two or more analytical methods, it is important to distinguish between (a) full validation, when developing and implementing an analytical method for the first time or if metabolites are added to an existing assay for quantification, (b) partial validation, which includes minor modifications of validated analytical methods that do not necessarily require full revalidations (e.g., analytical method transfer between laboratories and change in sample processing procedure), and (c) cross-validation [6, 7]. One main aspect common to most analytical validation methods is the attempt to maximize the number of known metabolites of interest that can be quantitatively measured in a single study [3].

MS-based plant metabolomics experiments deal with two approaches for quantitative analyses, namely, relative and absolute quantification. Relative quantification can be defined as the target analyte(s) instrument response relative to an internal standard or another analyte(s) level such as the sum of all analyte abundance, and is typically used in metabolite profiling experiments. Absolute quantification is the determination of the absolute concentration of a target analyte(s) through correlation of its instrument response to that of a known concentration series of the same analyte and is commonly used in targeted metabolite(s) analysis.

In this chapter, we present guidelines to key steps required for the analytical development and validation of MS-based methods for quantitative plant metabolomics experiments, namely the (a) establishment of the calibration curves and instrument linearity, (b) assessment of the limit of detection (LOD) and limit of quantification (LOQ), (c) method precision and accuracy, (d) method specificity and selectivity, (e) analytical recoveries during metabolite extraction, and finally, (f) matrix effect on metabolite quantification (Fig. 1) [3, 8].

2 Instrument Performance and Analytical Method Development

Developing an analytical MS-based method is an iterative process that can be very time consuming depending on the method requirements, namely the number of target analytes, analyte ionization,

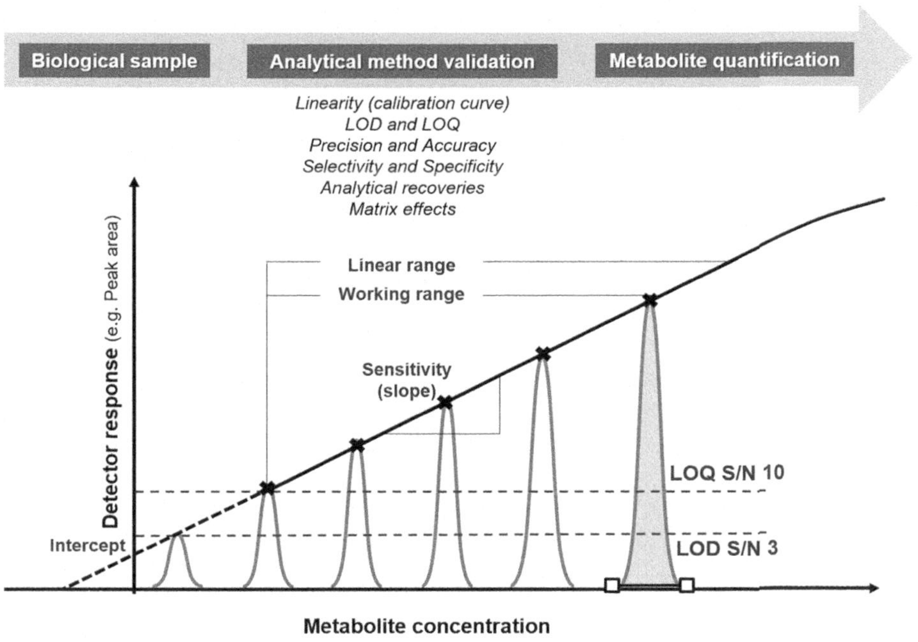

Fig. 1 Overview of the required parameters for the analytical validation of MS-based methods for quantitative plant metabolomics experiments, adapted from [3]

and matrix effects. Method development experiments should always start with a proper research planning in order to gather chemical and structural information of the target analyte(s). This planning stage sets the starting point for all subsequent experiments.

When developing an analytical method, the first experiment to assess instrument performance is the mass spectrometer initial tuning. Tuning is usually carried out by introducing a constant flow of analyte into the ionization source to allow to determine (a) whether the analyte(s) ionize in the mass spectrometer, (b) the best ionization mode (positive/negative), (c) the presence of adducts (very common when performing LC-MS), and finally, (d) the most informative fragments if performing tandem MS experiments. The next step is the chromatographic method development, where chromatographic parameters such as mobile phase, chromatographic gradient, and flow rate are optimized to achieve a good chromatographic separation of the target analyte(s). Coelution of different analytes can be acceptable as long as they do not interfere with the ionization process and can be distinguished with the mass spectrometer. Once these parameters are determined, the mass spectrometer can then be optimized because its performance is dependent on the flow rate and composition of the chromatographic mobile phase [9].

The final step in an analytical method development is the optimization of the extraction procedure. This step attempts to remove as much interferences as possible while maximizing the recovery of the target analyte(s). In quantitative plant metabolomics experiments, when dealing with low-abundant metabolites, the extraction procedure can include a concentration step, generally solid-phase extraction (SPE) [10].

2.1 Calibration Curve: Range and Linearity

Linearity is the study of the calibration line and its corresponding slope. For a first-order calibration function ($y = a + b{\cdot}x$), axis intercept a and slope b are characteristic of the sensitivity of the analytical method (Fig. 1). Calibration samples are prepared at different concentration levels, and the increasing analyte concentration is plotted against the corresponding detector response (peak height, peak area or peak intensity) [11]. The resulting graphical representation of the calibration data is then evaluated for the degree of linear association between two variables with the correlation coefficient (R^2) determination. However, it has been shown that a correlation coefficient very close to 1 might also be obtained for a curved relationship, typical of a second-order calibration function ($y = a + b{\cdot}x + c{\cdot}x^2$) [12, 13]. In this case, the Lack-of-fit and Mandel's fitting test are reported to be more appropriate for evaluating the linearity of the calibration curve during analytical method validation [3, 12–14].

Firstly, when establishing a new analytical method, the range of concentration used for the calibration curves must be defined for each analyte based on the amount present in the matrix and taking into consideration the changes that might occur during experiments. The calibration should always cover the working range in which the usual content of real samples is expected, thus a working range between 0 and 150% can be considered as applicable. Commonly, calibration samples should be prepared in blank matrix (i.e., a specimen of the matrix containing effectively none of the analyte), but due to the fact that analytical instruments have an inputted background signal or noise which is expected to be nonzero, a zero-point calibration (0,0) should not be used unless a real standard zero-point is available and can be reasonably interpreted as zero [14]. The working range should be defined by the LOQ and the upper limit of quantification, corresponding to the highest standard concentration used for calibration [6] (Fig. 1). Although information on specific requirements for the number of replicates at each concentration level is scarce, the calibration levels (calibration points) recommended for an appropriate calibration curve range from 5 to 8 over the whole working range, ideally evenly distributed or equidistant and each analyzed at least in triplicate [13–18]. The choice of an appropriate calibration model is also necessary for reliable quantification. In quantitative MS-based chromatographic methods, quantification of target analyte(s) can be performed using (a) external standard calibration plots, using standard solutions of the analyte; (b) internal standard calibration plots, where an internal standard is added to the sample and is analyzed together with the analyte; (c) standard addition calibration plots, where increasing amounts of standard are added to several sample matrix aliquots and the standard addition calibration curve is calculated between the measured analytical response and the known added concentrations; and (d) matrix-matched calibration plots, where calibration is obtained using a matrix reference material [11]. However, because there is no matrix reference material with known analyte concentrations available for high-throughput analyses in plant metabolomics studies, this approach has not yet been adopted [18]. Calibration samples can also be used to determine the correction factor (i.e., response factor) for each standard in comparison with its internal standard, as is often performed in plant metabolomics for phytohormone quantification [19, 20]. The response factor is determined comparing the slope of the calibration curve from the standard to that of its internal standard. If the relative response factor is significantly different from 1, a correction factor should be applied in the calculation, or the reported results will be grossly overestimated or underestimated [21].

2.2 Limit of Detection and Limit of Quantification

The limit of detection (LOD) represents the lowest analyte amount that can be detected in a sample, even if it cannot be quantified. In chromatographic methods, a quick estimation of the LOD can be simply defined as the concentration for which the signal-to-noise ratio (S/N) equals 3 (Fig. 1). S/N can be defined as the height of the analyte peak (signal) and the amplitude between the highest and lowest point of the baseline (noise) in a certain area around the analyte peak [22]. The limit of quantification (LOQ), also known as the lower limit of quantification (LLOQ) is the lowest analyte amount that can be quantitatively determined using chromatographic methods based on (a) precision and accuracy data based on precision and accuracy data (lowest concentration with a precision and accuracy better than 20%) (b) signal-to-noise or (c) a specific calibration curve in the range of LOQ, using the standard deviation (SD) on the lowest point of the calibration line ($\text{LOQ} = k.SD/b$), where k-factor is usually 10 [18, 22, 23]. Despite the difficulties in calculating the background noise and the significant equipment dependence, the S/N approach using of spiked samples with decreasing concentrations of analyte(s) is the most widely used [6, 22]. The LOQ is often calculated as the analyte concentration for which the S/N ratio equals 10 [3, 11, 15, 18].

2.3 Precision and Accuracy

The precision of an analytical method is defined as the closeness of agreement between a set of independent analytical measurements of the same sample, under the optimal developed conditions, expressed as relative standard deviation (RSD, %) or coefficient of variation (CV, %) of retention/migration times or peak areas [3, 6]. Precision expresses numerically the random error of an analytical method [15]. Within a laboratory, several levels of precision can be determined, namely, (a) instrument precision, or repeatability (b) intraday, or intrabatch, precision, and (c) interday precision, also known as intermediate precision or interbatch precision, reflecting small variations in the experimental or operational conditions of repetitive analysis. Repeatability is determined by the consecutive repeated injection of the same sample, whereas intraday and interday precision are determined by injecting a minimum of three replicates at three different concentration levels, representative of the validation range (low, medium, and high concentration). Interday precision is determined within different days, consecutive or not [3, 15, 16, 18]. Several experimental designs (replicates X days) are reported in the literature [15–18]. These samples can also be used for accuracy and trueness evaluation [6, 22]. Reproducibility is used to describe the precision between different laboratories but is only determined when the analytical method is intended to be implemented in different laboratories [23].

Accuracy is defined as the closeness of agreement between a quantity value obtained by measurement and the true value of the measurand. In plant metabolomics studies, where no reference materials with known metabolite concentrations are available, accuracy can be assessed by determining the analytical recoveries (*see* next section) of analytical standards or stable isotopically labeled standards spiked into the plant matrix [5, 18]. In analytical method validation, another terminology used to describe the quality of measurements is trueness, which corresponds to the closeness of agreement between the average value obtained from a large series of replicate measurements and an accepted reference value. The meaning of precision, accuracy and trueness terminology used to describe the quality of measurements has been focus of discussion among the scientific community, especially when the term accuracy is erroneously used instead of trueness causing confusion and misinterpretation of results [6, 15–17, 24]. Trueness is a quantitative measurement of the systematic error of an analytical method whereas accuracy is measured and reported as an uncertainty. Due to the high workload of analyzing large series of measurements needed to determine trueness, only accuracy is typically determined during method validation [23]. Nonetheless, accuracy is considered an overall concept related to trueness and precision [17, 24].

2.4 Analytical Recovery

In plant metabolomics studies, the analytical recovery measures the ability of the developed analytical method to extract an analyte from a biological matrix and; it is often used to simultaneously assess the influence of the analytical procedure (e.g., extraction) and of the plant matrix. It is generally assumed that if the analytical recoveries are 100% then the procedure in question is free from systematic errors (e.g., extraction). However, there is no minimum established value for analytical recoveries because they are dependent on the analyte; the recovery value is less crucial as long as high sensitivity, precision and accuracy are obtained for the analyte over the entire calibration range [6]. Similarly to precision and accuracy, analytical recovery can be measured in the same experiment using three different concentration levels (low, medium, and high) covering the whole analytical range under evaluation [15]. Recoveries can be expressed as the ratio between the analyte amount spiked before extraction procedure and the absolute spiked concentration of the analyte in solution [3, 6, 22, 23] or, alternatively, the ratio between the analyte amount spiked before extraction procedure and the analyte amount spiked after extraction procedure in the biological matrix [25].

As stated before, in plant metabolomics studies no reference materials with known analyte concentrations are available. In this case, the analyte recoveries can be estimated by using (a) surrogates (i.e., a pure compound or element added to the sample with the chemical and physical behavior taken to be representative of the native analyte) as stable isotopically labeled standards, where its recovery will be the same as that of the analyte, and (b) analytes added as a spike, a less costly and very commonly applied procedure. If a blank matrix is available, the analyte can be spiked into that and its recovery determined after the extraction procedure. If no blank matrix is available, the spike can be added to the sample matrix that is analyzed alongside an unspiked sample matrix [26].

2.5 Selectivity and Specificity

Selectivity is the ability of an analytical method to measure and discriminate several analytes of interest without interference from other components in the sample matrix, whereas specificity relates to how well a method is able to unequivocally measure one particular analyte in the presence of other components that might be present (e.g., endogenous or exogenous impurities, degradation products, or matrix components [27, 28]. An appropriate discrimination between these terms should be taken into consideration, as they are often wrongly used [6, 16, 22]. If a method is perfectly selective for an analyte or group of analytes, it can be considered specific [6, 27]. The most selective detection system is the mass spectrometer, particularly the triple quadrupole (QqQ)-MS mass spectrometer that allows for higher selectivity and sensitivity while minimizing chromatographic interferences. The QqQ-MS is typically operated in the multiple reaction monitoring (MRM) mode in which collision energies, dwell times, and resolution parameters for each individual target analyte are optimized using analytical standards to enhance sensitivity and selectivity [27, 29]. To establish the selectivity in an analytical method two approaches can be considered, namely (a) to prove the lack of response in blank matrix, i.e., that there are no signals interfering with the signal of the analyte(s) or the internal standard, and (b) consider that for merely quantitative procedures, small interferences can be accepted as long as accuracy and precision at the LOQ range remain within certain acceptance limits [22, 23]. In plant metabolomics studies, selectivity can be evaluated by standard addition method of the analyte (s) of interest using the plant matrix under study. A minimum of four standard additions within the established concentration range are advised, and the resulting calibration curve must be compared to the calibration with standard solutions [3, 16].

2.6 Matrix Effects

Matrix effect (ME) is defined as the effect of sample matrix and interferences from other metabolites in the matrix, on the quantification of the analyte of interest (e.g., ion suppression or enhancement, elevated baseline, impact on the extraction or the retention time) [3, 6]. These effects are usually caused by endogenous (e.g., conjugated forms of the target analyte, proteins or lipids) or exogenous (all substances introduced during sample extraction process and analysis) compounds [30]. ME are considered a serious problem in LC-MS methods, particularly when compounds from the matrix coelute with the target analyte(s) during chromatographic separation, thereby affecting ionization of the target analyte(s) and causing a change in the response of the analyte, either positive (ion enhancement effect) or negative (ion suppression effect) [6, 30]. ME can be quantitatively evaluated by the ratio between the signal response of a target analyte(s) in the absence of matrix (standard solution) and the signal response of the analyte(s) spiked (at the same concentration) in the matrix after extraction, where ME = 100% indicates no matrix effect, ME <100% indicates ion suppression, and ME >100% indicates ion enhancement [31]. Alternatively, ME can be assessed using (a) matrix-matched calibration curves and comparing the slope of a calibration curve prepared in the mobile phase and the slope of the same calibration curve prepared in the matrix [5, 32], and (b) stable isotopically labeled standards as surrogates [33].

ME issues in quantitative LC-MS can be especially complex as a high number of metabolites are usually profiled in metabolomics experiments in one single analytical run, and the absence of ME should be demonstrated for all individual analytes of interest. Overall, ME evaluation is an essential step in analytical method validation and could be eliminated or minimized by (a) improving the sample extraction procedure and by eliminating undetected matrix interferences, (b) optimizing the chromatographic conditions to separate analytes of interest from undetected endogenous compounds that might affect the efficiency of ionization, (c) employing corrective calibration methods such as the standard addition method [3, 30, 31], and (d) sample dilution [34], an effective method to reduce interfering compounds, thus, reducing ME. A flowchart describing typical strategies for a full MS-based method development and validation in plant metabolomics studies is presented in Fig. 2.

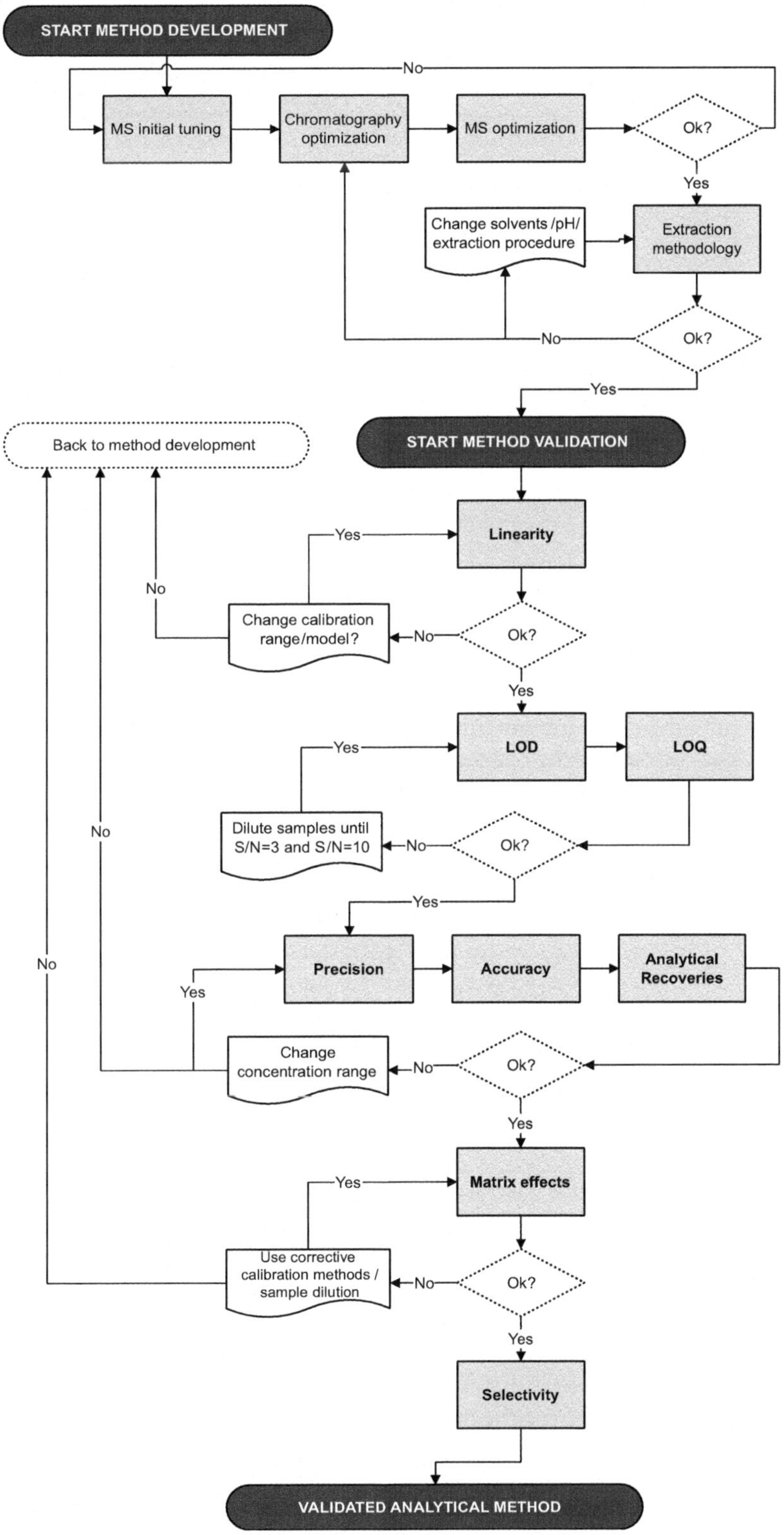

Fig. 2 Analytical method development and validation. Strategy flowchart for the full development and validation of chromatographic methods, suitable for quantitative plant metabolomics studies

3 Stable-Isotope-Labeled Internal Standards in Analytical Method Validation: Plant Metabolomics Application

Stable-isotope-assisted metabolomics is particularly useful in the study of plant metabolomes. Plant metabolomes are large and chemically diverse, and they are underrepresented in public database resources, which are more focused on clinical, pharmaceutical, or petrochemical-related compounds. Furthermore, the use of stable isotopes can be very useful when dealing with complex plant matrices for analyte quantification [35, 36]. While radioisotopes are detectable as a signal from their radiation without any contribution of noise from unlabeled compounds, stable-isotope-labeled standards are detected with the same sensitivity as their unlabeled counterparts [36]. The use of stable isotope and radioisotope labeling in plant metabolomics flux analysis studies, and respective advantages and disadvantages have been reviewed elsewhere [37] (*see* Chapters 4 and 5). Stable-isotope-labeled internal standards are commonly taken for having nearly identical chemical and physical properties to the unlabeled analyte, and are generally used in analytical method validation of low-abundant plant metabolites, such as phytohormones [19, 20, 25, 38, 39].

Due to their low abundance and diverse chemical structures, different classes of phytohormones are difficult to analyze simultaneously in one single analytical run or using one single analytical platform [39]. In this case, stable-isotope-labeled internal standards, with nearly identical chemical structure or eluting within the same time window of the target analyte, can provide correction for both matrix effects and phytohormone loss during the extraction procedure [20, 40]. Nevertheless, care must be taken as matrix effects might not be completely compensated by the use of these labeled internal standards [41].

4 Concluding Remarks

Despite the rapid advances in MS technology (i.e., mass resolution, mass accuracy, mass range, and sensitivity), quantitative plant metabolomics still faces many challenges and limitations. The major analytical challenges in plant metabolomics include (a) the chemical diversity that can be found in the plant metabolome, (b) the huge dynamic range of metabolite concentrations, and (c) the spatial and temporal distribution of these metabolites. The MS-based analytical approaches commonly used for quantitative plant metabolomics studies allow measuring a broad range of chemical classes. Nevertheless, when performing absolute quantification, the number of analytes that can be accurately measured is limited mainly due to the availability of authentic standards required to perform the calibration curves. In addition, when performing the validation of MS-based analytical methods in plant metabolomics studies, no reference materials or a blank matrix are available. To address these analytical challenges, high-throughput MS-based plant metabolomics methodologies are being developed at a rapid pace. One example is the development of new approaches using stable-isotope-labeled internal standards for absolute quantification. This methodology provides additional control for the metabolite extraction procedure, derivatization (if needed) and detection, thus improving metabolite coverage.

Acknowledgments

A.M. Rodrigues acknowledges Fundação para a Ciência e a Tecnologia (FCT) for the PhD fellowship (PD/BD/114417/2016) and the ITQB NOVA International PhD Program Plants for Life (PD/00035/2013). C. António gratefully acknowledges support from the FCT Investigator Program (IF/00376/2012/CP0165/CT0003), ITQB NOVA R&D Green-it: Bioresources for Sustainability (UID/Multi/04551/2013).

References

1. Fiehn O, Kopka J, Dormann P et al (2000) Metabolite profiling for plant functional genomics. Nat Biotechnol 18:1157–1161
2. Jorge TF, Rodrigues JA, Caldana C et al (2016) Mass spectrometry-based plant metabolomics: metabolite responses to abiotic stress. Mass Spectrom Rev 35:620–649
3. Jorge TF, Mata AT, António C (2016) Mass spectrometry as a quantitative tool in plant metabolomics. Phil Trans R Soc A 374:20150370
4. FDA (2001) .Guidance for industry Bioanalytical method validation https://www.fda.gov/downloads/Drugs/Guidance/ucm070107.pdf. Accessed 10 May 2017
5. Thompson M, Ellison SLR, Wood R (2002) Harmonized guidelines for single laboratory validation of methods of analyis (IUPAC Technical Report). Pure Appl Chem 74:835–855
6. González O, Blanco ME, Iriarte G et al (2014) Bioanalytical chromatographic method validation according to current regulations, with a special focus on the non-well defined parameters limit of quantification, robustness and matrix effect. J Chromatogr A 1353:10–27
7. Shah VP, Midha KK, Findlay JWA et al (2000) Bioanalytical method validation – a revisit with a decade of progress. Pharm Res 17:1551–1557
8. Sumner LW, Amberg A, Barret D et al (2007) Proposed minimum reporting standards for chemical analysis. Metabolomics 3:211–221
9. Sargent M (ed) (2013) Guide to achieving reliable quantitative LC-MS measurements. RSC Analytical Methods Committee, Teddington, London
10. Dettmer K, Aronov PA, Hammock BD (2007) Mass-spectrometry based metabolomics. Mass Spectrom Rev 26:51–78
11. Krull IS, Swartz M (1999) Analytical method development and validation for the academic researcher. Anal Lett 32:1067–1080
12. Loco JV, Elskens M, Croux C, Beernaert H (2002) Linearity of calibration curves: use and misuse of the correlation coefficient. Accred Qual Assur 7:281–285
13. Funk W, Dammann V, Donnevert G et al (2007) Phase I: establishing a new analytical process. In: Funk W, Dammnann V, Donnevert G (eds) Quality assurance in analytical chemistry: applications in environmental food, and materials analysis, biotechnology, and medical engineering. Wiley-VCH, Weinheim
14. Raposo F (2016) Evaluation of analytical calibration based on least squares linear regression for instrumental techniques: a tutorial review. Trends Anal Chem 77:167–185

15. Hartmann C, Smeyers-Verbeke J, Massart DL, McDowall RD (1998) Validation of bioanalytical chromatographic methods. J Pharm Biomed Anal 17:193–218
16. Araujo P (2009) Key aspects of analytical method validation and linearity evaluation. J Chromatogr B 877:2224–2234
17. Stöckl D, D'Hondt H, Thienpont LM (2009) Method validation across the disciplines–critical investigation of major validation criteria and associated experimental protocols. J Chromatogr B 877:2180–2190
18. Koek MM, Jellema RH, van der Greef J et al (2011) Quantitative metabolomics based on gas chromatography mass spectrometry: status and perspectives. Metabolomics 7:307–328
19. Pan X, Welti R, Wang X (2010) Quantitative analysis of major plant hormones in crude plant extracts by high-performance liquid chromatography–mass spectrometry. Nat Protoc 5:986–992
20. Delatorre C, Rodríguez A, Rodríguez L et al (2016) Hormonal profiling: Development of a simple method to extract and quantify phytohormones in complex matrices by UHPLC–MS/MS. J Chromatogr B 1040:239–249
21. Lee Y (2004) Method Validation for HPLC Analysis of related substances in pharmaceutical drug products. In: Chang CC, Lee YC, Lam H, Zhang XM (eds) Analytical method validation and instrument performance verification. Wiley-Interscience, New Jersey
22. Peters FT, Drummer OH, Musshoff F (2007) Validation of new methods. Forensic Sci Int 165:216–224
23. Peters FT, Maurer HH (2002) Bioanalytical method validation and its implications for forensic and clinical toxicology – A review. Accred Qual Assur 7:441–449
24. Menditto A, Patricarca M, Magnusson B (2007) Understanding the meaning of accuracy, trueness and precision. Accred Qual Assur 12:45–47
25. Trapp MA, de Souza GD, Rodrigues-Filho E et al (2014) Validated method for phytohormone quantification in plants. Front Plant Sci 5:417
26. Thompson M, Ellison SLR, Fajgelj A et al (1999) Harmonised guidelines for the use of recovery information in analytical measurement. Pure Appl Chem 71:332–348
27. Aboul-Enein HY (2000) Selectivity versus specificity in chromatographic analytical methods. Accred Qual Assur 5:180–181
28. Danzer K (2001) Selectivity and specificity in analytical chemistry. General considerations and attempt of a definition and quantification. Fresenius J Anal Chem 369:397–402
29. Sumner LW, Lei Z, Nikolau BJ, Saito K (2015) Modern plant metabolomics: advanced natural product gene discoveries, improved technologies, and future prospects. Nat Prod Rep 32:212–229
30. Bylda C, Thiele R, Kobold U, Volmer AD (2014) Recent advances in sample preparation techniques to overcome difficulties encountered during quantitative analysis of small molecules from biofluids using LC-MS/MS. Analyst 139:2265–2276
31. Matuszewski BK, Constanzer ML, Chavez-Eng CM (2003) Strategies for the assessment of matrix effect in quantitative bioanalytical methods based on HPLC-MS/MS. Anal Chem 75:3019–3030
32. Kmellár B, Fodor P, Pareja L et al (2008) Validation and uncertainty study of a comprehensive list of 160 pesticide residues in multi-class vegetables by liquid chromatography–tandem mass spectrometry. J Chromatogr A 1215:37–50
33. Marín JM, Gracia-Lor E, Sancho JV et al (2009) Application of ultra-high-pressure liquid chromatography–tandem mass spectrometry to the determination of multi-class pesticides in environmental and wastewater samples. Study of matrix effects. J Chromatogr A 1216:1410–1420
34. Ferrer C, Lozano A, Agüera A et al (2011) Overcoming matrix effects using the dilution approach in multiresidue methods for fruits and vegetables. J Chromatogr A 1218:7634–7639
35. Hegeman AD (2010) Plant metabolomics – meeting the analytical challenges of comprehensive metabolite analysis. Brief Funct Genomics 2:139–148
36. Freund DM, Hegeman AD (2017) Recent advances in stable isotope-enabled mass spectrometry-based plant metabolomics. Curr Opin Biotechnol 43:41–48
37. Silva WB, Daloso DM, Fernie AR et al (2017) Can stable isotope mass spectrometry replace radiolabeled approaches in metabolic studies? Plant Sci 249:59–69
38. Pan X, Wang X (2009) Profiling of plant hormones by mass spectrometry. J Chrom B 877:2806–2813
39. Cao D, Lutz A, Hill CB et al (2017) A quantitative profiling method of phytohormones and

other metabolites applied to barley roots subjected to salinity stress. Front Plant Sci 7:2070

40. Hoyerová K, Gaudinová A, Malbeck J et al (2006) Efficiency of different methods of extraction and purification of cytokinins. Phytochemistry 67:1151–1159

41. Wang S, Cyronak M, Yang C (2006) A matrix effect study on a LC/MS/MS method for the determination of carvedilol enantiomers in human plasma. J Pharm Biomed Anal 43:701–707

Chapter 3

Guidelines for Sample Normalization to Minimize Batch Variation for Large-Scale Metabolic Profiling of Plant Natural Genetic Variance

Saleh Alseekh, Si Wu, Yariv Brotman, and Alisdair R. Fernie

Abstract

Recent methodological advances in both liquid chromatography–mass spectrometry (LC-MS) and gas chromatography–mass spectrometry (GC-MS) have facilitated the profiling highly complex mixtures of primary and secondary metabolites in order to investigate a diverse range of biological questions. These techniques usually face a large number of potential sources of technical and biological variation. In this chapter we describe guidelines and normalization procedures to reduce the analytical variation, which are essential for the high-throughput evaluation of metabolic variance used in broad genetic populations which commonly entail the evaluation of hundreds or thousands of samples. This chapter specifically deals with handling of large-scale plant samples for metabolomics analysis of quantitative trait loci (mQTL) in order to reduce analytical error as well as batch-to-batch variation.

Key words Large-scale metabolomics, Batch normalization, Variation, Natural genetic variation, QTL mapping, LC-MS, GC-MS

1 Introduction

The metabolites of the plant kingdom are extremely diverse; a commonly quoted estimate is that plants produce somewhere in the order of 200,000 unique chemical structures [1]. Recently, there has been an increasing use the analytical technologies such as metabolomics for comprehensive profiling of metabolites in biological samples and its subsequent application in several related research areas such as human nutrition, drug discovery and plant breeding [2, 3]. Given the diversity of structural classes of metabolites, ranging from primary metabolites such as carbohydrates, amino acids, and organic acids to very complex secondary metabolites such as phenolics, alkaloids, and terpenoids, there is no single methodology that can measure the complete metabolome in one step. It is, therefore, often necessary to combine different techniques to detect (even a significant proportion of) all metabolites

Carla António (ed.), *Plant Metabolomics: Methods and Protocols*, Methods in Molecular Biology, vol. 1778, https://doi.org/10.1007/978-1-4939-7819-9_3,

within a complex mixture [4]. Both liquid chromatography–mass spectrometry (LC-MS) and gas chromatography–mass spectrometry (GC-MS) have been intensively used to profile a broad natural variance in the form of recombinant inbred lines (RILs), introgression lines (ILs) and, more recently, genome-wide association mapping panels in order to boost our understanding of the regulation of plant primary and secondary metabolite levels [5–7].

In all metabolomics applications, it is important to understand and control factors that contribute to sources of variation within the datasets. The variability between samples can arise from multiple sources including natural biological variation itself and that which occurs on sample collection and storage [8, 9]. In addition, analytical variation caused by suboptimal performance of the chosen apparatus, and instrument drift over time, are two major issues in large-scale metabolomics studies [10].

Batch-to-batch variation is a technical source of variation arising from the sum of both manual and robotic samples handling [11]. The presence of batch-to-batch variation makes it difficult to integrate data from independent batches of samples. This issue is particularly problematic when dealing with large number of samples such as is the case when analyzing structured plant populations.

To counter this, several normalization methods have been developed and suggested to overcome these issues and to minimize nonbiological variation [11–13]. For example normalizations by a single or multiple internal or external standard compounds based on empirical rules, such as specific regions of retention time have been used [2]. Similarly, isotope-labeled internal standard approaches were developed to monitor analytical error [14]. While there is no single best way to conduct metabolomics studies, there are a number of pitfalls and known problems that need to be carefully avoided. Detailed guidelines and normalization protocols have been previously published for this purpose [15–17].

In this chapter, we describe a workflow to minimize analytical errors and provide guidelines for handling large sample numbers for the specific purpose of metabolic quantitative trait loci (mQTL) approaches which utilize sources of broad natural genetic variance. We solely concentrate on aspects pertinent to the large-scale analysis of genetic populations and normalization aspects that need to be adopted to ensure proper cross-sample comparability as well as the downstream analysis of the data within the framework of quantitative loci and association mapping analyses.

2 Experimental Design for Large-Scale mQTL Approaches

In order to correctly evaluate such large sample sets it is important to manipulate variables under strictly controlled conditions while taking precise measurements. Therefore, the precision of an experiment critically depends on the size of the experiment and the homogeneity of experimental materials. In large genomics experiments, the next step after choosing the population is to determine the number of lines and associated biological replicates. This is then followed by choosing the statistical approach to link genotype with phenotype. Here we neither focusing on the choices of the number of lines nor the population structure needed to obtain a complete genotype-to-phenotype matrix to identify all possible QTL [18] but rather on how many biological replicates are required per line for acceptable statistical analysis and data normalization.

The key to minimize technical sources of variation involves designing an experiment whereby several samples are taken per plant with multiple independent plants per parental genotype per replicate. Multiple independent replicates are conducted and all samples independently analyzed via metabolomics. Analysis of variance for this experiment will allow one to estimate the variation due to spatial differences within a plant, from differences between plants, from differences between replicate experiments and from differences between genotypes as well as any interactions between these different features. The optimum result is that most of the variance is due to genetic factors with the rest of the error being split between within replicates from the same plant or replicates between plants of the same genotype. If this is the case, it is best to take one measurement per plant with each line being represented by two or more plants per replicate.

In QTL analysis the number of replicates profiled will have a major influence on the reliability and reproducibility of the data and consequently on the QTL mapping results. Therefore, the ability to make broad conclusions or identify causal genes using quantitative studies of metabolic variation is greatly influenced by the fact that metabolic abundances measured in these studies are highly dependent on the environmental, developmental, and genetic variations present within the experiment as well as the experimental error. For these reasons and based on our own experience, it is recommended to use at least six independent biological replicates for each line (genotype) and many more control plants in a completely randomized design to overcome unavoidable effects associated with variation in microenvironmental factors such as light intensities, temperature and air humidity. This should be planed carefully in advance and he population size and time needed for collecting the samples should also take into account in order to ensure that harvesting is carried out in as homogeneous a manner as possible.

2.1 Plant Material and Sampling

Plant sampling (harvesting) is a crucial step in sample preparation for metabolomics, and much care needs to be afforded to it (*see* Chapter 1). The total variation in the dataset is a function of different sources of variation including variation introduced by differences in sample collection. Large scale experiments with vast sample size and genotypes (e.g., ILs, RILs, or GWAS) which might slightly different in their developmental age adding yet another source of variation. However, the experimental design is key to any metabolomics experiment and having a large number of biological replicates is an essential means to minimize metabolite variation during sample preparation. In the case of introgression lines (ILs) a reasonable number of biological replicates is six independent plants the best strategy being to collect several different plant organs [3–5] per biological replicate pool them and treat them as a single sample. In the case of other population such as RILs, BILs and GWAs less replication is needed than in the ILs since in these populations genetic variance is represented in multiple lines, as opposed to a single line, within the population.

Most metabolomics studies are carried out in the laboratory under highly controlled conditions. However, most mQTL studies have been carried out for crop species such as maize, tomato, and rice have been conducted in the field. For this reason and in order to minimize the variation there are several crucial points to take into consideration during harvest.

Given that the levels of metabolites vary through the day, and that some experiments are too large to allow harvest in a single day it is essential to harvest control samples for each temporally separate harvest. Also as mentioned above plant metabolomics experiments are generally performed at the organ level (developing fruit, whole leaf, root, etc.), and it is recommended to have pooled samples per replicate to reduce the level of within genotype variation. These issues are especially important when the harvest sessions of a given experiment are numerous or when each session requests several people harvesting to limit its duration. The age, or preferably the developmental stage, of the plants or their organs needs to be defined relative to standardized growth conditions and/or phenology descriptors, by using dedicated ontology's (Plant Ontology at http://www.plantontology.org/ for phenology) or reference articles for Arabidopsis [19] or tomato [20] when available.

2.2 Sample Processing and Extraction

After harvesting, plant organs (e.g., leaves, flowers, or fruits) or dissected tissues, plant should be immediately frozen in liquid nitrogen and stored at −80 °C, or immediately ground to a powder and extracted. Sample grinding is usually required to optimize solvent extraction and additionally aids in the homogenization of the sample material [21]. It is recommended that all samples for a given experiment follow exactly the same procedure before, during, and after grinding. For further reading on extraction protocols available for plant metabolomics we suggest the work of Shimizu et al. (*see* Chapter 12) for LC-MS and the comprehensive work of Osorio and colleagues for GC-MS [22].

However, there are some important points at which these protocols should be adapted when handling the large number of plant samples required for QTL analysis. First, quality control (QC) is necessary throughout the entire sample preparation process, from the field to the sample storage location and through distribution to chemical analysts for data normalization to reduce the analytical errors. The quality control (QC) samples should qualitatively and quantitatively representative the entire collection of samples included in the study, providing an average of all of the metabolomes analyzed in the study. Sample prepared by pooling aliquots of individual study samples, either all or a subset representative for the study. The QC sample has (should have) an identical or a very similar (bio) chemical diversity as the study samples. The QC samples are evenly distributed over all the batches and are extracted, derivatized, and analyzed at the same time as the individual study samples as part of the total sequence order. The data from the QC samples is used to monitor drift, separate high- and low-quality data, equilibrate the analytical platform, correct for drift in the signal and allow the integration of multiple analytical experiments. The data analysis technique such as principal component analysis can be used to quickly assess the reproducibility of the QC samples in an analytical run. The QC samples are used to determine the variance of a metabolite feature.

Before extraction QC samples should prepared by pooling aliquots of individual study samples, the QC samples should then be distributed across all machine-batches and aliquots thereof should be extracted, derivatized, and analyzed at the same time as the individual study samples (Fig. 1).

2.3 MS-Based Metabolomics Analysis (LC-MS, GC-MS)

Once the extraction has been made, extracts must be subsequently prepared for MS-based analysis. In the case of LC-MS, once the samples are extracted aliquots of the extract can be directly introduced into the LC-MS apparatus (*see* Shimizu et al., Chapter 12). In GC-MS-based metabolomics, however, additional preparation steps are necessary to confer volatility to the metabolites via silylation and to simplify chromatography of sugars via methoxyamination [16, 17, 22]. We recommend dividing the samples in batches so that each batch contains 50–80 samples with ample QC samples distributed across the sequence run (Fig. 1). Metabolite profiling via GC-MS involves several general steps [23]. After derivatization, automated sample injection robotics and separated in GC in highly standardized conditions of gas flow, temperature programming, and standardized capillary column material. Electron impact (EI) is the most widely used ionization technique applied in GC-MS. Mass separation and detection is achieved preferably by TOF detectors that can be tuned to fast scanning rates, and finally acquisition and evaluation of GC–MS data files.

In the case of LC-MS-based metabolomics approaches, the most frequently used protocols use C18-based reversed phase columns coupled to soft ionization techniques, such as electrospray

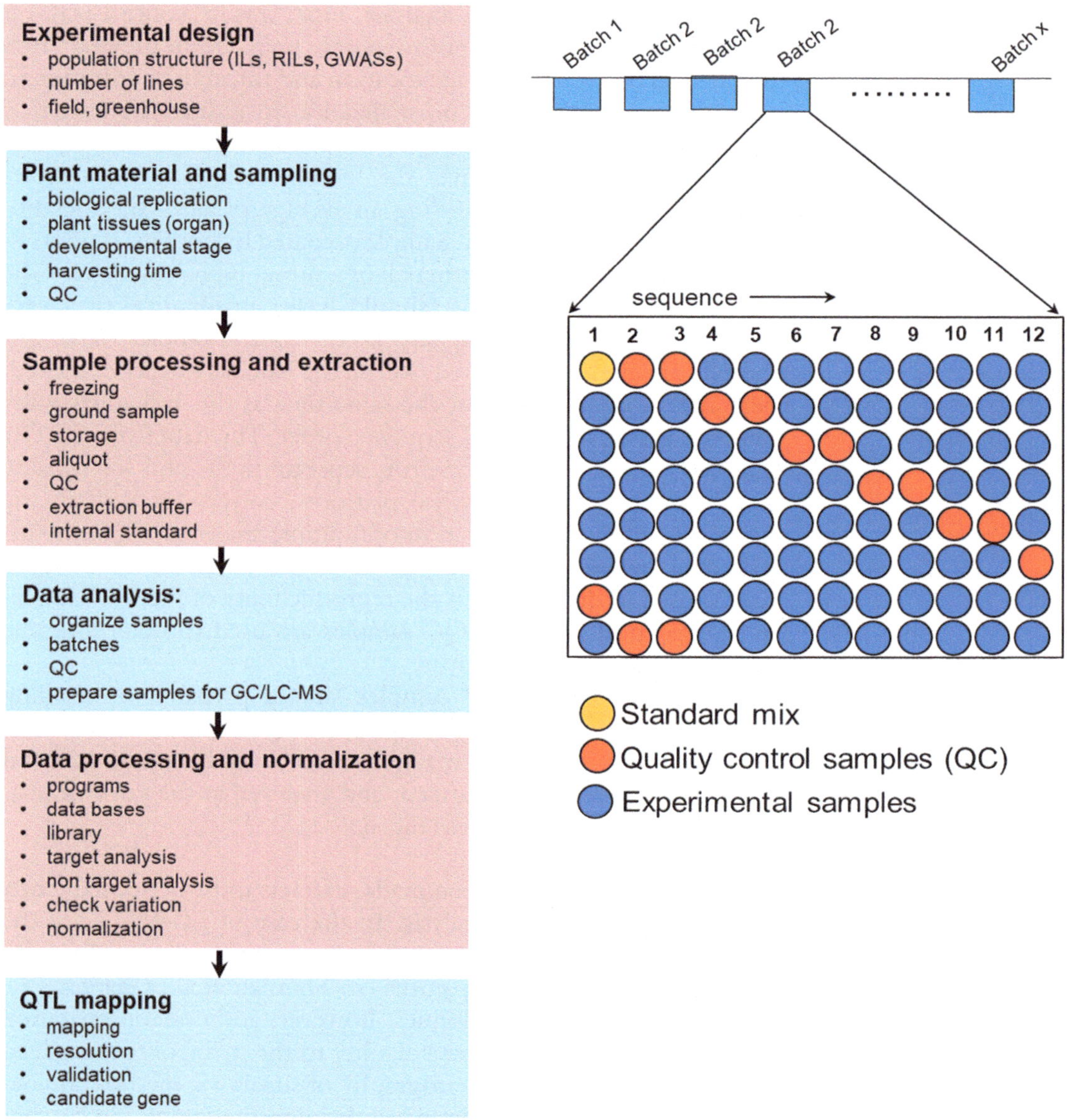

Fig. 1 Flowchart of the metabolomics study in plants. Left panel represent the different steps for experimental design, sample preparation and process for QTL experimental study. The left panel shows sample organization and suggested sequence running in GC-MS or LC-MS

ionization (ESI) or atmospheric pressure chemical ionization (APCI), resulting in protonated (in positive mode) or deprotonated (in negative mode) molecular ions. Modern high resolution instruments with exact mass detection, such as TOF-MS, ion cyclotron FT-MS, or Orbitrap FT-MS, nowadays enable the profiling of hundreds to thousands of compounds in plant extracts, combined with elemental formulae calculations of the detected masses [24, 25].

3 Data Processing

Once samples are analyzed, automatic data processing tools are required for peak picking and mass peak alignment. In GC-MS several tools, software and databases have been established and used for this purpose [23, 26, 27]. For further details on data processing from LC-MS metabolomics data see Shimizu et al. (Chapter 12). Chromatograms obtained from e.g., UPLC-FT-MS runs can be analyzed and processed with REFINER MS® 10.0 (GeneData, http://www.genedata.com), where molecular masses, retention time (RT), and associated peak intensities for each sample are extracted from the .raw files. The chemical noise is subtracted automatically. The chromatogram alignments are performed using a pairwise alignment-based tree using *m/z* windows of five points and RT windows of five scans within a sliding frame of 200 scans. Further processing of the MS data includes isotope clustering, adduct detection, and library searches. Resulting data matrices with peak ID, RT, and peak intensities in each sample are generated. However, for both LC- and GC-MS methods, manual checking of peaks is strongly recommended.

4 Data Normalization

The goal of metabolomics as a phenotyping platform depends on its ability to detect biologically related metabolite changes in complex biological samples. As with any high-throughput technology, systematic biases are often observed in LC-MS and GC-MS metabolomics data [26, 27]. As the number of samples in the dataset increases there is a corresponding time-dependent variation in the metabolite data. The variability in samples can arise from multiple sources including physiological differences and variability from the analytical method itself. Removing platform-specific sources of variability such as systematic errors is one of the top priorities in metabolomics data preprocessing. However, metabolite diversity leads to different responses to variations at given experimental conditions, making normalization a very demanding task [27]. For the effective elimination of different sources of analytical variation, preprocessing steps should follow a specific sequence.

The first step in data normalization is using an internal standard (IS); a compound added to the sample before a critical step in the analysis. An IS is not necessarily an isotope-labeled version of an analyte. However, it can be structurally related to one or more analytes, but not naturally occurring in the samples of interest. This normalization step reduces the differences in sample extraction (which can be caused by slight differences in the composition of the samples and also differences in the volumes injected). The second step is the removal of between-batch and within-batch variations and machine drifts. The final steps consist of the combination of data from replicate sample analysis and removal of noise and biomass correction. The biomass correction neutralizes differences in response due to sample weight or volume.

Here the QC samples are of key importance, and these are best prepared by pooling equal volumes of material from all of the biological samples to be analyzed. Alternatively, a chemically defined mixture of authenticated reference compounds [28] that mimics the metabolic composition of the investigated biological material can be employed. Both the synthetic mixtures and biological QC samples are then subjected to the same sample extraction, instrumental analyses (ideally distributed across the analytical run), and data processing, thus providing quality checks for technical and analytical error, and quantitative calibration to eliminate batch effects for the final processed data. This normalization is a crucial step for minimizing the batch-to-batch data variability across extended periods. As such this is a crucial requirement for large-scale phenotyping and facilitates interbatch data integration.

5 QTL Mapping

The principle of quantitative trait locus (QTL) mapping is based of detecting association of molecular genetic markers with the phenotype of interest in the resultant offspring [29]. Markers are used to partition the mapping population into different genotypic groups based on the presence or absence of a particular marker locus and to determine whether significant differences exist between groups with respect to the trait being measured [27]. If a QTL is linked to a marker locus, then the individuals with different marker locus genotypes will have different mean values of the quantitative trait. In plants, the use of such mapping populations is highly useful since the use of stable populations permits the growth of clonal replicates and, additionally, multiple analyses of genetically identical individuals across multiple harvests. There are several structural populations and methods have been used to detect the QTL and mapping. Therefore, choosing the proper population for such experiments is a key determinant in the success of any given project. There are several factors influencing the detection of QTL detections that should be considered in advance of planning such experiments.

Factors influencing QTL mapping: the genetic properties of QTL controlling traits include the magnitude of the effect of individual QTL. Only QTL with sufficiently large phenotypic effects will be detected; and the QTL with small effects may fall below the significance threshold of detection. Another genetic property is the distance between linked QTL; QTL that are closely linked will usually be detected as a single QTL in typical population sizes (<500) [30–32]. The environmental effects may have a large influence on the expression of quantitative traits. The size of the population used in the mapping study is also highly important; the larger the population, the more accurate the mapping study and the more likely it is to allow detection of QTL with smaller effects.

Owing to the factors and variations described above, QTL mapping studies should be independently confirmed or validated. Such confirmation studies (referred to as validation or replication studies) can be achieved by repeating the experiment and the QTL mapping at different sites, seasons, or years. The conserved detected QTL throughout several repeated experiment most likely the QTL that have strong genetic effect (high heritability) and that can be chosen as a region to focus on in further analysis. A second type of validation may involve independent populations constructed from the same parental genotypes or closely related genotypes used in the primary QTL mapping study. Once an association between a particular SNP and variation in a trait of interest has been established, a crucial but yet too often overlooked step is to replicate the association in an independent mapping population. As the number of studies documenting significant associations between SNPs and variation in quantitative traits of interest accumulates, increasing emphasis should be placed on replicating studies to validate effects of significant associations. In the following sections we briefly define some of the commonly used structural populations for the QTL mapping.

5.1 RIL Mapping

In plant species, the uses of immortal mapping populations consisting of homozygous individual have been used to map loci for complex traits. Recombinant inbred lines (RILs) (Fig. 2) can be obtained relatively easily and are produced by successively selfing the progeny of individual F2 plants (single seed descent method), from which the F8 generation and onward are practically homozygous lines that will produce further progeny that is essentially identical to the previous generation. Such a population can also be produced by induced chromosomal doubling of haploids, such as for doubled haploids (DHs) [33–35]. RILs are likely advantageous over DHs since they are characterized by a higher frequency of recombination within the population, resulting from multiple meiotic events occurred during repeated selfing [36]. Candidate mutations (such as a SNP, illustrated by the red dots) are then identified.

5.2 IL Mapping

Another type of immortal population consists of introgression lines (IL) (Fig. 2) which are obtained through repeated backcrossing and extensive genotyping. These can also be referred to as near isogenic lines (NILs) [37] or backcross inbred lines (BILs) [38, 39]; although the latter are slightly different in nature. These lines

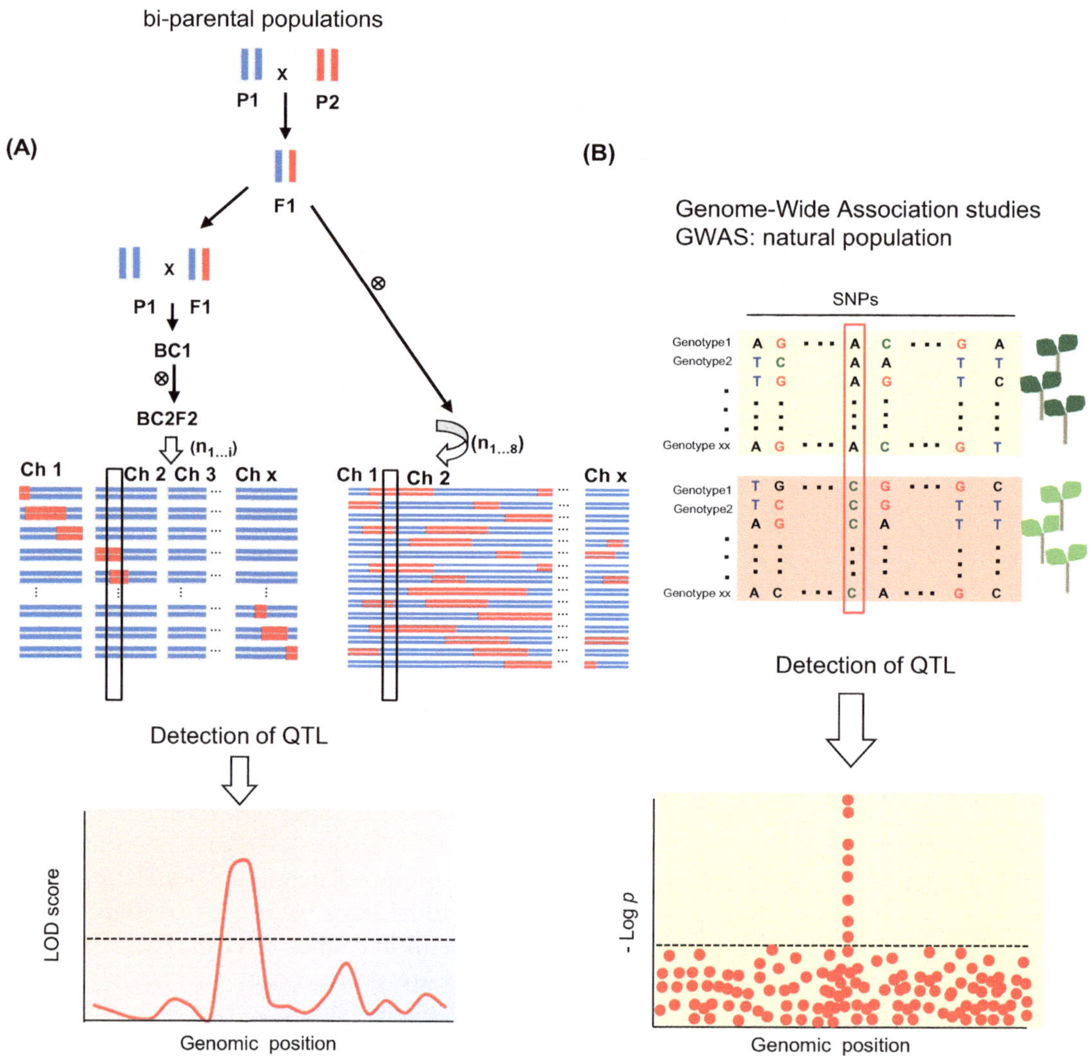

Fig. 2 Quantitative trait locus mapping. (**a**) Comparison of introgression and recombinant inbred lines; which created by backcrossing an F1 of a cross between two parental lines to a recurrent parent for several times. Recombinant inbred lines are generated by selfing an F1 for at least eight generations when full homozygosity is reached. (**b**) Genome-Wide Association Study (GWAS) for Identification of QTL using natural population; GWAS makes use of natural genotypic variation and enables the analysis of associations between hundreds of thousands of single nucleotide polymorphisms (SNPs) and specific traits

contain a single or a small number of genomic introgression fragments from a donor parent into an otherwise homogeneous genetic background.

5.3 GWAs Mapping

The IL and RILs have historically been the most common types of experimental populations used for the analysis of quantitative traits and powerful method to identify regions of the genome that cosegregate with a given trait. However, they suffer from some limitations; only allelic diversity that segregates between the parents of the particular F2 cross or within the RIL population can be assayed [27], and second, the amount of recombination that occurs during the creation of the RIL population places a limit on the mapping resolution [40]. The basic principle of genome-wide association studies (GWAS) (Fig. 2), which was initially developed for use in medical genetics, is that the incidence of nucleotide polymorphisms is associated with the presence of variance is overcome the limitations of using the IL and RILs. This approach has several major advantages over conventional QTL mapping. First, a much larger and more representative gene pool can be surveyed. Second, it bypasses the expense and time of mapping studies and enables the mapping of many traits in one set of genotypes. Third, a much finer mapping resolution can be achieved, resulting in small confidence intervals of the detected loci compared to classical mapping, where the identified loci need to be fine-mapped. Finally, it has the potential not only to identify and map QTLs but also to identify the causal polymorphism within a gene that is responsible for the difference in two alternative phenotypes [27]. A major issue with association studies is false positives, and the main sources of such false positives are linkage between causal and noncausal sites [41, 42].

QTL analysis is predicated by looking for associations between the quantitative trait and the marker alleles segregating in the population. Classical mapping approaches use segregating populations, such as recombinant inbred lines (RIL) and introgression lines (IL), to shed light on the genetic contributions to diverse phenotypes. Both are obtained from a cross (F1) of two parental accessions (P1 and P2), through repeated selfing (RIL) or backcrossing of the initial hybrid with one of its parents (IL) followed by a selfing until a homozygous state is reached. Alternate alleles at loci in homozygous lines with distinct genetic basis in such structured populations potentially influence the trait of interest and allow to map the genomic loci responsible for the observed intraspecific or interspecific variation (Fig. 2a).

GWAS makes use of natural genotypic variation and enables the analysis of associations between hundreds of thousands of single nucleotide polymorphisms (SNPs) and specific traits. For Identification of QTL; DNA obtained from hundreds of natural genetic accessions and tested for genetic variations, like SNPs. If certain SNPs are found significantly more frequently in group of genotypes (accessions) with a certain phenotype (trait) than in the general population, the mutations are said to be "associated" with the trait. The GWAS analysis, represented in a Manhattan plot with significance ($-\log_{10}$ (Pvalue) on the y-axis, and genomic position shown as chromosomes in the x-axis, is done to look for genetic variants that are associated with certain trait in a group of genotypes (leaf color for example) but not found among the other group. Significant variants (positive z-scores) are represented by the red dots. The dot that rises above background variation and is significantly associated with the phenotype is represented in the Manhattan plot at chromosomal position (Fig. 2b).

6 Conclusions

Both GC-MS and LC-MS are widely used analytical tools for profiling highly complex mixtures of primary and secondary metabolites. High-throughput use of these techniques is faced with a large number of potential sources of nonbiological variation that can compromise the interpretation of the results. However, by following several recommendations prior to and during the conductance of large-scale genomics and QTL mapping experiments such problems can be circumvented in a relatively facile manner.

Acknowledgments

This wok was in part supported by the PlantaSYST project by the European Union's Horizon 2020 Research and Innovation Programme (SGA-CSA Number 664621 and Number 739582 under FPA Number 664620).

References

1. Dixon RA, Strack D (2003) Phytochemistry meets genome analysis, and beyond. Phytochemistry 62:815–816
2. Bijlsma S, Bobeldijk I, Verheij ER et al (2006) Large-scale human metabolomics studies: a strategy for data (pre-)processing and validation. Anal Chem 78:567–574
3. Schauer N, Semel Y, Roessner U et al (2006) Comprehensive metabolic profiling and phenotyping of interspecific introgression lines for tomato improvement. Nat Biotechnol 24:447–454
4. Fiehn O (2002) Metabolomics - the link between genotypes and phenotypes. Plant Mol Biol 48:155–171
5. Fu J, Keurentjes JJ, Bouwmeester H et al (2009) System-wide molecular evidence for phenotypic buffering in Arabidopsis. Nat Genetics 41:166–167
6. Rowe HC, Hansen BG, Halkier BA et al (2008) Biochemical networks and epistasis shape the Arabidopsis thaliana metabolome. Plant Cell 20:1199–1216
7. Wentzell AM, Rowe HC, Hansen BG et al (2007) Linking metabolic QTLs with network and cis-eQTLs controlling biosynthetic pathways. PLoS Genet 3:1687–1701
8. Biais B, Bernillon S, Deborde C et al (2012) Precautions for harvest, sampling, storage, and transport of crop plant metabolomics samples. In: Hardy N, Hall R (eds) Plant Metabolomics, Methods in Molecular Biology (Methods and Protocols), vol 860. Humana Press, New York, pp 51–63
9. Gibon Y, Rolin D (2012) Aspects of experimental design for plant metabolomics experiments and guidelines for growth of plant material. Methods Mol Biol 860:13–30
10. Sysi-Aho M, Katajamaa M, Yetukuri L, Oresic M (2007) Normalization method for metabolomics data using optimal selection of multiple internal standards. BMC Bioinformatics 15:8–93
11. van der Kloet FM, Bobeldijk I, Verheij ER, Jellema RH (2009) Analytical error reduction using single point calibration for accurate and precise metabolomic phenotyping. J Proteome Res 8:5132–5141
12. van der Greef J, Martin S, Juhasz P et al (2007) The art and practice of systems biology in

medicine: Mapping patterns of relationships. J Proteome Res 6:1540–1559

13. Dunn WB, Broadhurst D, Brown M et al (2008) Metabolic profiling of serum using ultra performance liquid chromatography and the LTQ-orbitrap mass spectrometry system. J Chromatogr B Anal Technol Biomed Life Sci 871:288–298
14. Chen MJ, Rao RP, Zhang Y et al (2014) A modified data normalization method for GC-MS-based metabolomics to minimize batch variation. Spring 3:439
15. Fiehn O, Kopka J, Dörmann P et al (2001) Metabolite profiling for plant functional genomics. Nat Biotechnol 18:1157–1161
16. Lai Z, Fiehn O (2016) Mass spectral fragmentation of trimethylsilylated small molecules. Mass Spectrom Rev 9999:1–13
17. Lisec J, Schauer N, Kopka J et al (2006) Gas chromatography mass spectrometry-based metabolite profiling in plants. Nat Protocols 1:387–396
18. Joseph B, Corwin JA, Kliebenstein DJ (2015) Genetic variation in the nuclear and organellar genomes modulates stochastic variation in the metabolome, growth, and defense. PLoS Genet 11:e1004779
19. Boyes DC, Zayed AM, Ascenzi R et al (2001) Growth stage-based phenotypic analysis of arabidopsis: a model for high throughput functional genomics in plants. Plant Cell 13:1499–1510
20. Brukhin V, Hernould M, Gonzalez N et al (2003) Flower development schedule in tomato Lycopersicon esculentum cv. sweet cherry. Sex Plant Reprod 15:311–320
21. Markert B (1995) Sample preparation (cleaning, drying, homogenization) for trace element analysis in plant matrices. Science Total Environ 176:45–61
22. Osorio S, Do PT, Fernie AR (2012) Profiling primary metabolites of tomato fruit with gas chromatography-mass spectrometry. In: Hardy N, Hall R (eds) Plant Metabolomics, Methods in Molecular Biology (Methods and Protocols), vol 860. Humana Press, New York, pp 101–109
23. Kopka J, Fernie A, Weckwerth W et al (2004) Metabolite profiling in plant biology: platforms and destinations. Genome Biol 5:109
24. Allwood JW, De Vos RC, Moing A et al (2011) Plant metabolomics and its potential for systems biology research background concepts, technology, and methodology. In: Jameson D, Verma M, Westerhoff HV (eds) Methods in Enzymology, vol 500. Academic Press, Amsterdam, pp 299–33623
25. Allwood JW, Clarke A, Goodacre R, Mur LA (2010) Dual metabolomics: a novel approach to understanding plant-pathogen interactions. Phytochemistry 71:590–597
26. Karpievitch YV, Nikolic SB, Wilson R et al (2014) Metabolomics data normalization with EigenMS. PLoS One 9:e116221
27. Sehgal D, Singh R, Rajpal VR (2016) Quantitative trait loci mapping in plants: concepts and approaches. In: Rajpal V, Rao S, Raina S (eds) Molecular breeding for sustainable crop improvement. sustainable development and biodiversity, vol 11. Springer, Cham
28. Strehmel N, Hummel J, Erban A et al (2008) Retention index thresholds for compound matching in GC-MS metabolite profiling. J Chromatogr B AnalTechnol Biomed Life Sci 871:182–190
29. Broman KW (2001) Review of statistical methods for QTL mapping in experimental crosses. Lab Anim 30:44–52
30. Tanksley SD (1993) Mapping polygenes. Ann Rev Genetics 27:205–233
31. Collard BCY, Pang ECK, Taylor PWJ (2003) Selection of wild Cicer accessions for the generation of mapping populations segregating for resistance to ascochyta blight. Euphytica 130:1–9
32. Soltis NE, Kliebenstein DJ (2015) Natural variation of plant metabolism: genetic mechanisms, interpretive caveats, and evolutionary and mechanistic insights. Plant Physiol 169:1456–1468
33. Han F, Ullrich SE, Kleinhofs A et al (1997) Fine structure mapping of the barley chromosome-1 centromere region containing malting-quality QTLs. Theoretical Applied Genetics 95:903–910
34. Rae AM, Howell EC, Kearsey MJ (1999) More QTL for flowering time revealed by substitution lines in Brassica oleracea. Heredity 83:586–596
35. von Korff M, WJ LK, Pillen K (2004) Development of candidate introgression lines using an exotic barley accession (Hordeum vulgare ssp spontaneum) as donor. Theoretical Applied Genetics 109:1736–1745
36. Balding DJ, Bishop M, Cannings C, Jansen RC (2004) Quantitative Trait Loci in Inbred Lines. In: Balding DJ, Bishop M, Cannings C (eds) Handbook of Statistical Genetics, 3rd edn. John Wiley & Sons Ltd, Chichester, UK
37. Monforte AJ, Tanksley SD (2000) Development of a set of near isogenic and backcross recombinant inbred lines containing most of the Lycopersicon hirsutum genome in a L-esculentum genetic background: A tool for

gene mapping and gene discovery. Genome 43:803–813

38. Jeuken MJW, Lindhout P (2004) The development of lettuce backcross inbred lines (BILs) for exploitation of the Lactuca saligna (wild lettuce) germplasm. Theoretical Applied Genetics 109:394–401
39. Blanco A, Simeone R, Gadaleta A (2006) Detection of QTLs for grain protein content in durum wheat. Theoretical Applied Genetics 113:563–565
40. Jamann TM, Balint-Kurti PJ, Holland JB (2015) QTL mapping using high-throughput sequencing. In: Alonso J, Stepanova A (eds) Plant functional genomics, Methods in molecular biology, vol 1284. Humana Press, New York
41. Platt A, Vilhjalmsson BJ, Nordborg M (2010) Conditions under which genome-wide association studies will be positively misleading. Genetics 186:1045–1052
42. Larsson SJ, Lipka AE, Buckler ES (2013) Lessons from Dwarf8 on the strengths and weaknesses of structured association mapping. PLoS Genet 9:e1003246

Chapter 4

Gas Chromatography–Mass Spectrometry-Based ^{13}C-Labeling Studies in Plant Metabolomics

Valéria F. Lima, Leonardo Perez de Souza, Thomas C. R. Williams, Alisdair R. Fernie, and Danilo M. Daloso

Abstract

Stable-isotope labeling analysis has been used to discover new metabolic pathways and their key regulatory points in a wide range of organisms. Given the complexity of the plant metabolic network, this analysis provides information complementary to that obtained from metabolite profiling that can be used to understand how plants cope with adverse conditions, and how metabolism varies between different cells, tissues, and organs. Here we describe the experimental procedures from sample harvesting and extraction to mass spectral analysis and interpretation that allow the researcher to perform ^{13}C-labeling experiments. A wide range of plant material, from single cells to whole plants, can be used to investigate the metabolic fate of the ^{13}C from a predefined tracer. Thus, a key point of this analysis is to choose the correct biological system, the substrate and the condition to be investigated; all of which implicitly relies on the biological question to be investigated. Rapid sample quenching and a careful data analysis are also critical points in such studies. By contrast to other metabolomic approaches, stable-isotope labeling can provide information concerning the fluxes through metabolic networks, which is essential for understanding and manipulating metabolic phenotypes and therefore of pivotal importance for both systems biology and plant metabolic engineering.

Key words Fluxomics, Mass spectrometry, Metabolic flux analysis, Metabolic network, Stable isotope, Systems biology

1 Introduction

The chapters in Part II of this book provide an overview of the theoretical and technical issues that surround targeted and untargeted mass spectrometry (MS)-based plant metabolomics analysis, in for example liquid chromatography–mass spectrometry (LC-MS) and gas chromatography–mass spectrometry (GC-MS) laboratory-based protocols. Here we introduce the concept and features of stable-isotope-labeling experiments coupled to GC-MS analysis. In contrast to metabolite profiling, such analysis provides information regarding the metabolic fluxes through a metabolic network rather

Carla António (ed.), *Plant Metabolomics: Methods and Protocols*, Methods in Molecular Biology, vol. 1778,
https://doi.org/10.1007/978-1-4939-7819-9_4, © Springer Science+Business Media, LLC, part of Springer Nature 2018

than merely information concerning the content of a battery of specific metabolites under a given condition. Thus, the aim of isotope-labeling studies is to investigate the fate of a supplied isotopic label through the metabolic network in different organs, tissues, and cell types of plants (*see* **Note 1**). While metabolite profiling and metabolic flux analysis present a number of theoretical and practical differences, it is noteworthy that these are complementary approaches that are key components of both systems biology and metabolic engineering. Despite this, flux analysis has generally received less attention most probably due the comparatively more laborious experimental and data analysis required [1].

Radiolabeled substrates have a long and important history of use in isotopic-labeling studies and remain important tools in the investigation of plant metabolism. Over time, however, the use of stable isotopes has become more frequent due in part to their greater safety and ease of use, but principally as a consequence of the excellent metabolite level resolution, and accuracy in isotope enrichment measurements that can be obtained using nuclear magnetic resonance spectroscopy or mass spectrometry (MS)-based approaches (*see* **Note 2**) [2]. Stable isotopes such as ^{2}H, ^{15}N, ^{13}C, ^{34}S, and ^{18}O are commonly used separately in metabolic and physiological studies [3] but stable isotopes such as ^{15}N and ^{13}C have also be used in combination to provide important information regarding the interaction between C and N metabolism [4]. Here we focus on studies using ^{13}C-labeled substrates. This approach has been applied to investigate ^{13}C-redistribution and metabolic fluxes (*see* **Note 3**) in primary and secondary metabolism using previously isolated heterotrophic [5–10] or autotrophic [11–17] tissues but has been recently adapted to analyze metabolic fluxes [18–20] and protein synthesis and degradation [21] at the whole plant level.

In general, stable-isotope labeling experiments are carried out by combining in vivo with in silico analysis and in addition may involve higher-level mathematical modeling (*see* **Note 4**; Chapters 23 and 24) (Fig. 1).

In the case of a single-cell ^{13}C-labeling experiment, the steps performed in vivo are plant cultivation, plant cell isolation and incubation in ^{13}C-labeled substrate (Fig. 1a). The time that the plant material is exposed to ^{13}C-labeled tracer depends on the objective of the study, the type of plant material to be used and whether steady-state can, or even needs to, be assumed or not. An isotopic steady state is reached when the ^{13}C distribution is constant over time (Fig. 1b) while metabolic steady state occurs when all the fluxes in a network are invariant [22, 23]. Given that plants are organized in organs which undergo marked diurnal variations in metabolic activities, an isotopic or metabolic steady state seems unlikely to occur at the whole-plant level [23]. Thus, steady-state stable-isotope labeling has been often used for generating flux maps of compartmented networks of central metabolism in

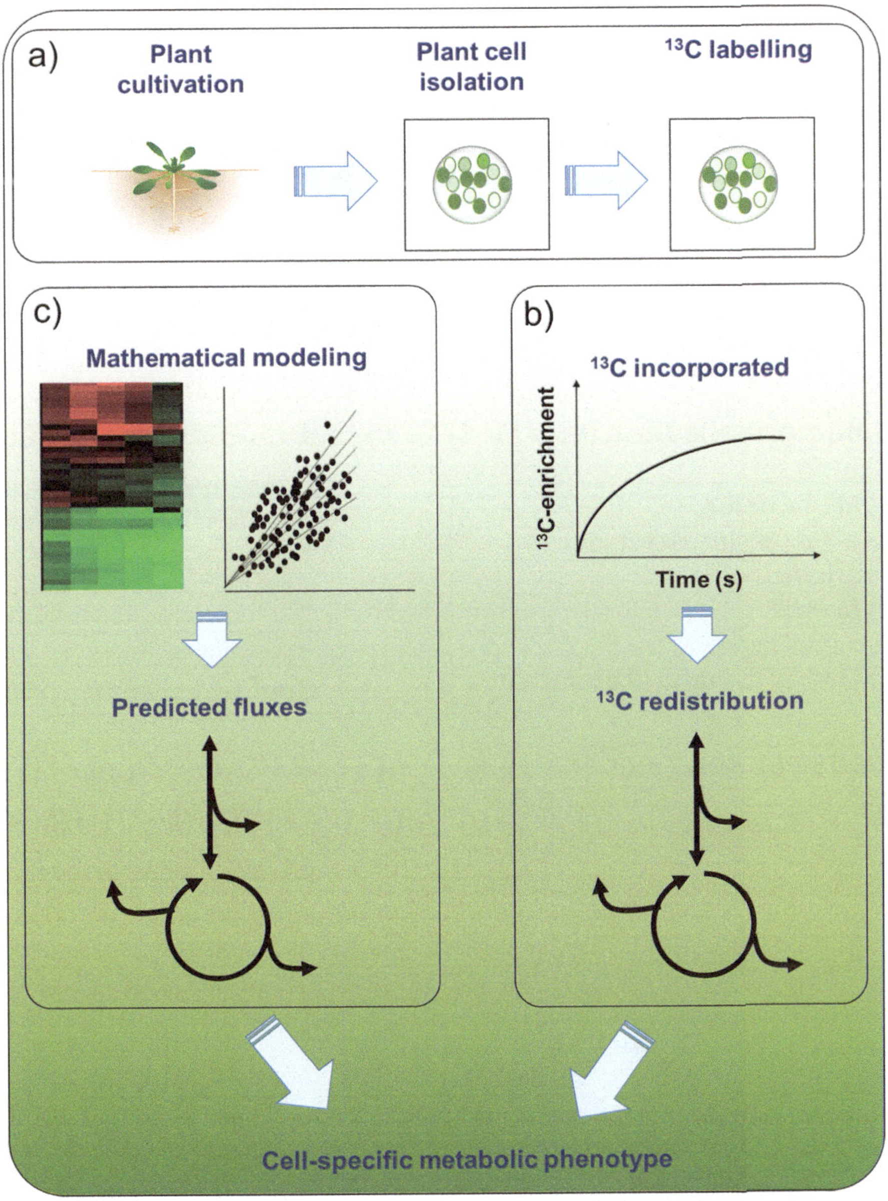

Fig. 1 Schematic representation of the procedures required for a MS-based ^{13}C-labeling study. (**a**) The experiment starts with the cultivation of the plant material to be used, which can range from intact plants to cell cultures. In case of single-cell based experiments, cell isolation has to be performed before the labeling period, which refers to the time that the plant material will be incubated with a specific ^{13}C-tracer. (**b**) After the mass spectral analysis (check Fig. 2 for details), the ^{13}C incorporated into the metabolites is obtained and a flux map of the ^{13}C-redistribution is generated. This allows a comprehensive analysis of the fate of the ^{13}C derived from the tracer. (**c**) Optionally, in silico analysis can be performed in parallel to predict the metabolic fluxes under the condition of the experiment. Several genome scale mathematical models are available that can also predict fluxes by using, for instance, previous published transcriptomic data of the same plant material under investigation. The combination of in silico with in vivo analyses aids to generate and understand the cell-specific metabolic phenotype

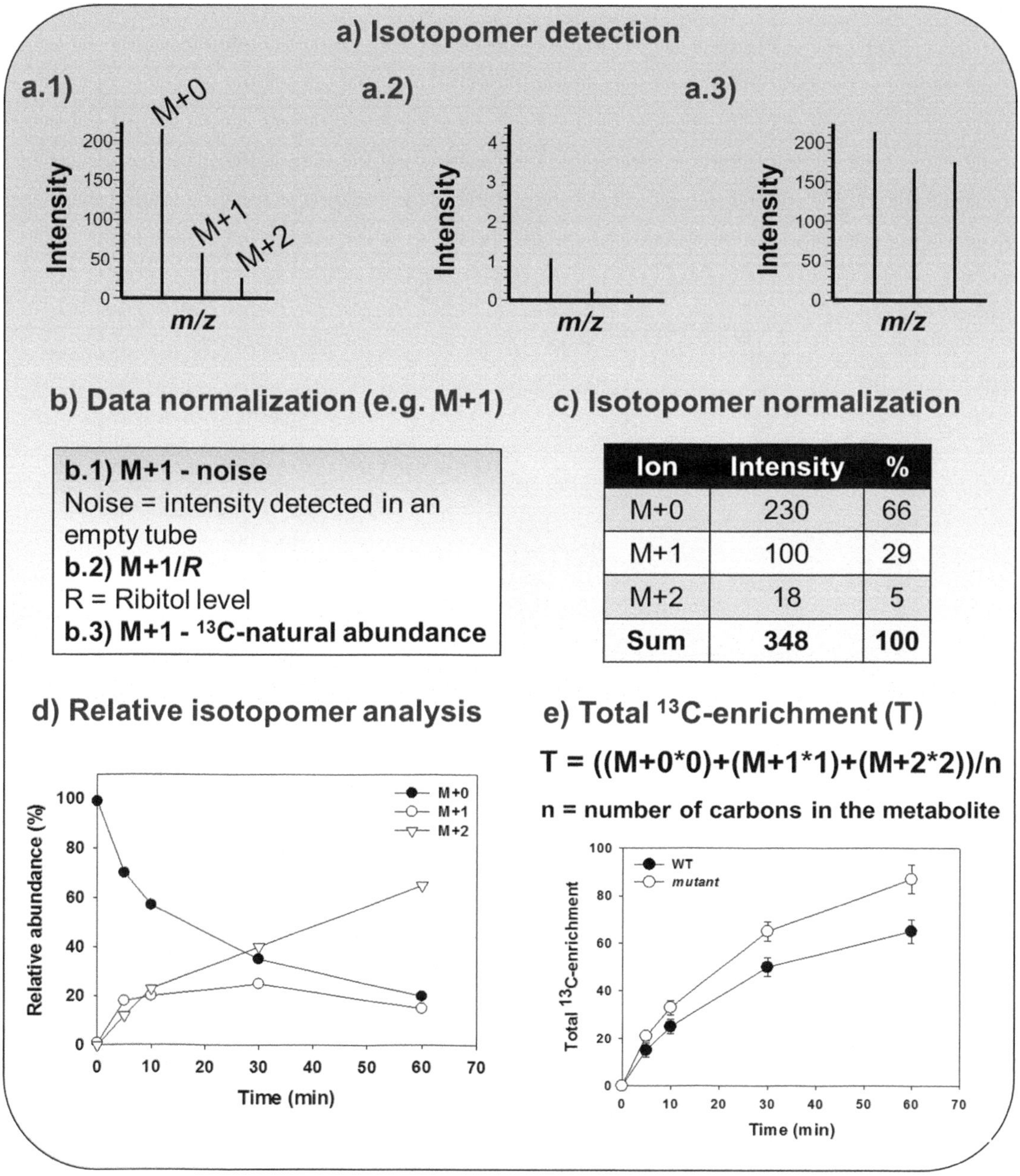

Fig. 2 Mass spectral data analysis of MS-based ^{13}C-labeling experiments. The figure shows an example of data analysis for a metabolite containing two carbon atoms. The metabolite identification and the detection of the mass isotopomers M+0, M+1, and M+2 follow the procedures used in metabolite profiling; *see* [30] for details. However, while metabolite profiling use the intensity of the mass isotopomer M+0 to determine the relative abundance of the metabolites, ^{13}C-labeling analysis involves measuring the abundance of the mass isotopomers M+1 and M+2 to determine how many ^{13}C atoms were incorporated into the metabolite (one or two) and the total amount of ^{13}C that was incorporated into this metabolite over time. (**a**) The isotopomer detection is carried out by analyzing the mass spectra. The aim is to detect all the different isotopic peaks of the fragment. The intensity of the mass isotopomers M+1 and M+2, which indicates the incorporation of one

heterotrophic plant tissues while dynamic stable-isotope-labeling approaches have been used to measure the time-course of the flow of label through the network in both heterotrophic and autotrophic tissues, and is furthermore particularly effective in situations when a short experimental time period is desirable [24–26].

After labeling the plant material, the quenching of metabolism, the extraction of metabolites and their analysis by GC-MS is identical to methods used in metabolite profiling found in the literature. However, data analysis and interpretation differ substantially. For instance, while metabolite profiling uses the intensity of the monoisotopic peak for a more quantitative comparison between different samples, stable-isotope-labeling analysis focuses on the shift in the mass-to-charge ratio (*m/z*) that occurs after the incorporation of a labeled isotope (Fig. 2a). A careful manual analysis of the mass spectra is recommended, although data can also be extracted via semiautomated tools [27–29].

Next, we describe the methods and procedures used to obtain and analyze the data from GC-MS-based ^{13}C-labeling experiments.

2 Methods

Detailed material and procedures needed to perform the extraction of polar metabolites, derivatization, and metabolite profile analysis using GC-MS are provided in Part II of this book and have also been reviewed elsewhere [30–32]. Nevertheless, in this chapter, we describe the methods needed for ^{13}C-labeling studies coupled to GC-MS analysis using the parameters described in [30].

3 Data Analysis

The data analysis is usually carried out following three basic steps: (a) mass spectra analysis for isotopomer detection and quantification; (b) data normalization and correction for the natural ^{13}C-abundance; and (c) determination of the ^{13}C-enrichment (Fig. 2).

Fig. 2 (continued) and two ^{13}C, respectively, is substantially higher in labeled (**a.3**) than nonlabeled samples (**a.1**). The abundance of these isotopomers in an empty tube sample is also recommended to assess the level of noise (**a.2**). (**b**) After detection of all isotopomers, the data is then normalized. The noise is used to normalize the values of the labeled samples (**b.1**). Then, the data is normalized by the level of ribitol (**b.2**) and the abundance of naturally occurring heavy isotopes (**b.3**). (**c**) The isotopomer normalization is a simple transformation of the intensity of all isotopomers of the fragment in percentage, in which the sum of the intensities of all isotopomers is equal to 100%. (**d**) This last normalization can then be plotted as the relative abundance of each ion (M+0, M+1, M+2) over time. In this hypothetical case, the labeling period was sufficient to fully label the metabolite, as evidenced by increases in M+2 intensity over time. (**e**) The intensity of each ion (M+0, M+1, M+2) is also used to obtain the total ^{13}C-enrichment of each metabolite, which is particularly useful when comparing two samples or two genotypes such as wild type (WT) vs mutant

The results can be presented in different manners such as the relative abundance of isotopomers (Fig. 2d) and the total ^{13}C-enrichment (Fig. 2e) of each metabolite of the network studied. With the help of systems-biology tools (Fig. 1c) flux maps can be generated and the metabolic phenotype of the plant, tissue or cell can be obtained (Fig. 1). However, even in the absence of these tools important information concerning the sequence of reactions within a biosynthetic or degradative pathway and likely points of metabolic regulation can be obtained.

The individual steps used to analyze mass spectra vary between different pieces of software and libraries (Fig. 2). Here we describe the procedures used to determine the ^{13}C-enrichment in a metabolite containing two carbon atoms as an example.

1. The first step aims to identify all isotopomers of a specific metabolite (Fig. 2a) and extract their intensities from the mass spectra for that metabolite (*see* **Note 5**). The mass isotopomers M+1 and M+2 indicate the incorporation of one and two heavy isotopes into the molecule, respectively (Fig. 2a.1). Thus, the researcher may check whether M+1 and M+2 were detected by the MS in all samples. The intensity of these ions is higher in samples harvested after a labeling period (Fig. 2a.3) compared to those from nonlabeled samples (Fig. 2a.1), in which the intensity is sometimes not detectable (*see* **Note 6**).
2. After detecting all mass isotopomers, the next step is data normalization (Fig. 2b), which can be initially carried out using a spreadsheet program. The first step of data normalization is to subtract the values of each isotopomers by the noise (Fig. 2b.1), which refers to the intensity of the respective mass isotopomer found in a blank tube used during the derivatization; this tube will be used in data analysis to discount the background noise of the GC-MS equipment (Fig. 2a.2). After that, the values should be normalized by the ribitol level (0.2 mg/mL ribitol in water) per sample; ribitol is used as an internal standard compound added to the extraction buffer (Fig. 2b.2). This normalization is performed to correct for errors from the extraction and derivatization. The last step of the normalization is to discount the effects of naturally occurring heavy isotopes of atoms found in the fragments (Fig. 2b.3), which can be carried out either by simple calculation or by using semiautomated platforms (*see* **Note 7**).
3. Following this normalization, semiautomated platforms can be used to perform the next steps (Fig. 2c–e) (*see* **Note 8**). These platforms usually requires a specific preparation of the spreadsheet file for input into the program; *see* [33] for an example. As outputs, different files related to isotopomer normalization (Fig. 2c), relative isotopomer analysis (Fig. 2d), and the

calculation of the total ^{13}C-enrichment are obtained (Fig. 2e). It is important to highlight that these steps can also be carried out manually using spreadsheet software (*see* **Note 9**).

4 Notes

1. A wide range of isolated organs, tissues, organelles, and cells have been used in ^{13}C-labeling studies to investigate the metabolic ^{13}C-redistribution under a wide range of conditions; these include plastids and mitochondria, root tips, potato tubers, seeds, embryos, leaf disks, detached leaves, cell suspensions, and guard cells. These analyses have provided important information concerning the structure and regulation of metabolic pathways, especially those of central metabolism. This approach has also recently been applied to secondary metabolism [34], albeit to a lesser extent given the complexity of the metabolites involved in these pathways as well as the lower number of commercially available labeled secondary metabolite precursors. The main advantage of working with isolated organelles/tissues/organs is that steady state can be reached faster, especially if they are heterotrophic. This is in general due the absence of assimilation of nonlabeled substrate from the atmosphere and other parts of the plant, which thus avoids dilution of the labeling. Therefore, if the aim of the work is to analyze the fluxes through a specific part of a metabolic network, for instance to compare control against treated tissues or to investigate the effect of a mutation, the use of isolated organelles/tissues/organs is recommended for such studies.
2. Isotope-labeling studies are carried out through the use of labeled substrates, also called tracers, which contain heavier or lighter isotopes of the most abundant chemical elements. A comparison highlighting the (dis)advantages of stable and radiolabeled precursors has been recently published [2]. ^{12}C is the most abundant isotope of C, corresponding to approximately 98.9% of the atmosphere, while ^{13}C represents 1.1% and the amount of ^{14}C is negligible. The labeled isotope can be inserted in different positions of the tracer or in all positions; for example different positionally labeled ^{14}C or ^{13}C glucose precursors are commercially available together with U-^{14}C or ^{13}C glucose (all carbons labeled). Given that different parts of the molecule of glucose are directed toward different pathways, such as glycolysis and the pentose phosphate pathway, the use of positional-labeled glucose allows the researcher to evaluate the fluxes through such pathways. This feature is especially useful in ^{14}C-labeling experiments, but it is also commonly exploited in ^{13}C-based experiments. The use of U-labeled

precursors on the other hand provides information on metabolic fluxes due to the mixing of labeled and unlabeled atoms within metabolites that occurs in central carbon metabolism and is extensively used in MS based metabolic flux analysis.

3. It is important to distinguish analysis of ^{13}C redistribution from complete ^{13}C-metabolic flux analysis. The principal difference is the scale of the study; whether the study allows the determination of fluxes through an entire network (or at least a compartmented subnetwork), or not. Both analyses provide information concerning which pathways are active in a given organ or tissue or are affected by a treatment or a mutation; however, only metabolic flux analysis allows for the exact determination of the fluxes through a set of biochemical reactions [31, 35–38]. In practical terms, metabolic flux analysis demands a massive computational effort, while a ^{13}C-redistribution analysis can be obtained following the more simple tasks described in Fig. 2.

4. Mathematical modeling has been coupled to stable-isotope-labeling studies, especially those carried out under steady state conditions. Therefore, another advantage of steady state isotope-labeling experiments is that diverse mathematical models such as flux balance analysis (FBA) are optimized to this condition. Interestingly, metabolic fluxes can be predicted by FBA using previous published transcriptomic data, in which the transcriptomic data is used to create genomic-scale models [39–42]. Thus, FBA, as other mathematical models, can create hypothesis, which helps the biologist to design isotope-labeling experiments to test such hypothesis. In turn, this experiment provide important information regarding the regulation of the metabolic network in vivo, which can be incorporated into the model to better predict the behavior of the metabolic fluxes. Although a wide range of models already exist for plants under steady state condition, it is important to highlight that a similar approach, called dynamic FBA, has also been used in tissues that are not under steady state condition. Furthermore, given the increased amount of metabolite profile data currently available, a new method to predict metabolic fluxes was recently proposed using these type of data [43]. The mathematical models commonly used for plant metabolism has been recently reviewed elsewhere [42, 44, 45] (*see* also Chapters 23 and 24).

5. Each metabolite detected by GC-MS is likely to yield more than one fragment with the potential to contain ^{13}C originating from the labeling experiment. Given that different fragments contain different parts of the carbon backbone of the original derivatized metabolite an analysis of a number of different fragments for the same metabolite can be worthwhile. During the extraction of isotopic data from GC-MS

chromatograms care must also be taken with mass isotopomer fractionation; mass isotopomers incorporating greater numbers of ^{13}C atoms typically elute earlier within the same chromatographic peak than those containing more ^{12}C. Thus, mass spectra should be averaged across the peak width before extraction of mass isotopomer intensities in order to avoid distortion of the data.

6. For other plant materials is may be necessary to dry different volumes and we recommend drying different volumes of the sample and checking the quality of the mass spectra data obtained. Drying different volumes is also recommended to ensure that all the mass isotopomers will be detected and potential problems with underestimation of enrichment are avoided. With quadrupole mass spectrometers single ion monitoring (SIM) methods can also prove useful when needing to detect low abundance mass isotopomers.
7. Raw mass isotopomer intensities reflect the presence of ^{13}C arising from the labeling experiment, ^{13}C at natural abundance in the metabolite of interest, and additional heavy isotopes present both within the metabolite and the groups added to the metabolite through reaction with MSTFA and methoxyamine hydrochloride during derivatization. In order to accurately calculate mass isotopomer abundances and ^{13}C enrichments the effect of heavy isotopes not arising from the labeling experiment must be discounted [46]. In this vein, several pieces of software are available that correct the mass spectral data for the distorting effect of naturally occurring stable isotopes [33, 47, 48], which can be used to facilitate the analysis. The information necessary to run these pieces of software (natural isotope abundance, chemical formula of metabolite fragments) can be found in the literature [49].
8. The time consumed using semiautomated platforms is considerably lower compared to a manual analysis of the mass spectra. However, given that the intensity of the ^{13}C-enriched mass isotopomers can be low, especially in single cell-based experiments, a careful manual analysis is recommended in order to detect all mass isotopomers of the metabolite as well as to make sure that the values obtained are not affected by errors in peak analysis introduced by the software.
9. The isotopomer normalization is made in order to obtain a relative value for the mass isotope abundances, in which the sum of the intensity of all mass isotopes of a given fragment is equal to 100% ($M+0+M+1+M+2 = 100$) (Fig. 2c) [32]. This is an important step in the analysis of the relative abundance of each isotopomer during the experiment (Fig. 2d). The relative abundance shifts from M+0 toward heavier isotopes (M+1, M

+2, etc.) over time, which is due to the ^{13}C incorporation. For example, if this metabolite has been fully labeled over time, the expectation would be to observe substantial increases in the M +2 in parallel of decreases in the intensity of the M+0, while M + 1 initially presents a small increase followed by a decrease over time (Fig. 2d). This information is crucial to determine the total ^{13}C enrichment and thereby to infer the fluxes through this metabolite. The total ^{13}C enrichment refers to the total ^{13}C that was incorporated into the metabolite. This value is obtained by summing the multiplication of the intensity of the ions by the number of ^{13}C atoms incorporated in each isotopomer (M+0*0+M+1*1+M+2*2) divided by the number of C of the molecule (Fig. 2e); check [32] for further details on ^{13}C-enrichment calculation. The total ^{13}C-enrichment is a quantitative parameter useful in the analysis of the effects of a treatment (control vs treated comparison) or a mutation (wild type vs mutant comparison) (Fig. 2e) on metabolic fluxes and ^{13}C-redistribution.

Acknowledgments

Scholarships granted by the Coordination for the Improvement of Higher Level Personnel (CAPES) to V.F.L. and the National Council for Scientific and Technological Development (CNPq-Brazil) to L.P.S. are gratefully acknowledged. The Foundation for Support of Scientific and Technological Development of Ceará (FUNCAP) grant number AEP-0128-00092.01.00/17 to D.M.D. is gratefully acknowledged.

References

1. Fernie AR, Geigenberger P, Stitt M (2005) Flux an important, but neglected, component of functional genomics. Curr Opin Plant Biol 8:174–182
2. Batista Silva W, Daloso DM, Fernie AR et al (2016) Can stable isotope mass spectrometry replace radiolabelled approaches in metabolic studies? Plant Sci 249:59–69
3. Salon C, Avice J-C, Colombié S et al (2017) Fluxomics links cellular functional analyses to whole-plant phenotyping. J Exp Bot 68:2083–2098
4. António C, Päpke C, Rocha M et al (2016) Regulation of primary metabolism in response to low oxygen availability as revealed by carbon and nitrogen isotope redistribution. Plant Physiol 170:43–56
5. Centeno DC, Osorio S, Nunes-Nesi A et al (2011) Malate plays a crucial role in starch metabolism, ripening, and soluble solid content of tomato fruit and affects postharvest softening. Plant Cell 23:162–184
6. Araújo WL, Nunes-Nesi A, Trenkamp S et al (2008) Inhibition of 2-oxoglutarate dehydrogenase in potato tuber suggests the enzyme is limiting for respiration and confirms its importance in nitrogen assimilation. Plant Physiol 148:1782–1796
7. Schwender J, Shachar-Hill Y, Ohlrogge JB (2006) Mitochondrial metabolism in developing embryos of Brassica napus. J Biol Chem 281:34040–34047
8. Sriram G, Fulton DB, Shanks JV (2007) Flux quantification in central carbon metabolism of Catharanthus roseus hairy roots by 13C

labeling and comprehensive bondomer balancing. Phytochemistry 68:2243–2257

9. Masakapalli SK, Bryant FM, Kruger NJ, Ratcliffe RG (2014) The metabolic flux phenotype of heterotrophic Arabidopsis cells reveals a flexible balance between the cytosolic and plastidic contributions to carbohydrate oxidation in response to phosphate limitation. Plant J 78:964–977
10. Alonso AP, Val DL, Shachar-Hill Y (2011) Central metabolic fluxes in the endosperm of developing maize seeds and their implications for metabolic engineering. Metab Eng 13:96–107
11. Daloso DM, Antunes WC, Pinheiro DP et al (2015) Tobacco guard cells fix CO2 by both Rubisco and PEPcase while sucrose acts as a substrate during light-induced stomatal opening. Plant Cell Environ 38:2353–2371
12. Daloso DM, Williams TCR, Antunes WC et al (2016) Guard cell-specific upregulation of sucrose synthase 3 reveals that the role of sucrose in stomatal function is primarily energetic. New Phytol 209:1470–1483
13. Daloso DM, Müller K, Obata T et al (2015) Thioredoxin, a master regulator of the tricarboxylic acid cycle in plant mitochondria. Proc Natl Acad Sci U S A 112:E1392–E1400
14. Tcherkez G, Mahe A, Gauthier P et al (2009) In folio respiratory fluxomics revealed by 13C isotopic labeling and H/D isotope effects highlight the noncyclic nature of the tricarboxylic acid "cycle" in illuminated leaves. Plant Physiol 151:620–630
15. Nargund S, Misra A, Zhang X et al (2014) Flux and reflux: metabolite reflux in plant suspension cells and its implications for isotope-assisted metabolic flux analysis. Mol BioSyst 10:1496–1508
16. Lonien J, Schwender J (2009) Analysis of metabolic flux phenotypes for two Arabidopsis mutants with severe impairment in seed storage lipid synthesis. Plant Physiol 151:1617–1634
17. Allen DK, Ohlrogge JB, Shachar-Hill Y (2009) The role of light in soybean seed filling metabolism. Plant J 58:220–234
18. Ma F, Jazmin LJ, Young JD, Allen DK (2014) Isotopically nonstationary 13C flux analysis of changes in Arabidopsis thaliana leaf metabolism due to high light acclimation. Proc Natl Acad Sci U S A 111:16967–16972
19. Szecowka M, Heise R, Tohge T et al (2013) Metabolic fluxes in an illuminated Arabidopsis rosette. Plant Cell 25:694–714
20. Arrivault S, Obata T, Szecówka M et al (2017) Metabolite pools and carbon flow during C4 photosynthesis in maize: 13CO 2 labeling kinetics and cell type fractionation. J Exp Bot 68:283–298
21. Ishihara H, Obata T, Sulpice R et al (2015) Quantifying protein synthesis and degradation in arabidopsis by dynamic 13CO2 labeling and analysis of enrichment in individual amino acids in their free pools and in protein. Plant Physiol 168:74–293
22. Antoniewicz MR (2015) Methods and advances in metabolic flux analysis: a mini-review. J Ind Microbiol Biotechnol 42:317–325
23. Schwender J (2009) Isotopic steady-state flux analysis. In: Plant Metab. Networks, pp 1–331
24. Kruger NJ, Le Lay P, Ratcliffe RG (2007) Vacuolar compartmentation complicates the steady-state analysis of glucose metabolism and forces reappraisal of sucrose cycling in plants. Phytochemistry 68:2189–2196
25. Sweetlove LJLJ, Fernie AR (2005) Regulation of metabolic networks:understanding metabolic complexity in the systems biology era. New Phytol 168:9–23
26. Ratcliffe RG, Shachar-Hill Y (2006) Measuring multiple fluxes through plant metabolic networks. Plant J 45:490–511
27. Luedemann A, Strassburg K, Erban A, Kopka J (2008) TagFinder for the quantitative analysis of gas chromatography-mass spectrometry (GC-MS)-based metabolite profiling experiments. Bioinformatics 24:732–737
28. Weindl D, Wegner A, Hiller K (2016) MIA: Non-targeted mass isotopolome analysis. Bioinformatics 32:2875–2876
29. Huang X, Chen Y-J, Cho K et al (2014) X13CMS: Global tracking of isotopic labels in untargeted metabolomics. Anal Chem 86:1632–1639
30. Lisec J, Schauer N, Kopka J et al (2006) Gas chromatography mass spectrometry–based metabolite profiling in plants. Nat Protoc 1:387–396
31. Zamboni N, Fendt S-M, Rühl M, Sauer U (2009) 13C-based metabolic flux analysis. Nat Protoc 4:878–892
32. Heise R, Arrivault S, Szecowka M et al (2014) Flux profiling of photosynthetic carbon metabolism in intact plants. Nat Protoc 9:1803–1824
33. Huege J, Goetze J, Dethloff F et al (2014) Quantification of stable isotope label in metabolites via mass spectrometry. In: Hicks GR, Robert S (eds) Plant Chem. Genomics Methods Protoc, Methods Mol. Biol, vol 1056. Humana Press, Totowa, NJ, pp 11–17

34. Fernie AR, Morgan JA (2013) Analysis of metabolic flux using dynamic labelling and metabolic modelling. Plant Cell Environ 36:1738–1750
35. Shachar-hill Y, Morgan JA, Grady JO (2012) Metabolic cartography: experimental quantification of metabolic fluxes from isotopic labelling studies. J Exp Bot 63:2293–2308
36. Kruger NJ, Ratcliffe RG (2009) Insights into plant metabolic networks from steady-state metabolic flux analysis. Biochimie 91:697–702
37. Schwender J, Ohlrogge J, Shachar-Hill Y (2004) Understanding flux in plant metabolic networks. Curr Opin Plant Biol 7:309–317
38. Shachar-Hill Y (2013) Metabolic network flux analysis for engineering plant systems. Curr Opin Biotechnol 24:247–255
39. Williams TCR, Miguet L, Masakapalli SK et al (2008) Metabolic network fluxes in heterotrophic Arabidopsis cells: stability of the flux distribution under different oxygenation conditions. Plant Physiol 148:704–718
40. Williams TCR, Poolman MG, Howden AJM et al (2010) A genome-scale metabolic model accurately predicts fluxes in central carbon metabolism under stress conditions. Plant Physiol 154:311–323
41. Cheung CYM, Poolman MG, Fell DA et al (2014) A diel flux balance model captures interactions between light and dark metabolism during day-night cycles in C3 and Crassulacean acid metabolism leaves. Plant Physiol 165:917–929
42. Nikoloski Z, Perez-Storey R, Sweetlove LJ (2015) Inference and prediction of metabolic network fluxes. Plant Physiol 169:1443–1455
43. Sajitz-Hermstein M, Töpfer N, Kleessen S et al (2016) IReMet-flux: Constraint-based approach for integrating relative metabolite levels into a stoichiometric metabolic models. Bioinformatics 32:i755–i762
44. Shi H, Schwender J (2016) Mathematical models of plant metabolism. Curr Opin Biotechnol 37:143–152
45. Nargund S, Sriram G (2014) Mathematical modeling of isotope labeling experiments for metabolic flux analysis, Plant Metab Methods Protoc Methods Mol Biol, pp 85–108
46. Van Winden WA, Wittmann C, Heinzle E, Heijnen JJ (2002) Correcting mass isotopomer distributions for naturally occurring isotopes. Biotechnol Bioeng 80:477–479
47. Jungreuthmayer C, Neubauer S, Mairinger T et al (2015) ICT: Isotope correction toolbox. Bioinformatics 32:154–156
48. Millard P, Letisse F, Sokol S, Portais JC (2012) IsoCor: Correcting MS data in isotope labeling experiments. Bioinformatics 28:1294–1296
49. Huege J, Sulpice R, Gibon Y et al (2007) GC-EI-TOF-MS analysis of in vivo carbon-partitioning into soluble metabolite pools of higher plants by monitoring isotope dilution after 13CO2 labelling. Phytochemistry 68:2258–2272

Chapter 5

Carbon Atomic Survey for Identification of Selected Metabolic Fluxes

Leonardo Perez de Souza, Alisdair R. Fernie, and Takayuki Tohge

Abstract

Stable-isotope-labeling experiments have become a frequently used tool to investigate different metabolic systems. They have been recently applied to several comprehensive studies in plant metabolomics providing interesting insights into metabolic dynamics and regulation. However, the complexity of mass spectrometry data originating from such experiments is rarely fully explored. Data analysis often considers metabolites in their entirety which obscures important information at the atomic level. Recently, the use of mass spectrometry fragmentation for obtaining positional-labeling information was described for a few specific metabolites. Here, we describe a general methodology that can be applied for extracting positional-labeling information based on the characterization of fragments that are inherent in gas chromatography electron impact mass spectrometry (GC-EI-MS) chromatograms.

Key words Stable-isotope labeling, Metabolic flux, Mass spectrometry, Isotopomer

1 Introduction

Isotopic-labeling techniques have long been an essential part of experiments for determining metabolic fluxes and pathway topology, from pioneering works defining primary metabolism [1] to more recent metabolomics-based analysis [2, 3]. Several recent reviews discuss in detail the different aspects of these experiments [4–6], which in general consist on feeding substrates containing heavier isotopes and measuring their incorporation into the samples. In this context, the use of stable isotopes, such as ^{13}C and ^{15}N, has become a prominent strategy by which to investigate the dynamics of metabolic networks. This is to a great extent due to its versatility (C and/or N are present in the vast majority of metabolites), and the possibility of using well-established metabolomics platforms for data acquisition. In these experiments, information is extracted by techniques such as mass spectrometry (MS) and nuclear magnetic resonance (NMR) and used for further modeling steps and flux calculations. Mass spectrometry is usually

Carla António (ed.), *Plant Metabolomics: Methods and Protocols*, Methods in Molecular Biology, vol. 1778, https://doi.org/10.1007/978-1-4939-7819-9_5,

preferred for metabolomics analysis due to its higher sensitivity and easy coupling with chromatographic techniques, allowing for the investigations of multiple metabolites in different concentration ranges at once. However, NMR has the significant advantage of providing fine details on positional enrichment, which can significantly improve further modeling steps. Hence, a few recent works dedicated extensive efforts in developing MS-based methods to obtain positional-enrichment data, most of them based on LC-MS/MS [7–10]. A particularly interesting work by Alves et al. [10] provides a practical example of how the atomic resolution obtained by analyzing different isotopomers of a metabolite can significantly improve the information extracted from labeling experiments, and compare two methods based on MS and NMR. In this work, the authors used different isotopomers of citrate and glutamate to obtain position-specific enrichments of mitochondrial acetyl-CoA, oxaloacetate, and citrate, being able to show the differences in fluxes through pyruvate carboxylase (PC) and pyruvate dehydrogenase (PDH) in the TCA cycle. Interestingly, data from harsher GC-EI-MS ionization, which yields highly fragmented spectra from which similar information could be extracted in a facile manner, is seldom analyzed in such way. This could be explained by the fact that most works trying to achieve MS-based positional-labeling information focus on very specific questions, and the relatively few metabolites for which tandem MS approaches offer higher selectivity. In this case, a direct link between parental and daughter ions, and a broader option of fragments that has greater chances to cover the whole molecule are used for modeling at the molecular level. Nevertheless, GC-EI-MS is a widespread and high-throughput platform frequently used to analyze stable-isotope-labeling experiments, and whilst this information is often neglected it is readily available. These two different approaches rely in fairly similar principles, whilst tandem MS uses fragmentation to investigate the enrichment over different portions of a parental ion, GC-EI-MS necessarily generates multiple fragments that often incorporate different parts of the original molecule. Independently of the MS system, the crucial information for obtaining reliable stable-isotope-enrichment data is the chemical identity of the ions analyzed. This information allows to correct mass isotopomer distributions (MID) for the natural abundance of heavier isotopes [11, 12] as well as to calculate the amount of incorporated label [6]. Additionally, multiple fragments containing different carbons carry the positional information described earlier, and proper characterization of these fragments is an essential step toward obtaining the atomic flux to individual parts of a metabolite. Data on mass spectra fragment structure is scarce and largely based on old fragmentation studies of a limited group of metabolites, including mainly amino acids, a few sugars, and organic acids [13, 14]. Even more scarce are thorough evaluations of MID measurement accuracy [15], which ultimately determines the quality of the data. It is important to note that these can be strongly affected by many factors, including different fragmentation mechanisms leading to the same mass-to-charge ratio (*m/z*), low-intensity signals, and partial overlapping with different fragments. Obtaining this information is a time-consuming procedure and requires several different labeled standards, and is therefore, rarely performed. Instead, most analysts rely solely on the available literature and, when there is lack of available data, simply infer the composition of fragments from fully labeled data. Here we describe how to characterize mass fragments obtained from GC-EI-MS and validate their use for stable-isotope-enrichment analysis, with the goal of improving the information that can be extracted from such experiments by the inclusion of positional-labeling information.

2 Materials

Unlabeled serine and ^{13}C serine labeled at carbon 1 are used as an example in this chapter, but the same procedure can be applied to any targeted metabolite, i.e., any unlabeled standards and ^{13}C labeled standards (full or point ^{13}C labeled). These standards are essential for characterizing mass spectra fragments and validating their use for calculating label enrichment.

3 Methods

3.1 Data Acquisition and Analysis

Sample preparation, data acquisition and data analysis of stable-isotope enrichment experiments coupled to GC-MS analysis have been previously described for different experimental designs [11, 16] as well as in the previous chapter of this book (*see* Chapter 4); hence, they will be only briefly summarized here.

1. Perform metabolite extraction, derivatization, and GC-EI-MS analysis according to the protocol described by Lisec et al. [17] (*see* **Note 1**).
2. Use the preferred software for extracting peak intensities (*see* **Note 2**).
3. Process the data as described by Heise et al. [6] and in Chapter 4 of this book.

3.2 Formula Assignment of Fragment Peaks

Knowledge about the structure of detected fragments is essential for the accurate determination of labeling enrichment and for selection of fragments carrying relevant information for flux predictions (*see* **Note 3**). Proper fragment assignment should provide the molecular formula of the number of labeled atoms and the positions of this label within the original molecular structure of

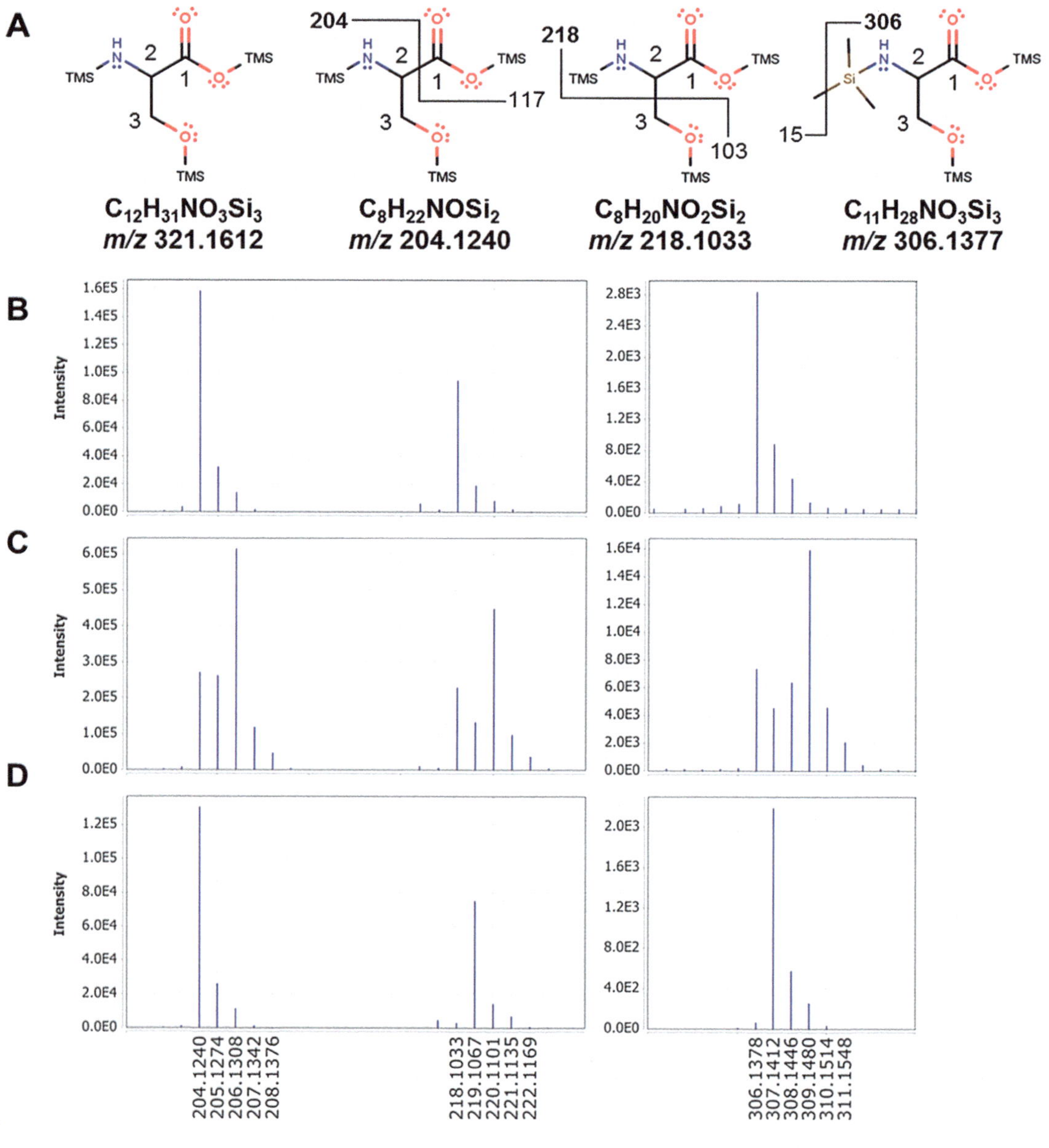

Fig. 1 (**a**) Characterized fragments of Serine (3TMS). (**b**) GC-EI-MS mass spectra of unlabeled Serine (3TMS) standard. (**c**) GC-EI-MS mass spectra of Serine (3TMS) from Arabidopsis rosette submitted to $^{13}CO_2$ feeding for 60 min. (**d**) GC-EI-MS mass spectra of Serine (3TMS) standard, labeled with ^{13}C at carbon 1

the metabolite (Fig. 1a). This information can be obtained from the analysis of mass spectra-features and the point ^{13}C–labeled standards.

1. Review the literature for previous information on the fragment composition of common analytes (*see* **Note 4**).

2. Use unlabeled samples to determine the structural formula of the fragments based on the accurate mass and natural isotope composition pattern (*see* **Note 5**).
3. Compare mass spectra of full labeled and unlabeled analytes to obtain the total possible number of labeled positions in the fragment which is equivalent to the mass shift observed between the fully labeled and unlabeled fragments (*see* **Note 6**) (Fig. 1b, c).
4. Analyze positionally labeled standards to confirm the identity of labeled positions in the metabolites (*see* **Note 7**) (Fig. 1d).

3.3 Evaluation of Fragment Accuracy for Enrichment Analysis

Previous work by Antoniewicz et al. [15] discussed in detail the importance in assessing potential errors in determining MID of tert-butyldimethylsilyl-derivatized amino acids to improve flux estimates. Factors that can affect accuracy in MID measurements include analytical error, alternative routes of fragmentation yielding the same *m/z* and the superimposition of different fragments. To validate the use of specific fragments, linearity and recovery of relevant labeled standards, under different concentrations, should be evaluated.

1. Prepare stock solutions of labeled and unlabeled standards (*see* **Note 8**).
2. Prepare working solutions by mixing labeled and unlabeled stock solutions at different proportions and concentrations (*see* **Note 9**).
3. Prepare and analyze samples and data as discussed in Subheading 3.1.
4. Calculate error, recovery and linear range for the total enrichment obtained for each fragment (Fig. 2).
5. Once the use of fragments containing different information is validated, they can be incorporated into further mathematical models as different fluxes. Additionally, multiple fragments can be used to estimate fluxes that otherwise cannot be individually measured.

4 Notes

1. In the original protocol for metabolite extraction, a 150 μL aliquot of the polar phase is used for further derivatization and GC-MS analysis. In this step, it is recommended to aliquot different volumes in case it is necessary to analyze low level fragments that may provide useful information.
2. GC-MS analysis for stable-isotope enrichment differs from the more commonly used metabolite profiling platforms in that not only the monoisotopic peak but all isotopic peaks of a

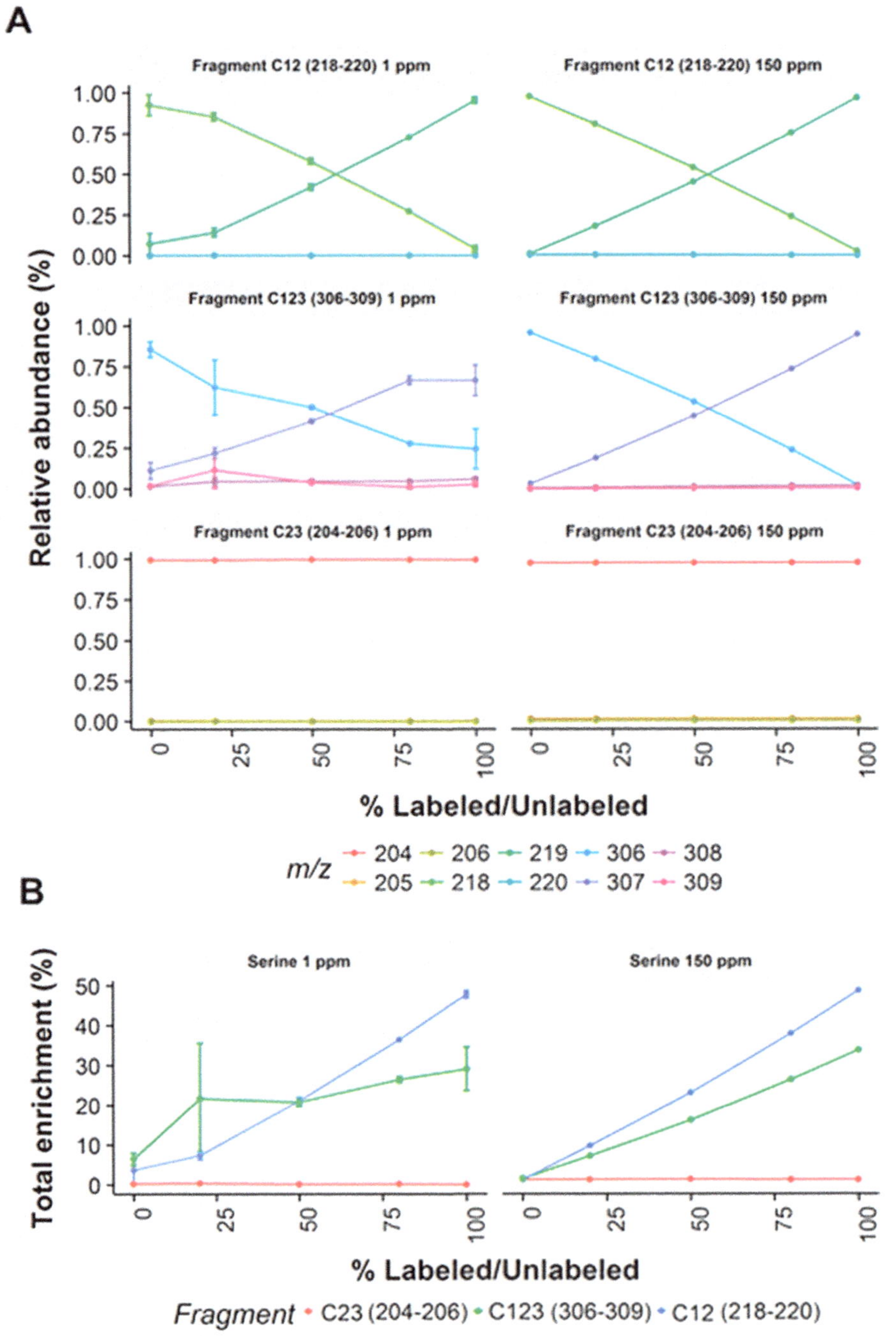

Fig. 2 Evaluation of error, recovery and linearity for ^{13}C enrichment of samples containing different proportions of Serine (3TMS), unlabeled and labeled at carbon 1, on two different concentrations. (**a**) Relative abundance of all isotopomers for each characterized fragment of Serine. (**b**) Total enrichment for fragments of Serine (3TMS)

specific fragment must be measured for further processing. More details on how to process this data are discussed by Heise et al. [6] and in Chapter 4 of this book.

3. This information is particularly important for proper correction of natural isotopic variation of metabolite derivatives with a significant portion of the structure not labeled and a high content of atoms with higher proportion of heavier isotopes [11] which is exactly the case of trimethylsilyl (TMS) derivatives analyzed by GC-EI-MS.
4. There are few comprehensive works describing the characterization of mass spectral fragments from GC-EI-MS, nevertheless several central metabolites including amino acids and sugars were extensively covered in the past [13–15]. It is important to highlight that analytes detected by GC-EI-MS are usually the derivatized form of the metabolites; therefore, the mass spectra and its features are also dependent on the derivatization process.
5. One of the great benefits of accurate mass measurements is its ability to significantly reduce the space of possible molecular formulas for a certain mass. It is usually assumed that mass accuracies with error below 5 ppm, achievable by most modern time-of-flight (TOF) equipment, can provide the elemental composition of many accurate masses. However, in certain high-mass regions, several molecular formulas are obtained even at higher mass accuracies bellow 1 ppm [18]. By imposing extra constraints such as limiting the atoms based on the original metabolite and derivatization protocol, and including evaluation of the natural mass isotopomer distribution can significantly improve these predictions [19].
6. This information can be obtained by directly analyzing unlabeled and fully labeled standards of specific compounds, or alternatively, by profiling unlabeled and fully labeled samples. The second approach has the significant advantage of providing information on all detected metabolites at once. However, it is not always feasible to achieve fully labeled samples. In the specific case of plants and other photosynthetic organisms, fully labeled samples can be obtained by long term $^{13}CO_2$ feeding. Additionally, partially labeled samples can often provide the necessary information, but this should be evaluated with extra care and considering each analyte individually. For this to be possible, the mass spectra obtained from analyte must contain at least one fragment with all possible labeling positions for which the fully labeled mass isotopomer should be the most intense (Fig. 1c). In this situation, it is reasonable to assume all other fragments will have the fully labeled mass isotopomer as the most intense within its MID.

7. By using all the information obtained in previous steps, together with common EI fragmentation mechanisms, one can propose putative structures for unknown fragments which can significantly reduce time and resources spent on analyzing point-labeled standards. Ultimately, the analysis of point-labeled standards will show the shift from M to M+1 in the MID of fragments containing the labeled position as presented in Fig. 1d.
8. Addition of ribitol as internal standard in the metabolite extraction solution can be used for further normalization (*see* Chapter 4).
9. It is important to make sure that all isotopomers of a fragment are properly quantified once they all contribute for the total enrichment. As an example, the quantification of isotopomer with *m/z* 307 at 1 ppm (Fig. 2a) is fairly good in the range between 0 and 75. The values obtained for the total enrichment (Fig. 2b); however, are strongly affected by the error in the quantification of the other isotopomers.

Acknowledgments

We thank the Max Planck Society and the National Council for Scientific and Technological Development CNPq-Brazil (LPS).

References

1. Bassham JA, Benson AA, Calvin M (1950) The path of carbon in photosynthesis: VIII. The role of malic acid J Biol Chem 185:781–787
2. Ma F, Jazmin LJ, Young JD, Allen DK (2014) Isotopically nonstationary 13C flux analysis of changes in Arabidopsis thaliana leaf metabolism due to high light acclimation. Proc Natl Acad Sci U S A 111:16967–16972
3. Szecowka M, Heise R, Tohge T et al (2013) Metabolic fluxes in an illuminated Arabidopsis rosette. Plant Cell 25:694–714
4. Antoniewicz MR (2015) Methods and advances in metabolic flux analysis: a mini-review. J Ind Microbiol Biotechnol 42:317–325
5. Batista Silva W, Daloso DM, Fernie AR et al (2016) Can stable isotope mass spectrometry replace radiolabelled approaches in metabolic studies? Plant Sci 249:59–69
6. Heise R, Arrivault S, Szecowka M et al (2014) Flux profiling of photosynthetic carbon metabolism in intact plants. Nat Protocols 9:1803–1824
7. Antoniewicz MR (2013) Tandem mass spectrometry for measuring stable-isotope labeling. Curr Opin Biotechnol 24:48–53
8. Choi J, Grossbach MT, Antoniewicz MR (2012) Measuring complete isotopomer distribution of aspartate using gas chromatography-tandem mass spectrometry. Anal Chem 84:4628–4632
9. Jeffrey FMH, Roach JS, Storey CJ et al (2002) 13C isotopomer analysis of glutamate by tandem mass spectrometry. Anal Biochem 300:192–205
10. Alves Tiago C, Pongratz Rebecca L, Zhao X et al (2015) Integrated, step-wise, mass-isotopomeric flux analysis of the TCA cycle. Cell Metabol 22:936–947
11. Huege J, Goetze J, Dethloff F et al (2014) Quantification of stable isotope label in metabolites via mass spectrometry. In: Hicks GR, Robert S (eds) Plant chemical genomics: methods and protocols. Humana Press, Totowa, NJ, pp 213–223
12. Van Winden WA, Wittmann C, Heinzle E, Heijnen JJ (2002) Correcting mass isotopomer

distributions for naturally occurring isotopes. Biotechnol Bioeng 80:477–479
13. Leimer KR, Rice RH, Gehrke CW (1977) Complete mass spectra of the per-trimethylsilylated amino acids. J Chromatogr A 141:355–375
14. DeJongh DC, Radford T, Hribar JD et al (1969) Analysis of trimethylsilyl derivatives of carbohydrates by gas chromatography and mass spectrometry. J Am Chem Soc 91:1728–1740
15. Antoniewicz MR, Kelleher JK, Stephanopoulos G (2007) Accurate assessment of amino acid mass isotopomer distributions for metabolic flux analysis. Anal Chem 79:7554–7559
16. Araújo WL, Tohge T, Nunes-Nesi A et al (2014) Analysis of kinetic labeling of amino Acids and organic acids by GC-MS. In: Dieuaide-Noubhani M, Alonso AP (eds) Plant metabolic flux analysis: methods and protocols. Humana Press, Totowa, NJ, pp 107–119
17. Lisec J, Schauer N, Kopka J et al (2006) Gas chromatography mass spectrometry-based metabolite profiling in plants. Nat Protocols 1:387–396
18. Kind T, Fiehn O (2006) Metabolomic database annotations via query of elemental compositions: Mass accuracy is insufficient even at less than 1 ppm. BMC Bioinformatics 7:234–234
19. Kind T, Fiehn O (2007) Seven golden rules for heuristic filtering of molecular formulas obtained by accurate mass spectrometry. BMC Bioinformatics 8:105

Part II

Mass Spectrometry-Based Methodology Chapters

Chapter 6

Quantification of Low-Abundant Phosphorylated Carbohydrates Using HILIC-QqQ-MS/MS

Tiago F. Jorge and Carla António

Abstract

Phosphorylated carbohydrates are central metabolites involved in key plant metabolic pathways, such as glycolysis and central carbon metabolism. Such pathways influence plant growth, development, and stress responses to environmental changes, and ultimately, reflect the plant's energy status. The high polarity of these metabolites, the variety of isomeric structures (e.g., glucose-1-phosphate (G1P)/fructose-6-phosphate (F6P)/mannose-6-phosphate (M6P)/G6P, sucrose-6-phosphate (S6P)/T6P), and rapid metabolic turnover makes their analysis particularly challenging. In this chapter, we describe the use of a set of known phosphorylated carbohydrates to develop and validate a hydrophilic interaction chromatography (HILIC) triple quadrupole (QqQ) tandem mass spectrometry (MS/MS) method in the highly sensitive and selective multiple reaction monitoring (MRM) mode for the target analysis of G1P, F6P, M6P, G6P, S6P, T6P, and the sugar nucleotide uridine 5-diphospho-glucose (UDPG). We present detailed information regarding HILIC column chemistry and practical considerations when coupling it with a QqQ-MS system.

Key words Phosphorylated carbohydrates, Trehalose-6-phosphate, HILIC, Triple quadrupole, Method development, Metabolite quantification

1 Introduction

Phosphorylated carbohydrates are key plant metabolites involved in different metabolic pathways, namely glycolytic and the central carbon metabolic pathways, the pentose phosphate cycle and starch and sucrose synthesis. Consequently, the plant's energy status, stress perception, and response to environmental changes largely depend on the availability of these key plant metabolites [1–3].

Trehalose-6-phosphate (T6P) is one such important phosphorylated disaccharide known to inhibit the sucrose nonfermenting-1-related protein kinase 1 (SnRK1), a key regulator of energy and carbon metabolism in plants [4]. In plants, T6P synthesis occurs via the trehalose biosynthetic pathway through a reaction between glucose-6-phosphate (G6P) and uridine 5-diphospho-glucose (UDPG) catalyzed by trehalose-6-phosphate synthase (TPS) [5, 6].

Carla António (ed.), *Plant Metabolomics: Methods and Protocols*, Methods in Molecular Biology, vol. 1778,
https://doi.org/10.1007/978-1-4939-7819-9_6, © Springer Science+Business Media, LLC, part of Springer Nature 2018

Over the last years, T6P has been reported to act as a signal of sucrose availability and a negative feedback regulator of sucrose levels in plants, as well as to promote carbon flux into organic and amino acids [7–10]. However, the quantitative analysis of T6P and other phosphorylated carbohydrates is difficult mainly due to the complexity of the plant matrix and the almost undetectable levels of these metabolites in plant tissues [11–13]. Therefore, advances in the development of high-throughput and sensitive methods able to measure such low metabolite levels is crucial for a better understanding of the molecular mechanisms by which T6P and its metabolic intermediates regulate sugar signaling and biological processes in plants [2, 12, 14–18].

Hydrophilic interaction chromatography (HILIC) is a suitable approach to separate highly polar analytes which performance is complementary to that of reversed-phase liquid chromatography (RP-LC), and similar to that of normal-phase liquid chromatography (NP-LC). Typical eluents for HILIC contain a high content of water-miscible organic solvent (typically acetonitrile in water or volatile buffer) to promote hydrophilic interactions between the analyte and a water-enriched hydrophilic stationary phase (Fig. 1). Therefore, HILIC is also well-suited for on-line coupling with electrospray mass spectrometry (ESI-MS) [19–24]. Under HILIC conditions, the retention of analytes is mainly influenced by (a) hydrophilic partitioning of analytes between the water-enriched and the organic-enriched layers and (b) weak electrostatic interactions between the charged HILIC stationary phase and the charged groups of the analytes (Fig. 1) [22, 23].

Triple quadrupole (QqQ)-MS methods are desirable in quantitative targeted approaches because this mass analyzer allows highly sensitive and selective multiple reaction monitoring (MRM) experiments that ultimately provide reliable absolute quantification of low-abundant metabolites [25].

In this chapter, we describe a detailed protocol for the development and validation of a HILIC-QqQ-MS method for the sensitive selective detection and quantification of highly polar phosphorylated carbohydrates, namely G1P, F6P, M6P, G6P, T6P, S6P, and UDPG from plant tissues. This protocol has been used to detect and quantify the fluctuations of S6P, T6P, and G6P in *Medicago truncatula* plants in response to environmental stress [18], and it is broadly applicable to different plant tissues and can be transferred to other QqQ-based LC-MS systems with minor modifications.

2 Materials

2.1 Plant Material and Sample Preparation

1. Leaves, roots, or other tissue from any plant species.
2. Screw-cap polypropylene Falcon tubes (15 or 50 mL depending on the amount of material) for storage of frozen plant tissues.

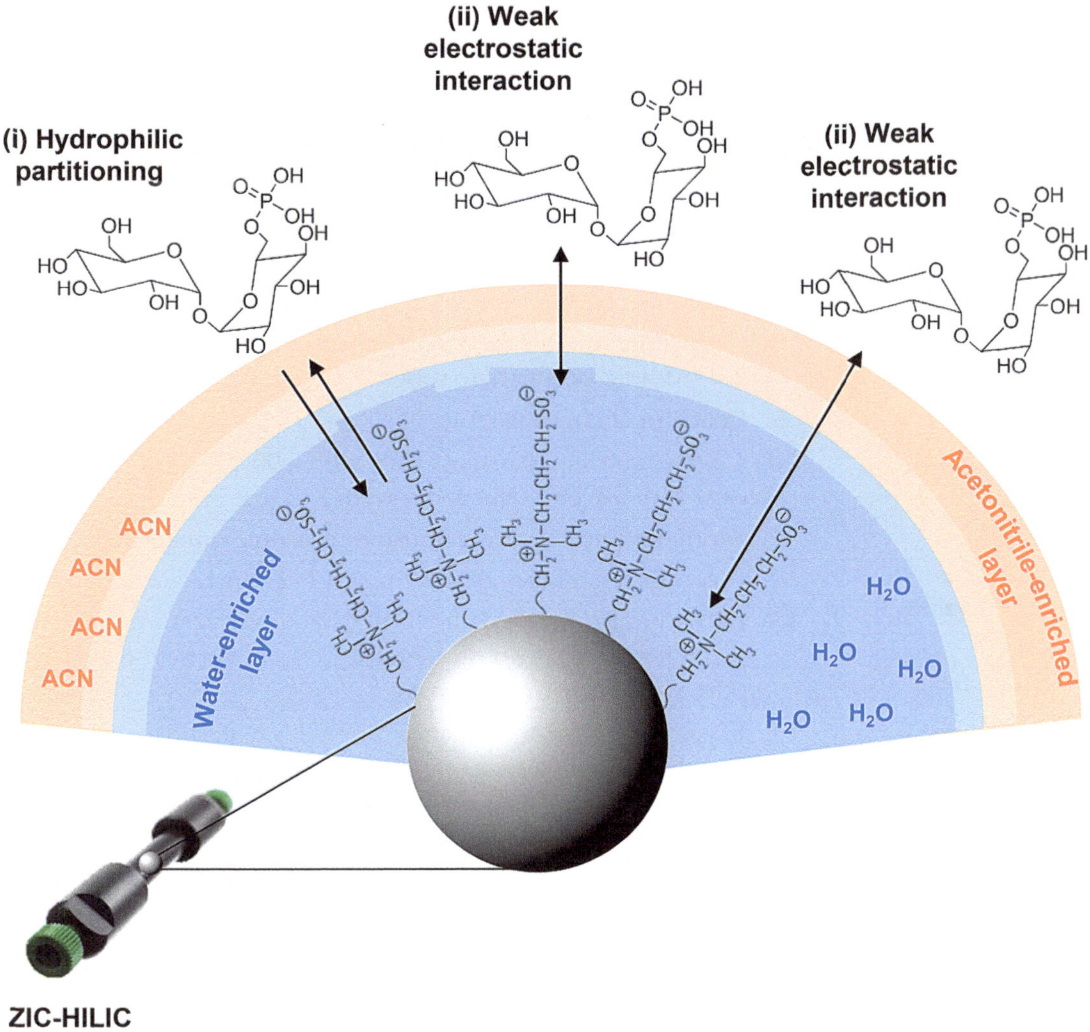

Fig. 1 Graphic representation of the retention mechanism in ZIC-HILIC. Polar analytes are retained through a combination of (1) hydrophilic partitioning between the water-enriched layer (blue) and the acetonitrile-enriched layer (orange) and (2) weak electrostatic interactions with the ionized functional groups (either positive or negative charges) of the stationary phase

3. 2.0-mL safelock microcentrifuge tubes (e.g., Eppendorf) for sample weighing and metabolite extraction.
4. Schott GL14 glass tubes for metabolite extraction.
5. Liquid nitrogen for sample harvest, quenching, grinding and weighing of plant samples (*see* **Note 1**).
6. Protective gloves for handling frozen objects.
7. Metal spatula or small spoon.
8. Ceramic pestle and mortar, or grinder with ball mill.

2.2 Equipment, Instrumentation, and Software

1. −80 °C freezer to store plant tissues.
2. Fridge-freezer.
3. Pipettes and tips suitable for handling organic solvents.
4. Balance for accurate weighing of standard chemical powders and plant tissues.
5. Vortex.
6. Thermoshaker with a 2.0-mL thermoblock.
7. Ultrasonic bath.
8. Centrifuge suitable for 15 mL tubes and centrifuge suitable for 2.0-mL microcentrifuge tubes.
9. Centrifugal concentrator.
10. 1.5-mL autosampler glass vials with 100 μL glass inserts and caps with a PTFE/silicon septum.
11. Syringe pump for direct infusion measurements.
12. Gas-tight Hamilton glass syringe 0.1–1.0 mL.
13. LC column: ZIC-HILIC HPLC column, 2.1 mm internal diameter, PEEK 150 mm length, 3.5 μm particle size, 100 Å pore size (Merck-Millipore).
14. LC system: Waters Alliance 2695 high performance liquid chromatography system or equivalent, equipped with an automatic sample injector, a sample cooler, an online solvent degasser and a column oven.
15. MS system: Triple quadrupole mass spectrometer (QqQ) Micromass® Quattro (Waters®) equipped with an electrospray ionization (ESI) or equivalent.
16. MS data acquisition software: MassLynx software (version 4.1, Waters®) or equivalent.

2.3 Solutions

Prepare all solutions in a fume hood following standard laboratory safety procedures as most of the organic solvents are very toxic, flammable, volatile, and/or corrosive.

1. Individual standard stock solutions: Prepare 1 mg/mL stock solution for each target compound (G1P, F6P, M6P, G6P, T6P, S6P, and UDPG) in HPLC-grade water, vortex until homogenization is completed, and store at −20 °C until further use (*see* **Note 2**).
2. Individual standard working solutions: Prepare 1 mM individual standard working solution for each target compound (G1P, F6P, M6P, G6P, T6P, S6P, and UDPG) in HPLC-grade water by dilution of the correspondent individual standard stock solution (1 mg/mL), vortex, and store at −20 °C before and after use.

3. HPLC mobile phase: 0.5 mM methylphosphonic acid in Milli-Q water containing 0.001% piperidine (v/v) (eluent A), and acetonitrile (eluent B) (*see* **Note 3**). Prepare eluents freshly in glass bottles and sonicate them in an ultrasonic bath for at least 15 min before use.
4. Cleaning and blank solution: acetonitrile/HPLC-grade water (90/10, v/v) (*see* **Note 4**).
5. Electrospray (ESI) solution: acetonitrile/HPLC-grade water (50/50, v/v).
6. Individual direct infusion standard solutions: Prepare a 50 μM standard working solution for each target compound (G1P, F6P, M6P, G6P, T6P, S6P, UDPG) in ESI solution by dilution of the correspondent individual standard working solution (1 mM), vortex, and store at −20 °C before and after use.
7. Method validation standard solution mixtures: Prepare a 50 and a 0.5 μM standard solution mixture containing all target compounds (G1P, F6P, M6P, G6P, T6P, S6P, and UDPG) in ESI solution by dilution of each individual standard working solution (1 mM), vortex and store at −20 °C before and after use (*see* **Note 5**).
8. Standard calibration curve mixtures: From the 1 mM individual standard working solutions, prepare a dilution series of standard calibration curve mixtures containing all target compounds (G1P, F6P, M6P, G6P, T6P, S6P, and UDPG) in ESI solution, vortex, and store at −20 °C before and after use. Prepare at least five calibration concentration points (ranging from 0.01 to 50 μM) (*see* **Note 6**).
9. Quality control (QC) mixture: Prepare a 50 μM QC mixture containing all target compounds (G1P, F6P, M6P, G6P, T6P, S6P, and UDPG) in ESI solution by dilution of the correspondent individual standard working solution (1 mM), vortex, and store at −20 °C before and after use (*see* **Note 7**).
10. Metabolite extraction solution: chloroform/acetonitrile (3/7, v/v), and store at 4 °C (*see* **Note 8**).

3 Methods

3.1 Plant Growth and Sampling Conditions

Plant material must grow under specific and controlled conditions to reduce biological variation between samples. In addition, the plant material must be representative of the total plant population under study, and a minimum of six biological replicates is required to normalize biological variation. Therefore, several samples are usually generated in a single metabolomics study, and it is recommended to prepare in advance an adequate sample identification file and prelabel all sample tubes. Plant harvest followed by quenching (i.e., rapidly freezing the plant tissue in liquid nitrogen) should be performed, if possible, in the field or in the greenhouse to avoid changes in metabolite levels. Metabolite extraction should be performed in finely homogenized fresh-frozen plant tissue to increase the extraction efficiency. During these steps, it is important to guarantee that the plant material does not thaw (*see* Chapter 1) [26–28].

3.2 Harvest and Tissue Sampling

1. Prelabel all sample screw-cap polypropylene Falcon tubes (15 or 50 mL depending on the amount of material) with freezer-compatible labels or pen markers.
2. Harvest a representative amount of the plant tissue of interest into a screw-cap polypropylene Falcon tube, and rapidly freeze it in liquid nitrogen (quenching).
3. Use a precooled pestle and mortar filled with liquid nitrogen (or a precooled ball mill) to finely homogenize the frozen plant tissue. While doing this, make sure that the plant material does not thaw by sequentially adding liquid nitrogen.
4. Once a fine homogenized powder is obtained, transfer it back to the precooled labeled screw-cap polypropylene Falcon tube with the help of a precooled metal spatula or small spoon.
5. Use precooled 2.0-mL safelock microcentrifuge tubes to weight aliquots of 50 mg of fresh weight (FW) powdered frozen tissue (*see* **Note 9**).
6. Prepare at least six biological replicates and make sure that they follow the same sample procedure to reduce biological variation.
7. Store all labeled and weighed 2.0-mL safelock sample tubes in a box (or plastic bag) at −80 °C until metabolite extraction and further LC-MS analysis.

3.3 Metabolite Extraction

Highly polar metabolites such as phosphorylated carbohydrates are extracted following a well-established protocol that uses 50 mg (FW) of plant tissue [18].

1. Add 500 μL of ice-cold metabolite extraction solution to each 2.0-mL safelock microcentrifuge tube containing 50 mg (FW) of frozen plant tissue. Vortex and keep tubes on ice (enzymatic activity stops and proteins precipitate in this step).
2. Incubate each sample tube for 2 h, at −20 °C, and vortex occasionally.
3. Add 400 μL of ice-cold HPLC-grade water, and place the sample tube in a thermoshaker during 15 min at 4 °C and 950 rpm. Attention! Turn on the thermoshaker in advance to cool it down (*see* **Note 10**).
4. Centrifuge for 10 min at 11,000 × *g* and at 4 °C.

5. Transfer the upper phase (aqueous-acetonitrile) to a new 2.0-mL safelock microcentrifuge tube and keep it on ice.
6. Reextract the chloroform phase with 400 μL of ice-cold HPLC-grade water and centrifuge as described above, in **steps 3** and **4**, respectively.
7. Add the second aqueous-acetonitrile phase to the first one.
8. Centrifuge again as described in **step 4**.
9. Transfer the combined upper phase (aqueous-acetonitrile) to a new 2.0-mL safelock microcentrifuge tube and keep it on ice.
10. Evaporate the combined aqueous-acetonitrile phase to dryness using a centrifugal concentrator set at 20 °C for a maximum time of 6 h (*see* **Note 11**).
11. Fill tubes with inert gas (Argon) to prevent the dry extract from oxidation and degradation reactions through components of atmospheric air.

3.4 HILIC-QqQ-MS/MS Method Development

The HILIC-QqQ-MS method here described is divided into two parts. The first part describes ESI-QqQ-MS direct infusion measurements of individual standard working solutions while the second part guides you through the development and validation of the HILIC-QqQ-MS method (*see* **Note 12**).

3.4.1 ESI-QqQ-MS Direct Infusion Measurements

1. Verify the MS signal and the presence of interfering background ions making use of the syringe pump to directly infuse ESI solution into the MS system at a flow rate of 20 μL/min, capillary voltage 3 V and capillary temperature 120 °C.
2. If the MS signal is low or a high presence of interfering background ions is detected, clean the MS system by continuously infusing ESI solution until the MS signal is improved.
3. Use the syringe pump to directly infuse each direct infusion individual standard working solution (50 μM) into the MS system at a flow rate of 20 μL/min, capillary voltage 3 V and capillary temperature 120 °C. Search for the mass-to-charge ratio (*m/z*) peak of interest for each target compound (G1P, F6P, M6P, G6P, T6P, S6P, and UDPG) in the negative ion mode. Clean the MS system between each measurement by directly infusing ESI solution.
4. When acquiring full MS and MS/MS data, set the mass range to *m/z* 60 to 1000, and the collision energy values for each target compound (G1P, F6P, M6P, G6P, T6P, S6P, and UDPG) as listed in Table 2. These values might need to be adjusted if a different QqQ-MS instrument is used.
5. Create one MS method file with the MRM transition for each individual target compound as listed in Table 2.
6. Clean the MS system at the end of each measurement by directly infusing ESI solution.

Table 1
ZIC-HILIC gradient method for the target analysis of phosphorylated carbohydrates. HPLC conditions: ZIC-HILIC column (3.5 μm, PEEK 150 × 2.1 mm, 100 Å); flow rate 300 μL/min; injection volume to 10 μL; mobile phase eluent (A) 0.5 mM methylphosphonic acid in Milli-Q water containing 0.001% piperidine (v/v) and eluent (B) acetonitrile

Time/min	A (%)	B (%)	Gradient stage
0	10	90	Initial conditions
19	90	10	Elution
20	10	90	Washing
30	10	90	Equilibration

3.4.2 HILIC-QqQ-MS Method Development and Validation

1. Switch on the HPLC pump and degas both eluents A and B for at least 5 min, or until there are no air bubbles within the HPLC lines.
2. Wash needle and flush HPLC syringe with cleaning solution, at least twice. Check if there are no air bubbles.
3. Plug the ZIC-HILIC column carefully into the HPLC system. Place the outlet from the column directly to a waste bottle and check for leaks.
4. Precondition the ZIC-HILIC column (*see* **Note 13**) by slowly increasing the percentage of the aqueous eluent A until the initial gradient conditions are reached (Table 1). For a good ZIC-HILIC column equilibration, we recommend keeping the initial gradient conditions (Table 1) for at least 1 h (i.e., running to waste at a flow rate of 200 μL/min).
5. Program the HPLC method according to the linear gradient described in Table 1.
6. Set the flow rate to 300 μL/min, and the sample injection volume to 10 μL.
7. After preconditioning the ZIC-HILIC column, connect the HPLC system to the ESI-QqQ-MS and check if there are no leaks in the system. It is recommended to regularly register the pressure values (*see* **Note 14**).
8. Connect the HPLC system to the ESI-QqQ-MS system. Check if there are no leaks.
9. Prepare at least 500 mL of each eluent A and eluent B (HPLC mobile phase).
10. Set a LC-MS sequence and run the blank solution, the method validation standard solution mixture (50 μM), and the QC mixture (50 μM). Analyze each solution in triplicate.
11. Create one MS method file with the MRM transitions for all target compounds of the standard mixture as listed in Table 2.

Table 2
MRM transitions of target phosphorylated carbohydrates and optimized analytical parameters for LC-MS/MS analysis. HPLC conditions: ZIC-HILIC column (3.5 μm, PEEK 150 × 2.1 mm, 100 Å); flow rate 300 μL/min; injection volume to 10 μL; mobile phase eluent (A) 0.5 mM methylphosphonic acid in Milli-Q water containing 0.001% piperidine (v/v) and eluent (B) acetonitrile

Standard compound	t_R (min)	Diagnostic ion (*m/z*)	Product ions (*m/z*)	Source potential (V)	MRM1	MRM2	Collision energy (eV)
S6P	8.36	421 $[M\text{-}H]^-$	97, 79	40	421 > 79	421 > 97	25
T6P	8.87	421 $[M\text{-}H]^-$	241, 97,79	40	421 > 79	421 > 241	25
G1P + F6P	8.97	259 $[M\text{-}H]^-$	97, 79	40	259 > 97	259 > 169	10
M6P	9.01	259 $[M\text{-}H]^-$	97, 79	40	259 > 97	259 > 169	10
G6P	9.11	259 $[M\text{-}H]^-$	97, 79	40	259 > 97	259 > 169	10
UDPG	9.27	565 $[M\text{-}H]^-$	323, 97, 79	50	565 > 323	565 > 79	20

12. For the method validation standard solution mixture (50 μM), evaluate for each target compound the chromatographic data (retention time, peak shape, peak area, and resolution), the MS data (ion current and peak intensity), and the LC-MS system repeatability (retention time and peak area) (Fig. 2).
13. Afterward, proceed with the HILIC-QqQ-MS method validation steps. Start by injecting the calibration curve mixtures containing all target compounds (G1P, F6P, M6P, G6P, T6P, S6P, and UDPG). The standard calibration curves are later used for the quantification of the target compounds in plant extracts.
14. Evaluate the intraday precision of the method by injecting six method validation standard solution mixtures (0.5 μM) within a single day. The intraday precision is calculated as the relative standard deviation (RSD) of peak areas for each target compound. Attention! According the literature, researchers have evaluated intraday precision in several different ways (*see* Chapter 2).
15. Evaluate the intermediate precision (interday precision) of the method by injecting six times the same method validation standard solution mixture (0.5 μM) in three different days. The intermediate precision is calculated as the relative standard deviation (RSD) of peak areas for each target compound.
16. Evaluate the instrument precision by injecting consecutively six times the method validation standard solution mixture (0.5 μM). The instrument precision is calculated as the relative standard deviation (RSD) of peak areas for each target compound.

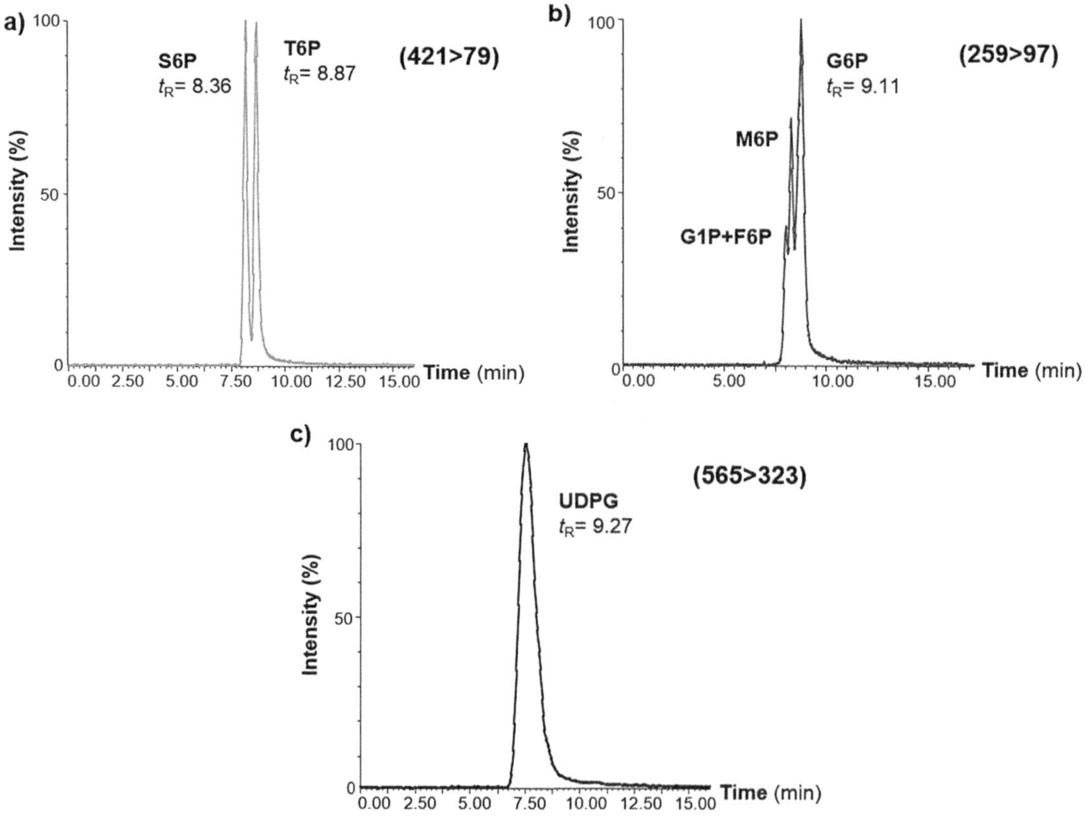

Fig. 2 Multiple reaction monitoring (MRM) chromatograms obtained for the HILIC-QqQ-MS/MS separation of a 50 μM method validation standard solution mixture containing all target compounds (G1P, F6P, M6P, G6P, T6P, S6P, UDPG): (**a**) S6P/T6P (421 > 79); (**b**) G1P/F6P/M6P/G6P (259 > 97); (**c**) UDPG (565 > 323). HPLC conditions: ZIC-HILIC column (3.5 μm, PEEK 150 × 2.1 mm, 100 Å); flow rate 300 μL/min; injection volume to 10 μL; mobile phase eluent (A) 0.5 mM methylphosphonic acid in Milli-Q water containing 0.001% piperidine (v/v) and eluent (B) acetonitrile

17. Determine the linearity range, limit of detection (LOD), and limit of quantification (LOQ) by making use of the five calibration curve mixtures (ranging from 0.01 to 50 μM containing all target compounds (G1P, F6P, M6P, G6P, T6P, S6P, and UDPG). LOD values for each target compound are estimated as the minimum concentration injected which gives a detector response higher than three times the signal-to-noise ratio (*S/N*). LOQ values are determined for each target compound at an *S/N* of 10. Each target compound LOQ value should be then confirmed by analyzing six times the correspondent individual standard working solution at the estimated LOQ concentration.
18. Determine the method accuracy, by adding known amounts of each target compound at three different spiking levels (1.0, 10, and 50 μM) to approximately 50 mg FW of plant tissue before sample extraction and HILIC-ESI-QqQ-MS/MS

analysis (*see* **Note 15**). The analytical recovery is then expressed as the percentage (%) of the known amount of target compound spiked into the plant matrix that remained after going through the extraction procedure.

19. Quantitatively assess the matrix effects (ME), by preparing four final concentration points of 0.5, 1.0, 10, and 50 μM of each target compound in 100 μL of extracted plant tissue. ME are evaluated by comparing the signal response of a target compound in the absence of matrix (individual standard working solution) to the signal response of the target compound spiked at the same concentration in the matrix after extraction (plant extract).
20. Reconstitute your biological samples in 100 μL acetonitrile--water (50:50, v/v) and inject 10 μL for HILIC-ESI-QqQ-MS/MS analysis in the MRM mode (*see* **Note 16**).

3.5 Data Analysis

Herein we present a brief procedure to analyze LC-MS raw data using the MassLynx software (version 4.1, Waters®), followed by the common use of Microsoft Excel® spreadsheet functions. Alternatively, the user can process the LC-MS data using the QuanLynx tool available in the MassLynx software (version 4.1, Waters®).

1. Open MassLynx to visualize LC-MS data (both individual target compounds and QC mixtures).
2. Click in Chromatogram to check the peak shape and ion intensity.
3. Click in Display followed by Mass and check for the presence of the peak for each transition (Fig. 3).
4. Check retention times and peak areas between analysis in different days or weeks.
5. Open the LC-MS raw file of the calibration curve mixtures and register the peak area values of the MRM1 transition for each target compound (Table 2) to construct the plot of peak area vs concentration, i.e., calibration curve, as well as the *S/N* values for LOD and LOQ evaluation for each target compound.
6. Once this is done, evaluate the linearity of each calibration curve by visual inspection and then, using the Mandel's Fitting test (*see* **Note 17**).
7. To evaluate intraday and interday precision for each compound, calculate the relative standard deviation (RSD, %) of retention times and peak areas.
8. To determine method accuracy, open the correspondent LC-MS raw files, register the peak areas, calculate the respective concentrations, and use Eq. 1:

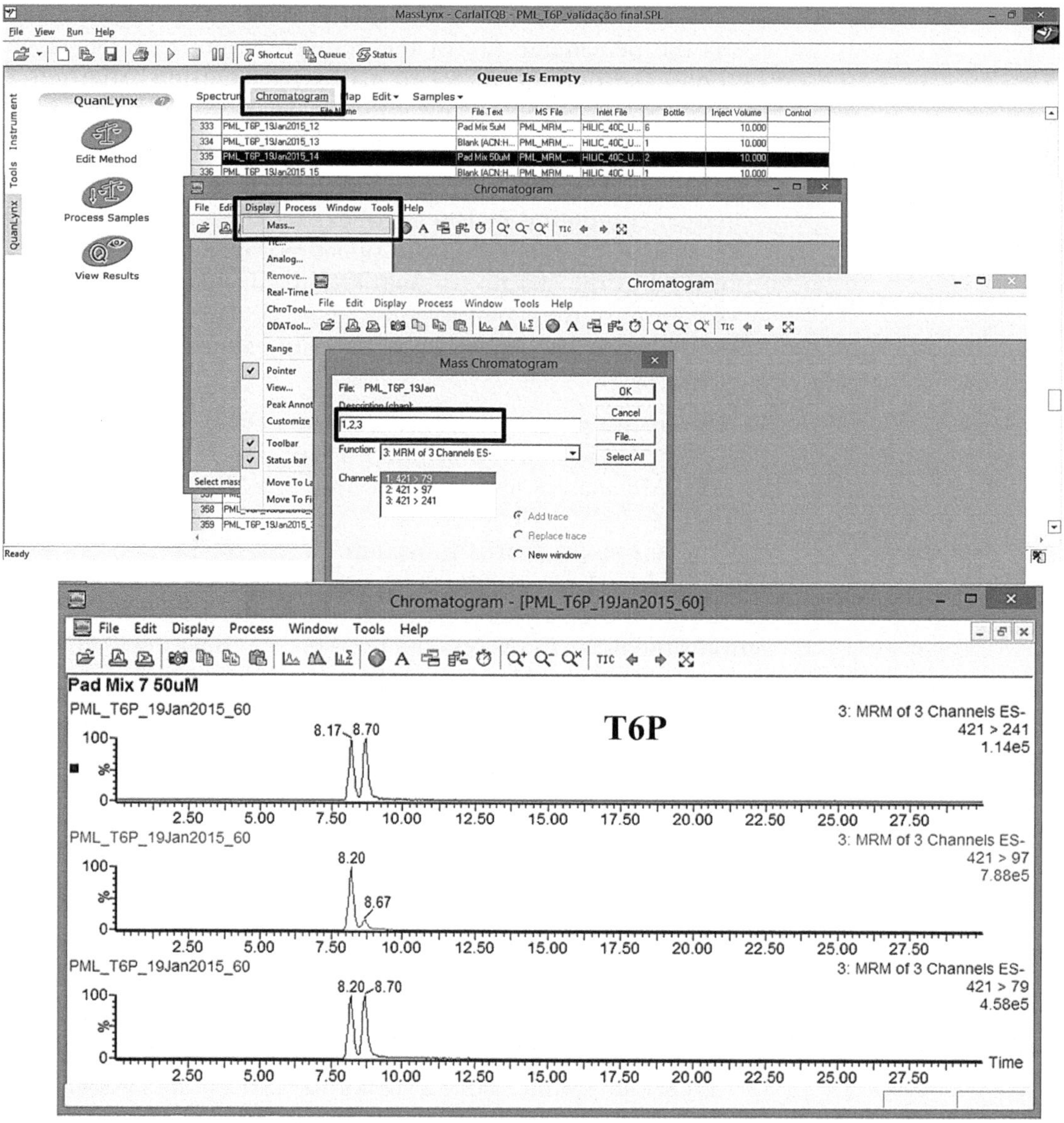

Fig. 3 MassLynx software (version 4.1, Waters®) windows displaying the settings to obtain T6P MRM transitions

$$\%\text{Analytical Recovery} = \frac{\text{mean observed concentration}}{\text{spiked concentration}} \times 100 \quad (1)$$

9. To determine percent ME (%), open the correspondent LC-MS raw files, register the analytes signal response, and use Eq. 2:

$$\%ME = \frac{\text{Analyte signal response in presence of matrix}}{\text{Analyte signal response in individual standard working solution}} \times 100 \qquad (2)$$

where ME = 100% indicates no matrix effect, ME<100% indicates ion suppression, and ME>100% indicates ion enhancement.

4 Notes

1. When handling and working with liquid nitrogen, all standard safety laboratory procedures must be followed.
2. Trehalose 6-phosphate dipotassium salt from Sigma-Aldrich (T4272, ~95% purity) should be used as the standard compound for T6P. Mass spectrometric and enzymatic assays performed [9] showed no detectable contaminants in this product, as previously discussed [12], and was considered appropriate for reliable measurement of T6P using LC-MS/MS methods.
3. Typical gradients used for HILIC consist of organic solvents (50–95%) in water or a volatile buffer, with acetonitrile being the most popular choice as it provides better retention of analytes. Buffers used for HILIC are typically ammonium salts of acetate and formate (5–20 mM) due to their solubility at such high percentages of organic solvent [19, 22]. However, the use of these volatile salts in HILIC has been reported to contribute to the low sensitivity required for the HILIC-ESIQqQ-MS/MS detection of low-abundant phosphorylated carbohydrates [29] and therefore, we strongly recommended replacing them by another volatile additive such as piperidine (Pip) that has high pKa value and requires very low concentrations for pH adjustments. Because our metabolites include a phosphate group, we also strongly advice to add a tail-sweeping reagent such as methylphosphonic acid to the aqueous eluent of the HILIC mobile phase. This addition will reduce the interaction of the phosphate group with stainless-steel components of the LC system, and consequently avoids peak tailing and improves the sensitivity of the HILIC-QqQ-MS method [30].
4. Always maintain a small percentage of water in the wash solution. A suitable wash solution for HILIC is 95% organic solution in water or a composition similar to the mobile phase but without buffer salts. If too much water is included in the wash solution, the polarity of the injection plug might increase, and therefore, cause undesired band-broadening. On the other hand, 100% organic solvents are also not suitable because

they are not polar enough to remove the hydrophilic analytes retained on the HILIC stationary phase.

5. The method validation standard solution mixture is used for method development and validation: evaluation and optimization of the retention times and MS signals as well as method precision (intraday and interday precision).
6. Our six calibration points were: 0.01, 0.05, 1, 5, 10, and 50 μM. We recommend using a minimum of five points for the calibration curve, each point analyzed in triplicate with LC-MS (*see* Chapter 2).
7. The QC mixture is prepared to assess the instrument performance throughout the LC-MS biological sample analyses.
8. Attention! Chloroform is a very toxic solvent. This step must be performed in the fume hood).
9. Avoid to weight small sample amounts (less than 20 mg FW) because it will produce a high weighing error.
10. Turn on the thermoshaker about 20 to 15 min before using. This will give enough time to the equipment to stabilize at the low temperature. We also recommend performing this step in a temperature control chamber at 4 °C.
11. We recommend not exceeding 6 h of evaporation time because it can lead to the degradation of phosphorylated carbohydrates. If not possible, compounds stability must be evaluated to reduce losses during this step. The same procedure must be followed if centrifugal concentrator temperatures higher than 20 °C have to be used.
12. It should be highlighted that LC-MS data is highly dependent on the type of mass spectrometer, ionization source and chromatographic system used. Therefore, the procedure and settings here described for a Micromass® Quattro QqQ with ESI source might need minor modifications when applied to other LC-MS systems.
13. ZIC-HILIC column use and care instructions should be read in advance. When using the ZIC-HILIC column for the first time, equilibrate it with a minimum of seven column volumes of test mobile phase (column certificate of analysis).
14. If the HPLC system does not record the pressure values, we recommend preparing in advance a file to register the pressure values of the column to evaluate the behavior throughout its life cycle.
15. In plant metabolomics studies, accuracy can be assessed by determining the analytical recoveries (*see* Chapter 2).
16. Attention! Do not reconstitute your sample in 100% of water. Aqueous samples should be diluted with organic solvent to at

least 50%. A suitable reconstitution solution for HILIC is a composition similar to the mobile phase but without buffer salts. If too much water is included in the wash solution, the polarity of the injection plug might increase and therefore cause undesired band-broadening. On the other hand, 100% organic solvents are also not suitable because they are not polar enough to remove the hydrophilic analytes retained with HILIC.

17. This test is based on a mathematical model and is used as a complementary test to evaluate the linearity of a calibration curve.

Acknowledgments

T.F. Jorge acknowledges Fundação para a Ciência e a Tecnologia (FCT) for the PhD fellowship (PD/BD/113475/2015) and the ITQB NOVA International PhD Program Plants for Life (PD/00035/2013). C. António gratefully acknowledges support from the FCT Investigator Program (IF/00376/2012/CP0165/CT0003), ITQB NOVA R&D Green-it: Bioresources for Sustainability (UID/Multi/04551/2013).

References

1. Rolland F, Baena-Gonzalez E, Sheen J (2006) Sugar sensing and signaling in plants: conserved and novel mechanisms. Annu Rev Plant Biol 57:675–709
2. Antonio C, Larson T, Gilday A et al (2007) Quantification of sugars and sugar phosphates from *Arabidopsis thaliana* tissues using porous graphitic carbon liquid chromatography-electrospray ionization mass spectrometry. J Chromatogr A 1172:170–178
3. Lastdrager J, Hanson J, Smeekens S (2014) Sugar signals and the control of plant growth and development. J Exp Bot 65:799–807
4. Martinez-Barajas E, Delatte T, Schluepmann H et al (2011) Wheat grain development is characterized by remarkable trehalose 6-phosphate accumulation pre-grain filling: tissue distribution and relationship to SNF1-related protein kinase 1 activity. Plant Physiol 156:373–381
5. Goddijn O, Smeekens S (1998) Sensing trehalose biosynthesis in plants. Plant J 14:143–146
6. Paul ML, Primavesi LF, Jhurreea D et al (2008) Trehalose metabolism and signalling. Annu Rev Plant Biol 59:417–441
7. Nunes C, O'Hara LE, Primavesi LF et al (2013) The trehalose6-phosphate/SnRK1 signaling pathway primes growth recovery following relief of sink limitation. Plant Physiol 162:1720–1732
8. O'Hara LE, Paul MJ, Wingler A (2013) How do sugars regulate plant growth and development? New insight into the role of trehalose-6-phosphate. Mol Plant 6:261–274
9. Yadav UP, Ivakov A, Feil R et al (2014) The sucrose-trehalose 6-phosphate (Tre6P) nexus: specificity and mechanisms of sucrose signalling by Tre6P. J Exp Bot 65:1051–1068
10. Figueroa CM, Feil R, Ishihara H et al (2016) Trehalose 6-phosphate coordinates organic and amino acid metabolism with carbon availability. Plant J 85:410–423
11. Schluepmann H, Pellny T, van Dijken A et al (2003) Trehalose 6-phosphate is indispensable for carbohydrate utilization and growth in Arabidopsis thaliana. Proc Natl Acad Sci U S A 100:6849–6854
12. Lunn JE, Feil R, Hendriks JH et al (2006) Sugar-induced increases in trehalose 6-phosphate are correlated with redox activation of ADPglucose pyrophosphorylase and

higher rates of starch synthesis in Arabidopsis thaliana. Biochem J 397:139–148

13. Lunn JE, Delorge I, Figueroa CM et al (2014) Trehalose metabolism in plants. Plant J 79:544–567
14. Antonio C, Larson T, Gilday A et al (2008) Hydrophilic interaction chromatography/electrospray mass spectrometry analysis of carbohydrate-related metabolites from *Arabidopsis thaliana* leaf tissue. Rapid Commun Mass Spectrom 22:1399–1407
15. Delatte TL, Selman MHJ, Schluepmann H et al (2009) Determination of trehalose-6-phosphate in Arabidopsis seedlings by successive extractions followed by anion exchange chromatography–mass spectrometry. Anal Biochem 389:12–17
16. Delatte TL, Schluepmann H, Smeekens SCM et al (2011) Capillary electrophoresis-mass spectrometry analysis of trehalose-6-phosphate in Arabidopsis thaliana seedlings. Anal Bioanal Chem 400:1137–1144
17. Toraño JS, Delatte TL, Schluepmann H et al (2012) Determination of trehalose-6-phosphate in Arabidopsis thaliana seedlings by hydrophilic-interaction liquid chromatography–mass spectrometry. Anal Bioanal Chem 403:1353–1360
18. Mata AT, Jorge TF, Ferreira J et al (2016) Analysis of low abundant trehalose-6-phosphate and related metabolites in Medicago truncatula by hydrophilic interaction liquid chromatography–triple quadrupole mass spectrometry. J Chromatogr A 1477:30–38
19. Alpert AJ (1990) Hydrophilic-interaction chromatography for the separation of peptides, nucleic acids and other polar compounds. J Chromatogr 499:177–196
20. Guo Y, Gaiki S (2005) Retention behavior of small polar compounds on polar stationary phases in hydrophilic interaction chromatography. J Chromatogr A 1074:71–80
21. Nguyen HP, Schug KA (2008) The advantages of ESI-MS detection in conjunction with HILIC mode separations: Fundamentals and applications. J Sep Sci 31:1465–1480
22. Cubbon S, Antonio C, Wilson J et al (2010) Metabolomic applications of HILIC-LC-MS. Mass Spectrom Rev 29:671–684
23. Buszewski B, Noga S (2012) Hydrophilic interaction liquid chromatography (HILIC)-a powerful separation technique. Anal Bioanal Chem 402:231–247
24. Tang DQ, Zou L, Yin XX et al (2016) HILIC-MS for metabolomics: an attractive and complementary approach to RPLC-MS. Mass Spectrom Rev 35:574–600
25. Jorge TF, Mata AT, António C (2016) Mass Spectrometry as a quantitative tool in plant metabolomics. Phyl Trans R Soc A 374:20150370
26. Lisec J, Schauer N, Kopka J et al (2006) Gas chromatography mass spectrometry-based metabolite profiling in plants. Nat Protoc 1:387–396
27. Kim HK, Verpoorte R (2010) Sample preparation for plant metabolomics. Phytochem Anal 21:4–13
28. Jorge TF, Rodrigues JA, Caldana C et al (2016) Mass spectrometry-based plant metabolomics: metabolite responses to abiotic stress. Mass Spectrom Rev 35:620–649
29. Han J, Tschernutter V, Yang J et al (2013) Analysis of selected sugars and sugar phosphates in mouse heart tissue by reductive amination and liquid chromatography-electrospray ionization mass spectrometry. Anal Chem 85:5965–5973
30. Wakamatsu A, Morimoto K, Shimizu M (2005) A severe peak tailing of phosphate compounds caused by interaction with stainless steel used for liquid chromatography and electrospray mass spectrometry. J Sep Sci 28:1823–1830

Chapter 7

Quantification of Soluble Sugars and Sugar Alcohols by LC-MS/MS

Regina Feil and John Edward Lunn

Abstract

Sugars are simple carbohydrates composed primarily of carbon, hydrogen, and oxygen. They play a central role in metabolism as sources of energy and as building blocks for synthesis of structural and nonstructural polymers. Many different techniques have been used to measure sugars, including refractometry, colorimetric and enzymatic assays, gas chromatography, high-performance liquid chromatography, and nuclear magnetic resonance spectroscopy. In this chapter we describe a method that combines an initial separation of sugars by high-performance anion-exchange chromatography (HPAEC) with detection and quantification by tandem mass spectrometry (MS/MS). This combination of techniques provides exquisite specificity, allowing measurement of a diverse range of high- and low-abundance sugars in biological samples. This method can also be used for isotopomer analysis in stable-isotope labeling experiments to measure metabolic fluxes.

Key words Disaccharide, Trisaccharide, Pentose, Hexose, High-performance anion-exchange chromatography (HPAEC), Isotopomer, Sugar alcohol, Tandem mass spectrometry (MS/MS)

1 Introduction

Sugars play a central role in the metabolism of most organisms as a source of carbon and energy for biosynthesis and growth [1]. They are widely used as short- and long-term carbon reserves, often after polymerization into oligosaccharides or polysaccharides (e.g., starch, glycogen, and fructans).

Sugars are classified according to the number of C atoms, with five-carbon (C_5, pentose) and six-carbon (C_6, hexose) sugars being the most common, although phosphorylated forms of C_3 (triose), C_4 (tetrose), and C_7 (heptose) sugars are also important as intermediates in photosynthetic and respiratory pathways. Reducing sugars contain either an aldehyde (aldose) or ketone (ketose) carbonyl group, while nonreducing sugar alcohols contain only hydroxyl groups. Isomeric forms differ in the position and orientation of hydroxyl groups (epimers and anomers), and in the size and

Carla António (ed.), *Plant Metabolomics: Methods and Protocols*, Methods in Molecular Biology, vol. 1778,
https://doi.org/10.1007/978-1-4939-7819-9_7,

configuration of the ring structure (furanose or pyranose) when the molecule is in a closed conformation. Condensation of simple sugars (monosaccharides) into disaccharides and oligosaccharides generates even more complexity and functional diversity. Structural polymers in prokaryotic, plant, and fungal cell walls are derived from five- (pentose) and six-carbon (hexose) sugars, and two pentoses, ribose and deoxyribose, are essential components of nucleic acids [1]. The functionality of many glycoproteins is dependent on the core polypeptide being decorated with sugars of the correct type and conformation.

Many secondary metabolites (e.g., flavonoids) are also modified by conjugation with sugars, with variation in the type and site of attachment increasing molecular diversity and the range of functions (pigments, abiotic and biotic stress protectants, etc.) [2]. Conjugation with sugars is also a common mechanism for detoxification of xenobiotics. As compatible solutes, nonreducing sugars (e.g., sucrose and trehalose) and sugar alcohols (e.g., glycerol, sorbitol, and mannitol) are widely used for osmoregulation and carbon transport. Trehalose is the most commonly transported sugar in fungal mycelia and in many invertebrates. In vascular plants, sucrose is the predominant transport sugar, although sucrose-derived oligosaccharides (e.g., raffinose and stachyose) or sugar alcohols (e.g., sorbitol and mannitol) share this function in some species [3].

Assays for sugars have a long history, being among the first scientific methods developed for analysis of foodstuffs. Their development was driven primarily by the winemaking and brewing industries, for measuring sugar content of developing grape berries to determine the optimal time for harvesting and to monitor sugar levels in cereal grains during the malting process. The earliest methods include refractometry for measuring total sugar content in Brix or Baumé units, and chemical assays using reagents (e.g., Fehling's solution, resorcinol, or anthrone) that react with reducing sugars to produce a colored product that can be quantified by its absorbance [4, 5]. Such methods do not require sophisticated equipment or expertise, and are still in use today for measuring total sugar levels in samples containing a predominant sugar (e.g., sucrose in sugarcane extracts) or sugars of the same type (e.g., glucose and fructose, which are both hexose reducing sugars). The development of enzymatic assays increased both the specificity and sensitivity of sugar measurements, allowing more complex samples to be analyzed and providing information about the sugar composition of samples [6, 7].

The most common approaches are based on spectrophotometric or fluorimetric measurements of glucose using either glucose oxidase (EC 1.1.3.4) and peroxidase (EC 1.11.1.7) or hexokinase (EC 2.7.1.1) and glucose 6-phosphate dehydrogenase (EC 1.1.1.49) to generate stoichiometric amounts of a quantifiable product (e.g., NADPH). The latter coupling system can be used to measure fructose by addition of phosphoglucose isomerase (EC 5.3.1.9). Glucose and fructose-containing disaccharides (e.g., sucrose, maltose, and trehalose) and oligosaccharides (e.g., maltotriose) can also be measured after hydrolysis to their constituent hexose sugars using specific hydrolytic enzymes, such as invertase (EC 3.2.1.26) to measure sucrose. Although enzymatic methods are widely used, they are limited by the availability of suitable coupling enzymes. Nuclear magnetic resonance (^{13}C-NMR) spectroscopy is sometimes used to measure sugars in complex mixtures, and even in living tissues [8, 9]. However, this technique has limited sensitivity and requires extremely expensive equipment.

Gas chromatography [10] offers high-resolution and sensitivity for profiling of sugars and sugar alcohols, especially if coupled to mass spectrometry (GC-MS) [11]. However, these techniques require the sugars and other metabolites in the sample to be derivatized before analysis, and variation in the efficiency of derivatization between standards and samples makes absolute quantification more challenging than measurement of relative amounts.

A broad spectrum of sugars can be analyzed by high-performance liquid chromatography (HPLC) using reverse phase or anion-exchange (HPAEC) columns to achieve separation of the various sugars in the sample [12]. The latter technique requires use of an alkaline eluent (e.g., KOH) to deprotonate one or more hydroxyl groups in the sugar molecule to generate anionic (i.e., charged) forms of the sugars ($R\text{–}OH + OH^- \longrightarrow R\text{–}O^- + H_2O$). Abundant sugars can be quantified by measuring the refractive index of the eluate as the sugars elute from the column, but greater sensitivity can be achieved using pulsed amperometric detection [13]. A disadvantage of both detection methods is that they do not distinguish between coeluting compounds, potentially leading to overestimation of individual sugars if the sample contains other compounds with the same retention time on the column as the sugar of interest [14].

The assay methods described in this chapter overcome this limitation by coupling an HPLC apparatus to a triple quadrupole mass spectrometer, allowing identification and quantification of sugars by tandem mass spectrometry (MS/MS) as they elute from the HPLC column [15]. HPAEC with an alkaline eluate (KOH/-methanol) is used for the initial separation of sugars in the liquid chromatography (LC) phase, using either a CarboPac™ PA1 or CarboPac™ SA10 (Dionex™; www.dionex.com) column. The choice of column depends on the type of sugar being analyzed, with the CarboPac™ SA10 column offering the best resolution of hexoses and inositols, while the CarboPac™ PA1 column is preferred for analysis of pentoses, disaccharides, trisaccharides, and the other sugar alcohols. However, it should be noted that there is considerable overlap in the range of sugars that can be adequately separated by these two columns, so for many applications either column would be suitable. The alkaline eluent (KOH) necessitates use of an electro-osmotic ion suppressor to remove excess K^+ ions from the eluate before entry into the mass spectrometer, which is operated in negative ion mode. Passage through the first quadrupole selects for parent ions with the expected mass-to-charge ratio (m/z) of the monovalent deprotonated form: $[M\text{-}H^+]^-$. The parent ion is then fragmented by collision with argon gas in the second quadrupole under optimized conditions, and then one of the characteristic product ions (usually the most abundant) is selected and quantified in the third quadrupole. The initial selection of parent ions allows coeluting compounds with different molecular masses to be separated (unlike refractive index or pulsed amperometric detection methods), while the detection of specific product ions often allows even isomeric compounds with overlapping retention times to be resolved. In the most advanced machines, greater specificity can be achieved by simultaneous measurement of multiple product ions in the third quadrupole, or by additional fragmentation analysis using a quadrupole ion trap (*see* **Note 1**).

Naturally occurring sugars predominantly contain the most abundant naturally occurring isotopes of hydrogen (^1H), carbon (^{12}C), and oxygen (^{16}O). The use of mass spectrometry allows quantification of other isotopomers, in which one or more of the naturally occurring isotopes in the sugar molecule are substituted by a heavier isotope, e.g., ^2H, ^{13}C, or ^{18}O. The presence of heavy isotopes increases the mass (and thus m/z) of the corresponding parent ion, so the heavier isotopomer is separated from the naturally occurring isotopomer in the first quadrupole. This capability is used in the described protocol to detect, and correct for, interference in the analysis of individual sugars by other components in the sample, by use of stable-isotope-labeled internal standards [16, 17] (*see* **Note 2**). The HPAEC-MS/MS method can also be used for isotopomer analysis in stable-isotope-labeling experiments, in which an isotopically labeled compound (e.g., $^{13}CO_2$) is supplied to living tissue to measure metabolic fluxes [18, 19] (*see* **Note 3**).

2 Materials

Prepare all solutions using analytical grade reagents. HPLC-grade methanol is supplied by Biosolve. Sugars suitable for use as calibration standards can be obtained from general biochemical suppliers (Sigma-Aldrich; Calbiochem) or from specialist carbohydrate suppliers (Carbosynth; Megazyme). Stable-isotope-labeled sugars are available from Omicron Biochemicals, Inc. and Sigma-Aldrich.

2.1 Sample Preparation

1. Extraction solution: Chloroform/methanol (3/7, v/v).
2. Ultrapure water: conductivity ≤18 MΩ cm at 25 °C.
3. 2-mL safelock microcentrifuge tubes (e.g., Eppendorf).
4. 1.5-mL screw-capped microcentrifuge tubes (e.g., Eppendorf).
5. Spatula suitable for weighing up to 20 mg; insulate the handle by insertion into a cylindrical block (approx. 5 cm × 2 cm diameter) of polystyrene.
6. Mortar and pestle.
7. Ball mill (e.g., Retsch).
8. MultiScreen-PCR$_{96}$ 96-well filter microplate (Millipore).
9. HPLC sample vials: 0.3-mL crimp-top fixed-insert vials fitted with polyethylene snap caps (e.g., Chromacol).
10. Centrifugal concentrator or lyophilizer.

2.2 LC-MS/MS: CarboPac™ SA10 Column (Hexoses and Inositols)

1. Eluent A: 5% (v/v) methanol degassed by sonication.
2. Dionex™ EGC III KOH eluent cartridge (Thermo Scientific).
3. 9 × 75 Dionex™ IonPac™ ATC-HC RFIC trap column (Thermo Scientific).
4. 2 × 50 mm 2-mm Dionex™ CarboPac™ SA10 guard column (Thermo Scientific) (*see* **Note 4**).
5. 2 × 250 mm 6-μm Dionex™ CarboPac™ SA10 column (Thermo Scientific) (*see* **Note 4**).

2.3 LC-MS/MS: CarboPac™ PA1 Column (Pentoses, Disaccharides, and Trisaccharides)

1. Eluent B: 2.5% (v/v) methanol degassed by sonication.
2. Dionex™ EGC III KOH eluent cartridge (Thermo Scientific).
3. 9 × 75 Dionex™ IonPac™ ATC-HC RFIC trap column (Thermo Scientific).
4. 2 × 50 mm 2-mm Dionex™ CarboPac™ PA1 guard column (Thermo Scientific) (*see* **Note 4**).
5. 2 × 250 mm 10-μm Dionex™ CarboPac™ PA1 column (Thermo Scientific) (*see* **Note 4**).

2.4 LC-MS/MS Apparatus

1. ICS-3000 chromatograph (Thermo Scientific) connected to an autosampler and an eluent generator fitted with Dionex EGC III KOH eluent cartridges (Thermo Scientific).
2. Finnigan TSQ Quantum Discovery MAX triple quadrupole mass spectrometer (Thermo Scientific).
3. Couple the Finnigan TSQ Quantum Discovery MAX mass spectrometer to the ICS-3000 chromatograph using DCMS-Link software (Thermo Scientific) and controlled using Xcalibur software (Thermo Scientific).

4. Operate the mass spectrometer in multiple reaction monitoring (MRM) mode using an electrospray ionization source (HESI-II) in the negative mode, with centroid data acquisition.
5. Nitrogen is used as the sheath and auxiliary gas.
6. Set the spray voltage to −3500 V, with a capillary voltage ranging from −1 to −56 V and a vaporizing temperature of 400 °C.
7. Set the ion transfer tube temperature to 300 °C, with a voltage range from −50 to −100 V.
8. Argon is used as the collision gas in the second quadrupole.

3 Methods

3.1 Sample Preparation

1. Grind frozen tissue to a fine powder at liquid nitrogen temperature using a mortar and pestle or a ball mill (e.g., Retsch).
2. Transfer 15–20 mg of fresh-frozen tissue powder to a precooled, 2.0-mL safelock microcentrifuge tube. It is very important not to let the tissue thaw during this step.
3. Use a small polystyrene box to hold the prechilled tube on the balance while weighing, and a spatula with an insulated handle.
4. Weigh out the tissue powder as quickly as possible to get an aliquot in the target range, note the exact weight, and then transfer the tube into liquid nitrogen or dry ice.
5. Add 175 μL of ice-cold extraction solution to the tissue while the tube is sitting in liquid nitrogen (or on dry ice). The mixture will freeze solid at this stage.
6. Holding the tube in the hand, use a plastic pestle to disperse the tissue in the solvent as it thaws. When thawed, place it on ice.
7. Rinse the pestle with 175 μL of ice-cold extraction solution, combining the rinses with the initial extract in the tube. Alternative option: add 350 μL of extraction solution to the frozen tissue, and vortex mix at room temperature until the mixture thaws and the tissue is evenly dispersed.
8. Place the tubes in a −20 °C freezer and leave for 2 h, mixing occasionally.
9. Add 350 μL of ice-cold water, and allow warming up to 4 °C with repeated shaking.
10. Centrifuge at 13,000 × *g*, 4 °C, for 10 min to separate the aqueous-methanol (upper) and chloroform (lower) phases.
11. Transfer the aqueous-methanol (upper) phase to a new 1.5-mL screw-capped tube and keep at 4 °C, taking care not to take any of the chloroform (lower) phase or the insoluble material at the interface.

12. Reextract the chloroform phase by adding 300 μL of ice-cold water to the chloroform phase, vortex mix, and centrifuge at 13,000 × *g*, 4 °C, for 10 min.
13. Take off the upper aqueous-methanol phase, and combine it with the first aqueous-methanol extract in the 1.5-mL screw-capped tube.
14. Evaporate the combined aqueous-methanol extract to dryness using a centrifugal concentrator at 35 °C, or a lyophilizer.
15. Redissolve the dried extract with 350 μL of water.
16. Remove the glycerol preservative from a 96-well filter microplate by adding 200 μL water to each well and centrifuging at 2500 × *g*, 15 °C, to dryness (approx. 30 min).
17. Discard the filtrate.
18. Repeat **step 16**.
19. Place a 200-μL aliquot of the extract in one well of the 96-well filter microplate. Centrifuge at 2500 × *g*, 15 °C, for 45 min, and keep the filtrate (filtered extract). This procedure removes high-molecular weight substances from the extract that might interfere with the chromatographic separation and reduce column longevity [16].
20. In a 0.3-mL HPLC vial, mix 45 μL of the filtered extract with 5 μL of a mixture containing appropriate stable-isotope-labeled internal standards (*see* **Note 2**).
21. Close the injection vial with a polyethylene snap cap.
22. The sample is now ready for LC-MS/MS analysis. Place the samples in the autosampler operated at 10 °C. Attention! If the extract contains a very high concentration of sugars, it might need to be diluted with water before analysis.

3.2 LC-MS/MS: CarboPac™ SA10 Column (Hexoses and Inositols)

1. Connect the 2 × 50 mm 2-mm Dionex CarboPac™ SA10 guard column and 2 × 250 mm 6-μm Dionex CarboPac™ SA10 column in series to the ICS -3000 chromatograph, operated with a constant column temperature of 15 °C (*see* **Note 4**).
2. Pass degassed Eluent A (5% (v/v) methanol) through a 9 × 75 Dionex IonPac™ ATC-HC RFIC™ trap column to remove anionic contaminants before it enters the eluent generator cartridges.
3. Equilibrate the column with 2 mM KOH/ 5% (v/v) methanol at a flow rate of 0.16 mL/min for 15 min.
4. Calibrate the LC-MS/MS by injecting 20-μL aliquots of sugar standards (spiked with appropriate internal standards) at five different dilutions (covering the expected range of metabolite abundance in the samples) (*see* **Note 5**).

5. Elute sugars isocratically with 2 mM KOH/ 5% (v/v) methanol for 25 min, followed by 50 mM KOH/5% (v/v) methanol for 10 min.
6. Pass the eluate through an AERS 500 ion suppressor (Thermo Scientific), operating in external water mode, before injection into the mass spectrometer.
7. Set the ion suppressor current to 5 mA for the first 25 min, and then to 50 mA for 10 min.
8. The parent ion/product ion transitions for the measured sugars are listed in Table 1, and the settings for the stable-isotope-labeled internal standards are shown in Table 2 (*see* **Note 1**).
9. For each sample, inject a 20-μL aliquot (spiked with appropriate internal standards) onto the columns.
10. Elute, ionize and detect the sugars as described in **steps 5–8**.
11. Use LCquan software (Thermo Scientific) for peak assignment, calculating calibration curves and quantification of sugars in the samples.
12. Normalize values automatically using the measured signal from the corresponding stable-isotope-labeled internal standard (*see* **Note 2**).

3.3 LC-MS/MS: CarboPac™ PA1 Column (Pentoses, Disaccharides, and Trisaccharides)

1. Connect the 2 × 50 mm 2-mm Dionex CarboPac™ PA1 guard column and 2 × 250 mm 10-μm Dionex CarboPac™ PA1 column in series to an ICS-3000 chromatograph (Thermo Scientific), operated with a constant column temperature of 22 °C (*see* **Note 4**).
2. Pass degassed Eluent B (2.5% (v/v) methanol) through a 9 × 75 Dionex IonPac™ ATC-HC RFIC™ trap column to remove anionic contaminants before it enters the eluent generator cartridges.
3. Equilibrate the column with 15 mM KOH/ 2.5% (v/v) methanol at a flow rate of 0.28 mL/min for 15 min.
4. Calibrate the LC-MS/MS by injecting 20-μL aliquots of sugar standards (spiked with appropriate internal standards) at five different dilutions (covering the expected range of metabolite abundance in the samples) (*see* **Note 5**).
5. Elute sugars using the following gradient, with all eluents containing 2.5% (v/v) methanol: (a) isocratic 15 mM KOH, 5 min; (b) linear gradient of 15–50 mM KOH, 7 min; (c) linear gradient of 50–100 mM KOH, 28 min; and (d) isocratic 100 mM KOH, 15 min.

Table 1
Chromatographic and mass-spectrometric parameters for LC-MS/MS analysis of sugars and sugar alcohols

Sugar	Retention time, min	Parent ion, $[M-H^+]^-$	Parent ion, *m/z*	Product ion, *m/z*
Pentoses				
Arabinose	9.64[a]	$C_5H_9O_5$	149.1	59.3
Lyxose	11.92[a]	$C_5H_9O_5$	149.1	59.3
Ribose	15.28[a]	$C_5H_9O_5$	149.1	59.3
Xylose	13.08[a]	$C_5H_9O_5$	149.1	59.3
Hexoses				
Allose	19.90[a]/15.20[b]	$C_6H_{11}O_6$	179.1	89.1
Altrose	14.79[a]/9.83[b]	$C_6H_{11}O_6$	179.1	89.1
2-deoxy-glucose	8.66[a]	$C_6H_{11}O_5$	163.08	85.0
Fructose	14.39[a]/12.36[b]	$C_6H_{11}O_6$	179.1	89.1
Fucose	5.27[a]	$C_6H_{11}O_5$	163.08	59.1
Galactose	11.80[a]/9.11[b]	$C_6H_{11}O_6$	179.1	89.1
Glucose	12.20[a]/10.49[b]	$C_6H_{11}O_6$	179.1	89.1
Gulose	13.55[b]	$C_6H_{11}O_6$	179.1	89.1
Mannose	12.59[a]/13.25[b]	$C_6H_{11}O_6$	179.1	89.1
Psicose	16.60[b]	$C_6H_{11}O_6$	179.1	89.1
Rhamnose	8.35[a]	$C_6H_{11}O_6$	163.08	59.1
Sorbose	12.36[b]	$C_6H_{11}O_6$	179.1	89.1
Tagatose	10.49[b]	$C_6H_{11}O_6$	179.1	89.1
Talose	22.54[a]/21.42[b]	$C_6H_{11}O_6$	179.1	89.1
Sugar alcohols				
Allo-inositol	2.10[a]/5.38[b]	$C_6H_{11}O_6$	179.1	87.2
Arabitol	2.70[a]	$C_5H_{11}O_5$	151.1	70.2
Galactinol	2.79[a]	$C_{12}H_{21}O_{11}$	341.1	179.1
Galactitol	3.03[a]	$C_6H_{13}O_6$	181.2	163.1
Glycerol	2.09[a]	$C_3H_7O_3$	91.1	59.1
Iso-erythritol	2.27[a]	$C_4H_9O_4$	121.1	89.1
Maltitol	9.10[a]	$C_{12}H_{23}O_{11}$	343.1	58.9
Mannitol	3.53[a]	$C_6H_{13}O_6$	181.1	101.1
Myo-inositol	2.10[a]/5.04[b]	$C_6H_{11}O_6$	179.1	161.1
Ribitol	3.06[a]	$C_5H_{11}O_5$	151.1	70.2
Sorbitol	3.12[a]	$C_6H_{13}O_6$	181.1	101.1
Xylitol	2.46[a]	$C_5H_{11}O_5$	151.1	70.2
Disaccharides				
Cellobiose	34.77[a]	$C_{12}H_{21}O_{11}$	341.1	161.1
Isomaltulose	35.80	$C_{12}H_{21}O_{11}$	341.1	221.1
Lactose	24.66[a]	$C_{12}H_{21}O_{11}$	341.1	161.1
Laminaribiose	53.51[a]	$C_{12}H_{21}O_{11}$	341.1	113.1
Maltose	49.59[a]	$C_{12}H_{21}O_{11}$	341.1	161.1
Melibiose	19.90[a]	$C_{12}H_{21}O_{11}$	341.1	221.1
Sucrose	22.16[a]	$C_{12}H_{21}O_{11}$	341.1	179.1
α,α-trehalose	3.94[a]	$C_{12}H_{21}O_{11}$	341.1	179.1
α,β-trehalose	8.27[a]	$C_{12}H_{21}O_{11}$	341.1	179.1
β,β-trehalose	6.78[a]	$C_{12}H_{21}O_{11}$	341.1	179.1
Turanose	28.94[a]	$C_{12}H_{21}O_{11}$	341.1	113.1

(continued)

Table 1 (continued)

Sugar	Retention time, min	Parent ion, $[M-H^+]^-$	Parent ion, *m/z*	Product ion, *m/z*
Trisaccharides				
1-Kestose	45.87[a]	$C_{18}H_{31}O_{16}$	503.25	323.1
Raffinose	38.96[a]	$C_{18}H_{31}O_{16}$	503.25	221.1

The second column shows the retention time(s) of sugars and sugar alcohols on the CarboPac™ SA10 and/or CarboPac™ PA1 columns under the described conditions. The third and fourth columns show the formula and mass-to-charge ratio (*m/z*) of the mono-deprotonated parent ion $[M-H^+]^-$, respectively, and column five shows the mass-to-charge ratio (*m/z*) of the product ion used for quantification (*see* **Note 1**)

[a]Dionex™ CarboPac™ PA1 column

[b]Dionex™ CarboPac™ SA10 column

6. Pass the eluate through an AERS 500 ion suppressor (Thermo Scientific), operating in external water mode with the current set to 65 mA, before injection into the mass spectrometer.
7. The parent ion/product ion transitions for the measured sugars are listed in Table 1, and the settings for the stable-isotope-labeled internal standards are shown in Table 2 (*see* **Note 1**).
8. For each sample, inject a 20-μL aliquot (spiked with appropriate internal standards) onto the columns.
9. Elute, ionize and detect the sugars as described in **steps 5–7**.
10. Use LCquan software (Thermo Scientific) for peak assignment, calculating calibration curves and quantification of sugars in the samples.
11. Normalize values automatically using the measured signal from the corresponding stable-isotope-labeled internal standard (*see* **Note 2**).

4 Notes

1. To determine suitable parent ion/product ion transitions for sugar analysis, individual sugars are initially injected directly into the mass spectrometer to determine the *m/z* of the major fragmentation products. The most abundant fragmentation product is usually chosen for quantification as this gives the greatest sensitivity. If two compounds have overlapping retention times and the same parent ion mass (*m/z*), they will not be separated by the column chromatography or during passage through the first quadrupole. However, such coeluting compounds can sometimes be resolved and quantified separately by selecting less abundant, but compound specific, product ions in the third quadrupole.

Table 2
Stable-isotope-labeled internal standards for sugars and sugar alcohols. The isotopically labelled atoms are highlighted in bold

Sugar	Parent ion, $[M-H^{+}]^{-}$	Parent ion, *m/z*	Product ion, *m/z*
Pentoses			
$[^{13}C_5]$arabinose	$\mathbf{^{13}C_5}H_9O_5$	154.1	61.3
$[^{13}C_5]$xylose	$\mathbf{^{13}C_5}H_9O_5$	154.1	61.3
Hexoses			
$[^{13}C_6]$fructose	$\mathbf{^{13}C_6}H_{11}O_6$	185.1	92.1
$[^{13}C_6]$galactose	$\mathbf{^{13}C_6}H_{11}O_6$	185.1	92.1
$[^{13}C_6]$glucose	$\mathbf{^{13}C_6}H_{11}O_6$	185.1	92.1
$[^{13}C_6]$mannose	$\mathbf{^{13}C_6}H_{11}O_6$	185.1	92.1
Sugar alcohols			
$[^{13}C_6]$mannitol	$\mathbf{^{13}C_6}H_{13}O_6$	187.1	104.1
$[^{2}H_6]$myo-inositol	$C_6H_5\mathbf{^{2}H_6}O_6$	185.1	161.1
$[^{13}C_6]$sorbitol	$\mathbf{^{13}C_6}H_{13}O_6$	187.1	105.1
Disaccharides			
$[^{13}C_{12}]$maltose	$\mathbf{^{13}C_{12}}H_{21}O_{11}$	353.1	167.1
[*Fru*-$^{13}C_6$]sucrose	$C_6\mathbf{^{13}C_6}H_{21}O_{11}$	347.1	179.1
$[^{13}C_{12}]$α,α-trehalose	$\mathbf{^{13}C_{12}}H_{21}O_{11}$	353.1	185.1
$[^{13}C_{12}]$α,β-trehalose	$\mathbf{^{13}C_{12}}H_{21}O_{11}$	353.1	185.1

2. An important feature of the described method is the use of stable-isotope-labeled internal standards for the compounds of interest [16, 17]. In the labeled standards, at least two of the ^{1}H, ^{12}C, or ^{16}O atoms are substituted by heavy stable isotopes—^{2}H, ^{13}C, or ^{18}O, respectively. This increases the molecular mass (and thus m/z) of the molecule, but has no significant effect on the physicochemical properties that determine its behavior during chromatography and ionization. Thus, the heavier isotopomer coelutes and is ionized along with the unlabeled compound, but is then separated in the first quadrupole as its parent ion has a slightly higher m/z ratio. (n.b. the natural abundance of ^{13}C is about 1.1%, therefore, to reduce the overlap from naturally occurring isotopomers, internal standards containing at least two substituted atoms are preferred). Ideally an internal standard should be used for each compound of interest. Known amounts of the isotopically labeled internal standards are spiked into each sample before analysis, and the mass spectrometer is set up to scan for the parent ion/product ion transition of the internal standard at the same time as it scans for the corresponding transition of the unlabeled compound. Quantification of the internal standard then allows the measurements of the naturally occurring compound to be corrected for any ion suppression or other matrix

effects. The choice of stable-isotope-labeled standard depends on (a) the type of sample being analyzed, (b) the main sugars of interest, and (c) the availability of suitably labeled isotopomers. If deuterium (^{2}H) labeled isotopomers are used, the deuterium atom(s) should be covalently linked to one of the C-atoms or located within a hydroxyl group that is not readily deprotonated. The amount of each internal standard added depends on the composition of the sample. Too little or too much of the internal standard can lead to problems with overlapping signals from the internal standard and naturally occurring isotopomers in the sample, especially those containing ^{13}C, which has a natural abundance of about 1.1%. Taking glucose as an example, 6–7% of the glucose molecules in an unlabeled sample will, on average, contain one ^{13}C atom, and about 2% of molecules will contain two or more ^{13}C atoms. If a glucose isotopomer containing two heavy isotope atoms (e.g., [$^{13}C_2$]glucose or [$^{2}H_2$]glucose) were used as an internal standard, the naturally occurring [$^{13}C_2$]glucose isotopomers in the sample would overlap with the internal standard and potentially interfere with its quantification if the amount of the internal standard were too low. Conversely, as commercially available isotopically labeled compounds are typically contaminated with 1–5% of the unlabeled isotopomer, quantification of the unlabeled compound in the sample could be compromised if a large (>10-fold) excess of the internal standard is used. The isotopic purity of commercially supplied labeled compounds should therefore also be taken into account when deciding how much to use as an internal standard. A useful online tool for estimating the natural occurrence of isotopomers for a compound of interest is available from the following website: http://www.chemcalc.org/main [20].

3. Product ions that are suitable for quantification can be determined empirically for unlabeled sugars (when the sugar is available) and for isotopically labeled standards, and it is not necessary to know the elemental composition of the fragmentation products. However, samples from stable-isotope labeling experiments typically contain a complex mixture of all possible isotopomers, and standards are not usually available to determine the fragmentation profile for each isotopomer. Therefore, to scan for appropriate parent ion/product ion transitions for isotopomer analysis of labeled samples, it is necessary to know the elemental composition of the fragment being detected in the third quadrupole. Elimination of a water (H_2O, $M_r = 18$) or CO_2 ($M_r = 44$) molecule is a common occurrence during fragmentation, and can be inferred if the *m/z* values of the parent and product ions differ by 18 (e.g., galactitol and myo-inositol) (Table 1) or 44 units, so the composition of

the residual fragment can be calculated by subtracting the eliminated moiety. Comparison of the product ions from unlabeled and fully labeled isotopomers can also give clues to the fragmentation product. For example, the mass of the product ion from the fully labeled isotopomer of glucose—[$^{13}C_6$]glucose—is 3 units greater than that of the product ion from unlabeled glucose, implying that the product ion contains three C atoms, and that the elemental composition of the product ion from unlabeled glucose is $C_3H_5O_3^-$ (*m/z* 89.1), while the product ion from [$^{13}C_6$]glucose is $^{13}C_3H_5O_3^-$ (*m/z* 92.1). With this knowledge, potential product ion masses and parent ion/product ion transitions can be calculated for all of the potential ^{13}C-isotopomers of glucose. Several online tools are available for identifying potential fragmentation products, e.g., http://calistry.org/calculate/massFragment.

4. Regular replacement of the guard columns and cleaning of the CarboPac™ PA1 and SA10 columns according to the manufacturer's instructions helps to maintain performance and maximize the working life of the columns.
5. When running a large batch of samples, monitor the performance of the column and ion-suppressor by intercalating a set of the calibration standards (one dilution is sufficient) after every ten samples. Any problems with the column or ion suppressor will usually be apparent by a decrease in the sugar peak areas relative to the initial calibration run.

Acknowledgments

This work was supported by the Max Planck Gesellshaft.

References

1. Nelson DL, Cox M (2017) Lehninger principles of biochemistry. W.H. Freeman & Company, New York
2. Saito K, Yonekura-Sakakibara K, Nakabayashi R et al (2013) The flavonoid biosynthetic pathway in Arabidopsis: structural and genetic diversity. Plant Physiol Biochem 72:21–34
3. Lunn JE (2016) Sucrose metabolism. In: Encyclopedia of life science (ELS). John Wiley & Sons, Chichester
4. Fehling H (1849) Die quantitative Bestimmung von Zucker und Stärkmehl mittelst Kupfervitriol. Ann Chem Pharm 72:106–113
5. Benedict SR (1909) A reagent for the detection of reducing sugars. J Biol Chem 5:485–487
6. Outlaw WH, Tarczynski MC (1984) Sucrose. In: Bergmeyer HU, Bergmeyer J, Graßl M (eds) Methods of enzymatic analysis. VCH, Weinheim, pp 96–103
7. Stitt M, Lilley RM, Gerhardt R et al (1989) Metabolite levels in specific cells and subcellular compartments of plant leaves. Meth Enzymol 174:518–552
8. Deslauriers R, Jarrell HC, Byrd RA et al (1980) Observation by ^{13}C NMR of metabolites in differentiating amoeba. Trehalose storage in encysted Acanthamoeba castellanii. FEBS Lett 118:185–190
9. Ratcliffe RG (1994) *In vivo* NMR studies of higher plants and algae. Adv Bot Res 20:43–123

10. Streeter JG, Bosler ME (1976) Carbohydrates in soybean nodules: identification of compounds and possible relationships to nitrogen fixation. Plant Sci Lett 7:321–329
11. Roessner U, Wagner C, Kopka J et al (2000) Simultaneous analysis of metabolites in potato tuber by gas chromatography-mass spectrometry. Plant J 23:131–142
12. Rocklin RD, Pohl CA (1983) Determination of carbohydrates by anion exchange chromatography with pulsed amperometric detection. J Liq Chromatogr 6:1577–1590
13. Hanko VP, Rohrer JS (2000) Determination of carbohydrates, sugar alcohols, and glycols in cell cultures and fermentation broths using high-performance anion-exchange chromatography with pulsed amperometric detection. Anal Biochem 283:192–199
14. Quéro A, Béthencourt L, Pilard S et al (2013) Trehalose determination in linseed subjected to osmotic stress. HPAEC-PAD analysis: an inappropriate method. Physiol Plant 147:261–269
15. Fichtner F, Barbier FF, Feil R et al (2017) Trehalose 6-phosphate is involved in triggering axillary bud outgrowth in garden pea (*Pisum sativum* L.) Plant J. https://doi.org/10.1111/tpj.13705
16. Lunn JE, Feil R, Hendriks JH et al (2006) Sugar-induced increases in trehalose 6-phosphate are correlated with redox activation of ADPglucose pyrophosphorylase and higher rates of starch synthesis in *Arabidopsis thaliana*. Biochem J 397:139–148
17. Arrivault S, Guenther M, Fry SC et al (2015) Synthesis and use of stable isotope labelled internal standards for quantification of phosphorylated metabolites by LC-MS/MS. Anal Chem 87:6896–6904
18. Ishihara H, Obata T, Sulpice R et al (2015) Quantifying protein synthesis and degradation in Arabidopsis by dynamic $^{13}CO_2$ labeling and analysis of enrichment in individual amino acids in their free pools and in protein. Plant Physiol 168:74–93
19. Fernandez O, Ishihara H, George GM et al (2017) Leaf starch turnover occurs in long days and in falling light at the end of the day. Plant Physiol 174:2199–2212
20. Patiny L, Borel A (2013) ChemCalc: a building block for tomorrow's chemical infrastructure. J Chem Inf Model 53:1223–1228

Chapter 8

Untargeted Analysis of Semipolar Compounds by LC-MS and Targeted Analysis of Fatty Acids by GC-MS/GC-FID: From Plant Cultivation to Extract Preparation

Camille Bénard, Sébastien Acket, Yannick Rossez, Olivier Fernandez, Thierry Berton, Yves Gibon, and Cécile Cabasson

Abstract

The way plants are grown and samples are harvested, prepared, and extracted has a profound impact on the output of a metabolomics experiment. In this chapter, we detail the experimental procedures from plant cultivation to extract preparation, in order to avoid difficulties that could result in contamination or undesired changes of the analytes. Two plant organs are mentioned as examples: tomato fruits (*Solanum lycopersicum*) and flax seeds (*Linum usitatissimum*). Extractions designed for the untargeted analysis of semipolar compounds by liquid chromatography–mass spectrometry (LC-MS) and targeted analysis of fatty acids by gas chromatography–mass spectrometry (GC-MS) or gas chromatography with flame ionization detector (GC-FID) are described.

Key words Plant cultivation, Sample preparation, Sample harvest, Fatty acids, Specialized compounds, Tomato, Flax seeds

1 Introduction

Plant metabolomics analysis consists in several steps [1, 2] that can be divided as follows: (1) experimental design; (2) plant growth; (3) plant harvest; (4) sample preparation; (5) extract preparation; (6) data acquisition; (7) data processing; (8) data mining. In this chapter, particular focus will be given to steps (2–5); the other steps are addressed in other chapters of this book.

The way plants are cultivated and samples are harvested, processed, and extracted has a profound impact on the outcome of a metabolomics experiment. Indeed the samples can be contaminated during the plant growth; its chemical integrity can be altered during the harvest if the quenching is not appropriate or during the

---

Camille Bénard and Sébastien Acket contributed equally to this work.

Carla António (ed.), *Plant Metabolomics: Methods and Protocols*, Methods in Molecular Biology, vol. 1778, https://doi.org/10.1007/978-1-4939-7819-9_8,

extract preparation. Additionally, a proper grinding is required to achieve a comprehensive extraction of a homogeneous material, and therefore limit technical error.

Necessary precautions for the collection and preparation of samples ahead of their biochemical analyses were previously published [1]. Moreover, all these steps may be carried out by scientists who may not be familiar with all requirements to get a relevant metabolomics study. Plant metabolomics has a broad field of applications [3–5] and is now routinely applied to middle and large scale experiments generating hundreds of samples. To deal with such studies, the whole process must be optimized: from plant harvesting to data processing including sample and extract preparation. Thus, whenever possible, bottlenecks should be removed from the analytical pipeline, and handling made easier. For example, the use of appropriate vials, such as the so-called "scintillation vials," which are made out of high-density polyethylene and thus very resistant to liquid nitrogen and shocks, for harvesting, transportation, storage, and grinding samples may result in substantial time saving. The use of pipetting robots represents a further way to increase the quality and throughput of the experiments.

One of the purposes of plant metabolomics is to better characterize plant quality in terms of food health properties, consumer and producer expectations. It is then relevant to grow the plants in conditions that are close to production practices. This implies considering extra features like climatic conditions and may require specific equipment and strategies to limit plant heterogeneity and other source of errors. We assume that the experimental design will provide a relevant setup and enough material for a statistically significant sampling. In particular, depending on the purpose of the experiment, it will be important to pool material gathered from several individuals [6]. It is also crucial to document the minimum information to describe an experiment as detailed in the Metabolomics Standards Initiative (MSI) [7, 8] to enable comparisons across metabolomics studies. Indeed the biological source (genotype and organ), climatic conditions (temperature, light, and humidity), and any crop management events (plant-health treatments, fertilization, plant pruning, and harvest conditions) should be recorded for further analysis as genotype and environmental conditions will impact the biochemical status of the plant [9, 10]. In addition, whatever the growth conditions (field, controlled, or semicontrolled), the growth environment is not uniform [11] and even at the organ scale, microclimate can alter the plant biochemical composition [12]. Such metadata, i.e., data about the experiment are highly relevant for data mining to improve the understanding of plant metabolism and biological phenomena and avoid erroneous conclusion. The management of all these datasets is a real challenge. The metadata and the data from the biochemical analyses should be comprehensible and well structured to be available for further data mining. For instance it is encouraged to use standard terms and ontologies for metadata annotation. A range of bioinformatics tools including LIMS, databases, web applications, and further software are available; however, more standardization and more bioinformatics tools are required to improve data sharing within the metabolomics community [13–15].

We introduce here general recommendations to get good quality samples for a metabolomics analysis, from the plant cultivation focused on long-term experiments, to the extract preparation through the sample preparation. As example, we use representative studies aiming at the untargeted analysis of semipolar compounds in tomato fruits by liquid chromatography–mass spectrometry (LC-MS) and targeted analysis of fatty acids in flax seeds by gas chromatography–mass spectrometry (GC-MS) or gas chromatography with flame ionization detector (GC-FID).

2 Materials

The vials used will highly depend on the lab equipment, and the sample type, consumable standardization and optimization are key points when handling large sets of samples. Prepare all solutions using glass flasks, ultrapure water (prepared by purifying deionized water, to reach a conductivity of 0.55 μS/cm, 18 MΩ cm at 25 °C), and HPLC or LC-MS grade reagents. All reagents used here are available from commercial suppliers. Prepare reagents at room temperature and store well closed at 4 °C for 1 month (maximum). Always handle toxic solvents and reagents, or solutions containing them, in a fume hood with the appropriate personal protection equipment. Diligently follow the safety datasheet recommendations and all the waste disposal regulations.

2.1 Plant Cultivation

1. Climate control devices: PAR Quantum Sensor (LI-190; LICOR, Lincoln, NE, USA) to measure photosynthetically active radiation (PAR) above the plant or similar tool to measure at least the global radiation; resistance temperature detector (PT-100) to measure temperature and relative hygrometry (RH).
2. Surface-sterilized seeds with a high germination potential (*see* **Note 1**).
3. Safety equipment in experimental facility: disinfectant foot bath, foot mats or overshoes, clean lab coat, hat, and mask.
4. Material to grow, water, stake, prune, and label plants.

2.2 Harvest, Quenching, and Sample Preparation

1. Liquid nitrogen and suitable containers (e.g., Dewar, PVC, or polyurethane ice pan), personal protective equipment (gloves and glasses).
2. Metal spatula, metal tweezers, stainless steel scalpel, knives, and scissors.

3. Permanent marker (Pentel® N50) resistant to long-term storage at −80 °C (*see* **Note 2**).
4. Analytical balance.
5. 2.0 mL tube with screw cap and 20 mL high-density polyethylene (HDPE) vials (Zinsser Analytic®) with screw cap or similar.
6. 4.0 mL glass tubes 4.0 mL with Teflon cap.
7. 2.0 mL safelock microcentrifuge tubes (e.g., Eppendorf).
8. Nitrogen dry shipper (Voyageur, Air Liquide®) (*see* **Note 3**).
9. Ball-mill grinder (Retsch MM400 or SPEX SamplePrep 2010 Geno/Grinder®) or equivalent with adjustable frequency and duration, and stainless steel balls (e.g., 5, 8, or 10 mm ϕ ball).
10. Cryo-station (SPEX SamplePrep®) to safely hold samples under cryogenic conditions.
11. Freeze-dryer (FTS Systems® Dura Dry Dura Stop MP Lyophilizer).
12. Precellys® 24 homogenizer (Bertin Technologies®) and glass beads 6 mm.
13. Cooling unit compatible with Precellys® 24 homogenizer (Cryolys®).
14. 1.0 mL plastic syringes, stainless steel needles and 0.2 μm pore size polyvinylidene difluoride membrane filters.
15. Ultrasonic bath.
16. Micronic 1.10 mL tubes.
17. 96-well format Micronic roborack.
18. Pierceable TPE Capcluster for capping 96 individual tubes.
19. Filter plate 0.22 μm, MultiScreenHTS-GV.
20. Greiner V-bottom Collection plate 300 μL.
21. Nitrogen evaporator system.
22. Centrifuge (capable of 14,000 × *g*).
23. Glass vials suitable for LC-MS, GC-MS, or GC-FID analysis.

2.3 Solutions to Extract Semipolar Compounds from Tomato Fruits

1. Extraction mix 1 (200 mL): 70% methanol, 30% water, 0.1% formic acid, 1.5 mM methyl vanillate as internal standard (*see* **Note 4**). Measure 140 mL of methanol in a graduated cylinder and transfer to a glass bottle. Use another graduated cylinder to add 60 mL of water in the bottle. Weight 50 mg of methyl vanillate, add it to the solution and mix. Use a 200 μL glass Hamilton syringe or similar to add 200 μL of formic acid (98%). Degas the solution by ultrasonication for 15 min.
2. Extraction mix 2 (200 mL): 80% ethanol, 20% water, 0.1% formic acid, 1.5 mM methyl vanillate (internal standard).

Measure 160 mL of ethanol in a graduated cylinder and transfer to a glass bottle. Use another graduated cylinder to add 40 mL of water in the bottle. Weight 50 mg of methyl vanillate, add it to the solution and mix. Use a 200 μL glass Hamilton syringe or similar to add 200 μL of formic acid (98%). Degas the solution by ultrasonication for 15 min.

2.4 Solutions to Extract Fatty Acids from Flax Seeds

1. Extraction mixture 1 (15 mL): mix 10 mL of chloroform into 5 mL of methanol.
2. Extraction mixture 2 (500 mL): KCl 1 M–methanol (4:1, v/v) with 0.034% of $MgCl_2$. Firstly prepare 1 M KCl. The molecular weight of KCl is 74.551 g/mol. To make 500 mL, weight 37.3 g and transfer to a cylinder with 500 mL of MilliQ water (solution 1). Secondly, prepare 0.034% of $MgCl_2$ into methanol. Dissolve 340 mg of $MgCl_2$ into 100 mL of methanol under the fume hood (solution 2). Mix 100 mL of solution 2 into 400 mL of solution 1.

2.5 Separation of Lipid Classes by Thin-Layer Chromatography (TLC)

1. TLC glass plate 20 × 20 cm, silica gel.
2. TLC developing tanks (L × H × W; 27.0 cm × 26.5 cm × 7.0 cm).
3. Prepare standard solutions of lipid classes (PS, PC, PE, PG, MAG, DAG, and TAG) in chloroform at a final concentration of 20 mg/mL (*see* **Note 5**).
4. Developing solvent 1: chloroform–methanol–acetic acid–water (90/15/10/3, v/v/v/v). To prepare 118 mL, mix 90 mL of chloroform with 15 mL of methanol and 3 mL of water. Add carefully 10 mL of acetic acid. Prepare this solution carefully under the fume hood.
5. Developing solvent 2: hexane–diethyl ether–acetic acid (70:30:1, v/v/v). To make 101 mL, combine, under the fume hood, 70 mL of hexane with 30 mL of diethyl ether. Add carefully 1 mL of acetic acid.
6. Iodine crystals.

2.6 Derivatization Prior to GC-MS/GC-FID Analysis

1. Butylated hydroxytoluene (BHT) stock solution (2 mL): dissolve 100 mg of BHT into 2 mL of methanol to a final concentration of 50 μg/μL. BHT prevents unsaturated lipids from being damaged through oxidation.
2. Derivatization mix (100 mL): 5% H_2SO_4 in methanol (v/v). Add carefully 13.5 mL of H_2SO_4 into 86.5 mL of methanol, prepared freshly. For each sample, add 1 mL of derivatization mix and 1 μL of BHT.
3. NaCl 0.9% solution (100 mL): dissolve 900 mg of NaCl into 100 mL of ultrapure water.

3 Methods

The accuracy and reproducibility of a metabolomics study is greatly dependent on the quality of the samples analyzed. Standardization of the plant cultivation, sample harvest, and sample and extract preparation are crucial. During the cultivation step, the biological material might suffer chemical or biological contamination. To date, the cultivation procedure is not standardized. In attempts to improve the documentation of experiments, standardization initiative was created (http://www.biowes.org/).

3.1 Plant Growth

Many general cultivation guidelines are available that explain how to grow a particular species for experimental purposes (e.g., tomato [16], flax [17], *Arabidopsis* [18]. So we only stress specific points that impact plant metabolomics analyses. Short-term experiments (several days up to 6 weeks for tomato) are simple in comparison with long-term experiments (several months, seeds to seeds for tomato). For this reason, we will focus our recommendations on advanced requirements for growing fruit and producing seeds (Fig. 1).

1. *Plant protection strategy*: The most important element for the cultivation step should be to harvest the samples from a healthy plant culture without plant protection treatment. So use hygiene procedures and personal safety equipment in order to avoid external contaminations. A fallow period, without crops in the growth chamber and greenhouse, is a good strategy to obtain pest elimination. In addition, start the culture with contamination-free or surface-sterilized seeds (*see* **Note 1**).

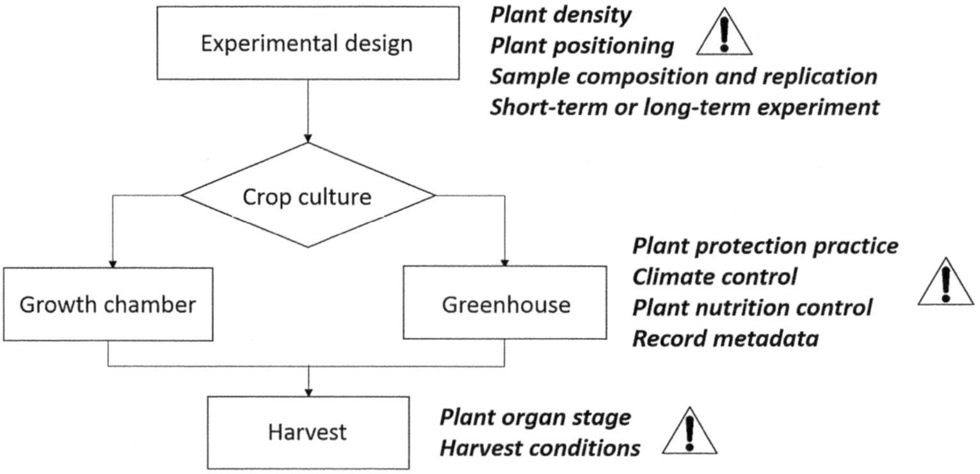

Fig. 1 Schematic representation of the different critical steps for plant cultivation and harvest for a metabolomics study, adapted from [1]

2. *Climate control strategy*: Because plant composition is influenced by the environment, related variables (PAR, temperature, and RH) should be recorded at the levels of the plants and at several places in the culture. In most experiments performed in growth chambers, conventional greenhouses or fully automated phenotyping greenhouses (Smarthouses), these variables are usually controlled but unfortunately, microclimatic variations are frequently observed (e.g., vertical and horizontal gradients of light, temperature, and air humidity), and should be taken into account when sampling (*see* Chapter 1). In the growth chamber, the climate is constant whereas in the greenhouse climate is modulated by the outside meteorological conditions (temperature, light, and relative air humidity). However techniques can be used to avoid extreme disparity (*see* **Note 6**).
3. *Crop management and nutrition control*: Commercial cultivation practices tend to maximize yield whereas for scientific purposes plant development and architecture homogeneity are essential. For example, when the radiation received is supposed to be sufficient and equal for all plants, experimenters should perform measurements at different places in order to check for intensity and homogeneity. Note that in growth chambers using artificial light, the distance of the canopy from the light source is critical. This is especially important when studying genotypes with contrasted heights. Tie the plants to a stake to support them may also be required, as well as the pruning of aerial or adventitious shoots or trusses, when a balance between vegetative and reproductive growth is wanted. Watering and supply of nutrients should be adapted throughout plant development to prevent water stress and nutrient deficiencies. The most important criteria to prepare a nutrient solution are (1) water quality, (2) optimal pH, and (3) recommended electrical conductivity (EC) (*see* **Note 7**). Irrigation and fertilization mismanagement might cause physiological disorders that might lead to misinterpretations. In addition, in experiments comparing genotypes and/or growth conditions, differences in growth rates are expected, thus requiring adjustments in the supply of water and nutrients.
4. *Metadata recording and management*: A number of events occurring during experiments might significantly influence the biochemical composition of the plants and be an important source of variation regarding the metabolome. Thus, it is recommended to record every possible data about the samples and the experiment, in particular climatic conditions and occurrences of planned or unplanned events (e.g., pest and disease control). In our study on tomato we used XEML-Lab and XEO to capture such metadata with controlled vocabularies and ontologies [13].

3.2 Harvest and Quenching

At each step, from the plant to the extract, identification has to be clear and referred to with a unique identifier to ensure sample traceability. Labeling can be achieved with a simple numbering, although QR coding and the associated informatics tools might dramatically improve traceability and documentation of experiments. Quenching, which stops metabolic processes, is achieved by plunging the plant material into liquid nitrogen. Note that whereas for small and thin leaves, freezing is quasi instantaneous, larger organs such as fruits should be cut into small pieces prior to freezing. Also, when photosynthesizing tissues are harvested during the day, quenching should be performed in the light and at the same distance of the source light. The turnover of many metabolites derived from photosynthesis is indeed very fast and even a short variation in light intensity may have big consequences on the metabolome. For safety reason when you close a vial make sure that there is no liquid nitrogen remaining, as it may explode when the nitrogen vaporizes. All harvests and sample preparation (different treatments, developmental stages, and/or organs) should be performed using the same methodology to limit variability within an experiment. A guide of precautions for crop plant emphasizes recommendations about field experiments, sample preparation and sample storage for metabolomics studies has been published previously [1]. The harvest and preparation steps for the tomato fruit pericarps and flax seed samples are detailed in the following paragraphs. According to the sample (tissues, species) adaptations may be undertaken. However, we show here the general steps and recommendations for the preparation of sample which includes grinding, drying and aliquoting necessary before extraction of fatty acid from different classes of lipid and/or semipolar compounds. Please note that fresh samples should stay frozen and stored at −80 °C.

The objective of the harvest is to collect a set of representative samples of the material under study, in sufficient amounts to be able to perform all the analyses, without contamination or alteration. To ensure homogeneity, samples should be chosen according to defined criteria that can be specific to the plant studied, and related to its phenological stage (e.g., days post anthesis, fruit index color, visual aspect, degree-days). Note that mutants or plants under stress may exhibit a delay in growth development compared to the control. We recommend to collect around 3 g of fresh material and not to exceed 2/3 of the vial volume. It can also be considered to harvest extra samples for setup analyses or grinding development. In addition harvest should be performed according to the planned analyses and tissues of interest. As the metabolic composition of plant tissues/organs may vary along the day [19, 20], the time of harvest during the day has to be in line with the questioning. For example, if the goal of the experiment is to study the accumulation of products of photosynthesis, the end of the day is probably the best period [19]. The duration of the harvest should also be evaluated as well as sample composition (part of the plant/organ to be representative, pool), number of replicates, and strategy of harvest (*see* **Note 8**).

Harvest and quenching should be achieved as quickly as possible to limit sample oxidation and degradation, so prepare carefully the bench with all the tools ready, the container filled with liquid nitrogen, the vials already labeled and filled with liquid nitrogen. In our study, tomato fruits were chosen according to days postanthesis, which were known thanks to the tagging of the flowers. Side plants and first fruit of a truss were avoided. The day, time, harvest duration, position of the plant in the greenhouse, truss number, and fruit position were recorded. Total fruit weight and size were measured. Fruit pericarp was cut into quarters (or less), and the placenta, jelly and seeds removed; 0.05 cm^3 of the median part was sliced and frozen. We accumulated at least four fruits to constitute a biological sample, and made four replicates per condition, all in duplicate. Flax seeds can be collected during the embryogenesis step but also at maturity. The stages of harvesting and conservation are not the same according to the date of harvest (*see* **Note 9**).

1. Wear gloves to avoid human skin metabolite contamination.
2. Collect representative parts of the plant. Plants growing on the boundaries of the experimental space may exhibit a different phenotype and should be avoided. Also avoid proximal and distal parts and wounded organs. Optional: clean the plant material with water (distilled or MilliQ) and wipe gently with disposal paper. The cleaning steps should be constant and consistent with the metabolites of interest [21].
3. Cut into small pieces the plant/organ, approximately 0.05 cm^3.
4. Immediately freeze pieces by plunging them into vials full of liquid nitrogen. The freezing step will be faster if plant/organ slices are thin, and the grinding easier if sample pieces are small and do not stick to each other.
5. Duplicate the sample into a second vial in case of sample loss (backup sample).
6. Record metadata (climatic condition, time of harvest, weight and size of entire organs, and fruit position on the truss).
7. Clean tools.
8. Repeat steps to prepare independent replicates.

3.3 Sample Grinding

Grinding allows tissue disruption and sample homogenization [21, 22]. An effective extraction is a key step to maximize the repeatability of the analysis. There is a large choice of grinders (equipped with knives, discs, tooth-rotors, balls, etc.), which allow dealing with the diversity of the plant matrix characteristics (e.g., hardness, fibrous tissue, particle size, and physicochemical properties), the sample volume, the type of chemical analyses planned and the final fineness desired. The particle size and shape can be measured using a microscope and a micrometer scale. In our study, we used ball-mill grinders to get a particle size around 5 μm, and stainless steel (or glass beads) to reduce the contamination of the sample through abrasion.

1. Perform grinding under cryogenic conditions, at a minimum temperature of −50 °C to avoid metabolite degradation due to enzyme activities.
2. Store the samples in liquid nitrogen or dry ice just before and after the grinding. Store the samples at −80 °C for long-term storage. The potential degradation duration storage should be evaluated.
3. It is recommended to observe in the grinding bowl the proportion of 1/3 of sample, 1/3 of beads, and 1/3 of empty space. However, there is no universal method to achieve a good grinding.
4. Set the following parameters: number of beads (from 1 to 3), its diameter (5, 8, 10, or 25 mm), frequency (up to 30 Hz) and the duration of the grinding according to the vials, the quantity, and nature of the matrix under study (e.g., species, organ, and phenology).
5. For each new matrix, we recommend to first try a method validated for a similar organ (Table 1), and then to modulate the parameters given above (**step 4**) to find the best compromise between duration, number, and bead diameter (*see* **Note 10**).

3.3.1 Grinding Fruit Pericarp Samples (e.g., Red Tomato)

1. Prepare the vials in liquid nitrogen, dry ice or in a precooled cryo-station.
2. In each sample vial add three balls of 8 mm diameter; close the vials with a metallic cap first and its lid.
3. Proceed the grinding for 1 min at 1750 rpm (30 Hz).
4. Visually inspect the quality of the powder (with a frozen spatula). It should be finely ground, powdery, and not melted; otherwise the settings are not appropriate.
5. Remove the balls with a magnet, close the vial with the lid, and store it at −80 °C.
6. Wash the balls and the cap with ethanol and water, wipe it well or let dry.

3.3.2 Grinding Seed Samples (e.g., Flax Seeds)

1. Place liquid nitrogen or dry ice into the cooling unit (Cryolys®).
2. Connect the cooling unit (Cryolys®) to the Precellys 24-Dual homogenizer.

Table 1
Example of grinding parameters for several matrices

Tissue	Apparatus	Bowl	Bead number	Bead diameter (mm)	Frequency (rpm or Hz)	Duration (s)
Tomato red fruit	Genogrinder	20 mL HDPE	3	8	30 Hz	1 × 90
Tomato green fruit	Genogrinder	20 mL HDPE	3	8	30 Hz	1 × 60
Grappe fruit	Genogrinder	20 mL HDPE	2	10	30 Hz	1 × 45
Tomato leaf	Genogrinder	20 mL HDPE	3	8	30 Hz	1 × 35
Arabidopsis leaf	Genogrinder	20 mL HDPE	1	5	30 Hz	2 × 30
Tomato stem	Genogrinder	20 mL HDPE	3	8	30 Hz	1 × 45
Tomato root	Retsch mill	50 mL stainless	1	25	30 Hz	1 × 60
Flax seed	Precellys	2 mL tubes	3	6	6000 rpm	2 × 30

Apparatus refers to ball-mill grinder, bowls are either vials used for the harvest (high-density polyethylene (HDPE) or polypropylene), or specific ones in stainless steel. The stroke frequency is expressed in Hertz (Hz) or rotation per minute (rpm), and the duration in second (s)

3. Place 200 mg of flax seeds (fresh material) with two glass beads (6 mm) into a 2.0 mL safelock microcentrifuge tube.
4. Place the sample tube into a Precellys 24-Dual homogenizer and agitate twice for 30 s at 6000 rpm. Control the quality of the powder.
5. Remove glass beads and close the sample tube.
6. Store all sample tubes at −80 °C or lyophilize the samples.

3.4 Sample Drying

Depending on the matrix under study, the stabilities and concentrations of metabolites, metabolomics analyses are carried out based on fresh or dry matter. Samples can be dried by several methods; however, freeze-drying is the only suitable method because free water is removed by sublimation and not by heating the material. Freeze-drying samples is risky and time-consuming (can last 2–5 days), but it facilitates storage and weighing of aliquots, concentrates metabolites, and to a certain extent, normalizes the data as water is removed [2]. Lyophilization is improved by using ground samples. Once samples have been lyophilized it is important to prevent any reabsorption of water from the surrounding environment to prevent their degradation.

1. Use a cryo-station or cold conditions to make an aliquot of the sample. Mind to aliquot enough material.
2. Adapt a drilled lid on the vials and store at −80 °C prior to freeze-drying.
3. Freeze-dry samples under appropriate conditions; samples should not melt (*see* **Note 11**).
4. Close the vials with a regular lid, and store. Vials can be stored prior to weighing for 2 weeks in the dark, at room temperature under vacuum. Alternatively, vials can be stored for longer periods with desiccant-like silica gel in sealed bags, at least at −20 °C.

3.5 Sample Aliquoting

The objective of this step is (1) to weight a precise amount of material and (2) to transfer the sample to a new vial (named extraction vial) suitable for the extract preparation. Weighing can be achieved manually or automated (*see* **Note 12**). For fresh weight (FW) material, proceed under cryogenic conditions, and for dry weight (DW) material, follow the same procedure at room temperature.

1. Precool the new labeled vials, tweezers, and spatula in liquid nitrogen or dry ice.
2. Use the frozen spatula to homogenize the sample.
3. Tare the extraction vial, move it with the cold tweezers.
4. Transfer a precise amount of fresh material in the extraction vial, the weight should be the same for all the samples ±10% (2 mg if you need 20 mg).
5. Store the vial in liquid nitrogen or dry ice and at −80 °C prior to analysis.
6. Let the spatula come up to room temperature and clean it with ethanol.

3.6 Sample Extraction

The next step in a plant metabolomics study, after sample harvest, grinding, and aliquoting, is the sample extraction. The objective of this step is to retrieve a maximum range of metabolites from the plant matrix by using a robust and reproducible method (*see* Chapter 2), and in the most efficient way, i.e., as simple and as fast as possible, with a good extraction yield, and without chemically modifying compounds. Sample extraction could be achieved using several techniques like solvent extraction, microwave-assisted extraction (MAE), supercritical fluid extraction (SFE), vapor phase extraction (headspace analysis), solid-phase extraction, or Soxhlet extraction; each one presents different levels of difficulty to implement and can be more selective for a specific compound class [2, 4, 22–24]. Among these techniques, for large-scale plant metabolomics studies, solvent extraction methods are frequently used. With plant samples, major difficulties are the wide variety of compounds that can be found and the fact that metabolite concentrations can vary over several orders of magnitude [5]. To detect as many metabolites as possible, several extractions and analytical techniques are commonly combined. Indeed, even if a one-step extraction method using MTBE that allows the detection of several metabolites, including polar, semipolar metabolites and lipids, has been previously described [25] to date there is no analytical technique that allows simultaneous detection of all metabolites in a single run. An appropriate internal standard (labeled analyte, deuterated or chemically close) should be implemented for semi or quantitative bioanalysis. Here, we will first detail two extraction procedures that are used to characterize semipolar compounds by untargeted LC-MS analysis (Subheadings 3.6.1 and 3.6.2). We then describe sample preparation details to extract the total lipid content from plant material, followed by the characterization of the fatty acid composition of different classes of lipids by TLC coupled to GC-MS or GC-FID analysis (Subheading 3.6.3).

3.6.1 Manual Methanolic Extraction of Semipolar Compounds

1. Weight 20 mg dry weight (DW) into a 2.0 mL safelock microcentrifuge tube and add 1 mL of extraction mix 1 (*see* **Notes 4** and **13**).
2. Vortex.
3. Extract the sample by ultrasonication for 15 min and at 4 °C (use crushed-ice in the ultrasonic bath) (*see* **Note 14**).
4. Centrifuge at 14,000 × *g* for 5 min.
5. Transfer and filter the supernatant using the combination of syringe-needle-filter to suitable analytical glass vials for LC-MS analysis (e.g., 2.0 mL glass vials with 200 μL glass inserts).
6. A quality control (QC) sample can be prepared by pipetting 5–10 μL of each sample in a new vial. Mix well.
7. Store the extract prior to analysis at 4 °C for a few days or at −20 °C for longer storage. The duration of the storage of the extract at 4 or −20 °C without degradation should be evaluated depending on the analyzed metabolites and plant materials.
8. The extract is ready to be analyzed by LC-MS using a previously described method [20].

3.6.2 Automated Ethanolic Extraction of Semipolar Compounds

The previously described methanolic extraction is extremely common to perform metabolomics studies. However, it requires specific procedures due to methanol toxicity and thus limits the possibility to use pipetting robots to achieve high-throughput performance, which is necessary to analyze large sample sets (>1000 samples). To adapt the previous protocol to our robotized platform, we have evaluated the replacement of 70% methanol with

80% ethanol and found that ethanol could partially replace methanol, at least for comparative approaches (*see* **Note 15**). The following version has been tested with a MicroLab star Hamilton pipetting robot but simpler, semiautomated robots can also be used.

1. Scale 10 mg DW of your sample in 1.10 mL Micronic tubes grouped in 96-well racks. In each rack, allow one tube for the QC sample (*see* **step 10**) and three tubes for blanks.
2. Program your pipetting robot to transfer 300 μL of extraction mix 2 in each rack and mix (*see* **Notes 4** and **13**).
3. Cap tubes and transfer racks to an ultrasonication bath for 15 min in crushed-ice.
4. Centrifuge at 4000 × *g* for 5 min.
5. Program your pipetting robot to transfer supernatants to a fresh 96-well rack.
6. Repeat **steps 2–5**.
7. Transfer your extract to a 96-well filter plate (0.22 μm) placed over a 96-well collection plate.
8. Centrifuge at 1000 × *g* for 2 min to perform filtration.
9. Transfer the filtrated extract to suitable analytical glass vials for LC-MS analysis (e.g., 2.0 mL glass vials with 200 μL glass inserts).
10. A QC sample can be easily prepared by pipetting 5–10 μL of each sample in a clean tank. Transfer to a 50 mL falcon and stir well.
11. The extract is ready to be analyzed by LC-MS (*see* **Note 16**).

3.6.3 Total Lipid Content Extraction, Separation of Lipid Classes by TLC Followed by GC-MS/GC-FID

Lipids from plant materials are complex mixtures of different lipid classes. These lipids can be separated in polar and neutral fractions and their acyl groups quantified. Basically, rigorous analysis of glycerolipids needs a freshly collected or properly stored tissue to minimize oxidation, or lipolysis. The cheapest and most convenient option to analyze purified polar and neutral glycerolipids is TLC. Once separated, the fatty acid content and the composition of each glycerolipids can be determined. The fatty acids are extracted from glycolipids, converted into fatty acid methyl esters (derivatization), and finally, separated and quantified by either GC-MS or GC-FID. A previously described acid-based method was used and modified to improve the accuracy for tissues with high triacylglycerol content [26]. This simple procedure can be used (1) for the detection of genotypes accumulating unusual fatty acids and to define in which lipid classes they are accumulated, and (2) to evaluate the effect of environmental conditions (or mutation) on the fatty acid content and their incorporation in different lipid classes.

Several sample preparation steps are necessary to determine the composition of fatty acids of the different lipid classes and/or quantify them (Fig. 2). These steps consist in the extraction of the total lipid content (**step 1**), followed by the separation of the different lipid classes by TLC (**step 2**), and finally, the derivatization of the fatty acids before GC-MS or GC-FID analysis (**step 3**). Although optional, lipid extracts can be stored prior to **step 2** (*see* **Note 17**).

1. Place 200 mg of lyophilized plant material (i.e., flax seeds) with two glass beads (6 mm) into 2.0 mL safelock microcentrifuge tubes (Fig. 2, **step 1**).
2. Add 1 mL of lipid extraction mix 1, in a fume hood.
3. Place the sample tubes into a Precellys 24-Dual homogenizer and agitate twice for 30 s at 6000 rpm.
4. Centrifuge sample tubes at 5000 × *g* during 5 min.
5. Collect the supernatant into a 4.0 mL glass tube with glass Pasteur pipettes, in a fume hood.
6. Add 1.0 mL lipid extraction mix 1.
7. Add 2.0 mL lipid extraction mix 2 and homogenize during 2 min.
8. Centrifuge the tubes at 5000 × *g* during 7 min.
9. Collect the lipid phase (lower-phase) and place into 4.0 mL glass tubes, for which the weight has previously been measured using an analytical balance (*see* **Note 18**).
10. Evaporate to dryness using a nitrogen evaporator system during 30 min at 40 °C.
11. Calculate the weight of the lipid as follows:

$$\text{Amount of lipid (mg)} = (\text{weight of 4.0 mL glass tube} + \text{extracted lipid}) - (\text{weight of 4.0 mL glass tube}).$$

12. Calculate the lipid content in the sample as follows:

$$\text{Lipid content (\%)} = \text{amount of lipid extracted (mg)/weight of original sample (mg)} \times 100.$$

13. Dissolve the lipid extracts into 200 μL of chloroform just before loading them on the TLC plate.
14. Activate the TLC plate by baking in an oven at 105 °C for 1 h (Fig. 2, **step 2**).
15. Let the activated plates cool down to room temperature.
16. Prepare the TLC developing chamber in a fume hood as follows: add 118 mL of the developing solvent 1 into a clean TLC developing chamber n° 1; add 101 mL of the

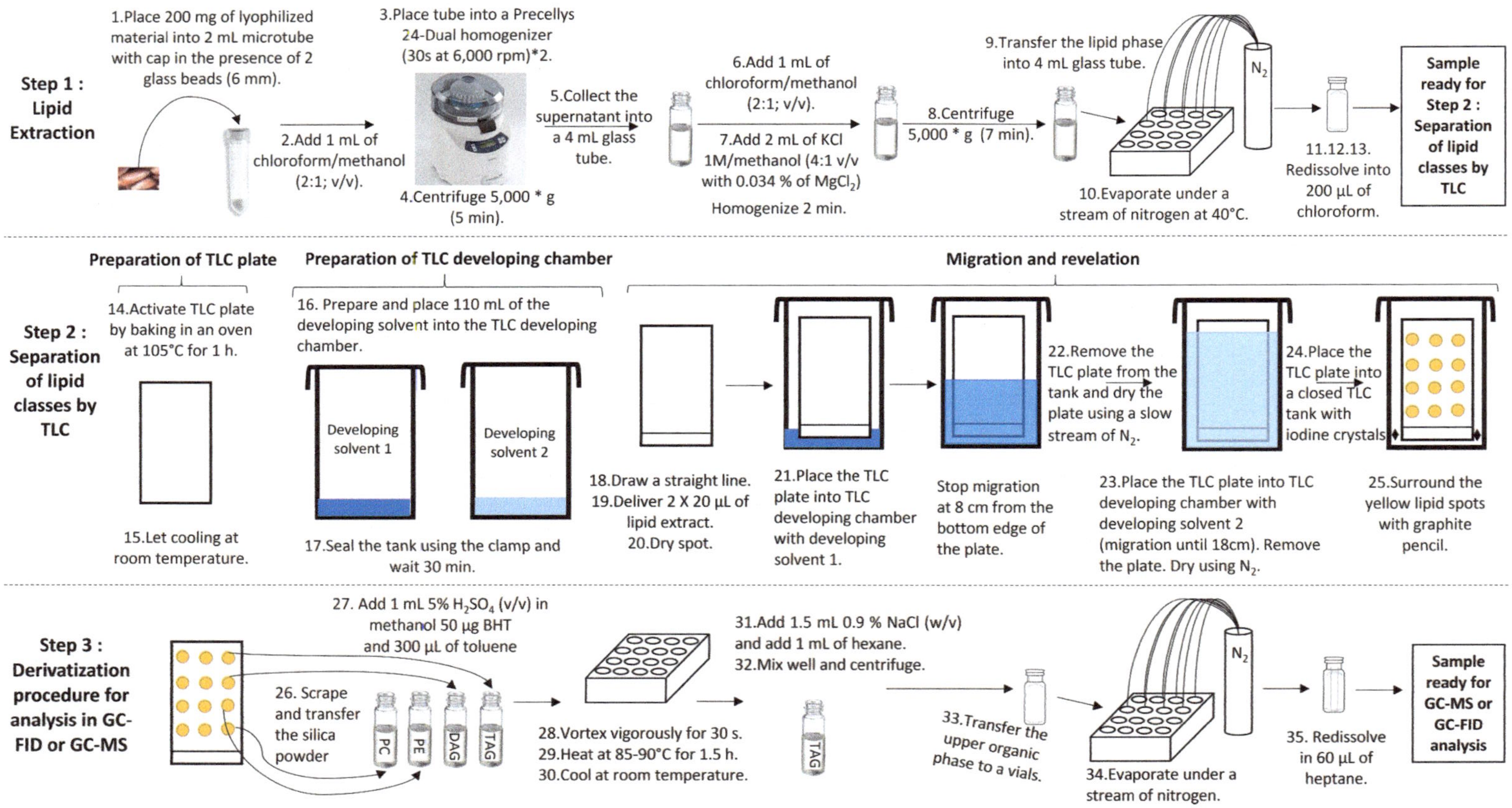

Fig. 2 Schematic representation of the principal sample preparation steps for fatty acid analysis of different lipid classes extracted from plant samples (Subheading 3.6.3, **steps 1–35**). For more details and recommendations, refer to this section. *TLC* thin-layer chromatography, N_2 nitrogen, *PC* phosphatidylcholine, *PE* phosphatidylethanolamine, *DAG* diacylglycerol, *TAG* triacylglycerol

developing solvent 2 into a clean TLC developing chamber n° 2 (*see* **Note 19**).

17. Seal the tank using the clamp and wait for 30 min.
18. Use a pencil to draw a straight line (1.5 cm from the edge of the plate) across the TLC plate at the origin of the chromatogram.
19. Prepare the standards (PS, PC, PE, PG, MAG, DAG, and TAG) at 20 mg/mL. In a fume hood, apply 2 × 20 μL of lipid extract and 8 μL of each standard on the TLC plate (Fig. 3) (*see* **Note 20**).
20. Dry spots with the help of a nitrogen flow.
21. Place the TLC plate into the elution tank n° 1. The solvent line should be below the baseline marked on the TLC plate.

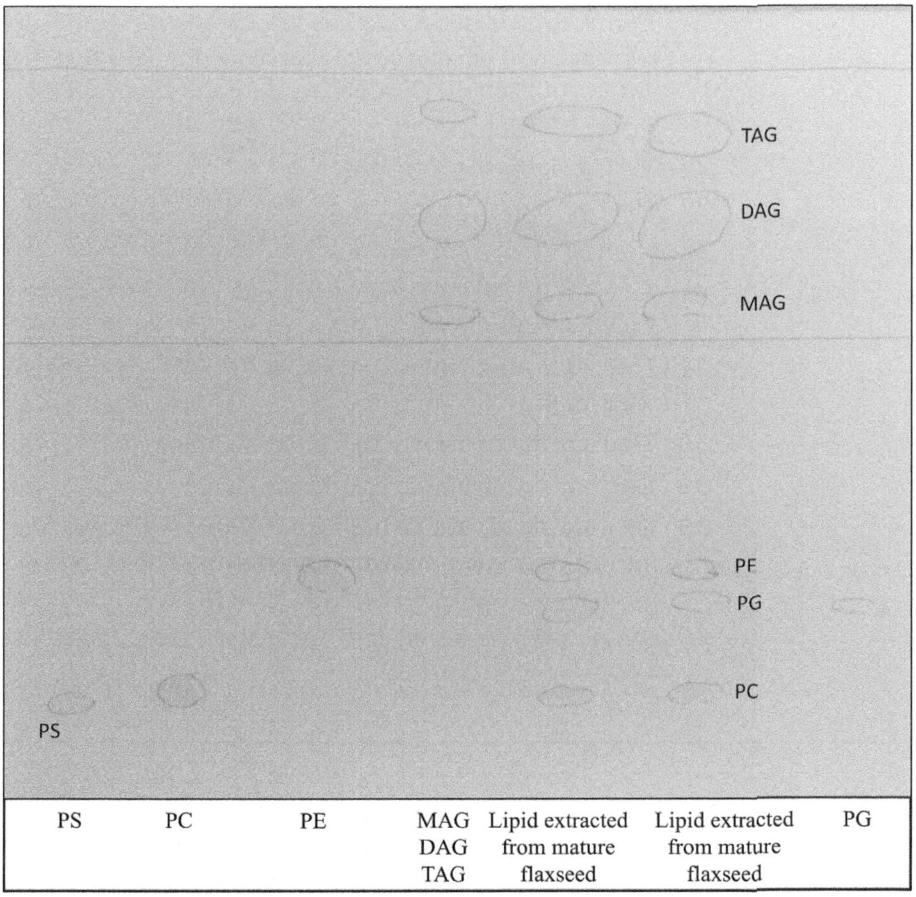

Fig. 3 Separation of lipid classes extracted from mature flax seeds. Eight microliters of standards (PS, PC, PE, PG, MAG, DAG, and TAG) at 20 mg/mL was deposited on the TLC plate. *PS* phosphatidylserine, *PC* phosphatidylcholine, *PG* phosphatidylglycerol, *PE* phosphatidylethanolamine, *MAG* monoacylglycerol, *DAG* diacylglycerol, *TAG* triacylglycerol

Then allow migration until the solvent front has reached 8.5 cm from the top of the plate.

22. Remove the TLC plate from the tank and let it dry under the fume hood.
23. Place the TLC plate into the developing chamber n° 2 and allow migration until the migration solvent has reached 18.5 cm from the top of the plate. Remove the plate, immediately mark the front of the solvent with a pencil. Dry completely the plate under the fume hood.
24. To reveal the lipids, place the dry TLC plate into a closed TLC tank with iodine crystals still in fume hood (*see* **Note 21**).
25. Surround the yellow lipid spots with a pencil.
26. Scrape the lipid containing silica and transfer the silica powder using a funnel into a 4.0 mL glass tube with a Teflon (PTFE)-lined screw cap (*see* **Note 22**).
27. Add 1 mL of derivatization mix, 1 μL BHT and 300 μL of toluene. Add an appropriate internal standard that is not present in your sample. For flax seed samples, add 20 μg of C17:0 TAG (triheptadecanoin) (Fig. 2, **step 3**).
28. Vortex vigorously for 30 s.
29. Heat at 85–90 °C for 1.5 h.
30. Cool at room temperature.
31. Add 1.5 mL 0.9% NaCl (w/v) and 1 mL of hexane to extract FAMEs.
32. Mix well and then centrifuge briefly to facilitate phase separation.
33. Transfer the upper organic phase to a new vial.
34. Evaporate under a stream of nitrogen.
35. Dissolve in 60 μL of heptane (vortex well), transfer samples into suitable glass vials and cap tightly. The extract is ready to be analyzed by GC-MS or GC-FID [27].
36. Samples can be stored at 4 °C for short term and at −20 °C for a few days (*see* **Note 17**).

4 Notes

1. To prevent disease transmission, tomato seeds should be surface sterilized in a calcium hypochlorite solution (4%) with a drop of surfactant (Tween 20), for 20 min under stirring, and then rinsed with sterile and distilled water three times. The number of seeds must be larger than the desired amount (20% more) to select uniform plants for transplantation.

2. QR codes or stickers can also be used. Check whether they withstand cryogenic conditions, water and solvents used.
3. Nitrogen dry shipper allows the safe transportation of frozen samples. Cryogenic temperatures can be kept for several days to weeks. Note that the transportation of samples involving liquid nitrogen or dry ice requires ventilated cars and might be subject to restrictions or local regulation when using public transportation.
4. Prepare a large amount of extraction mix (solvent) in order to use the same solution for a full set of samples. Formic acid is added to prevent the degradation of secondary metabolites like phenolic compounds. Internal standard is added to avoid intensity variation during the analysis. During data processing sample compound areas are normalized with the internal standard area and with the compound area mean of QC samples providing relative quantification data.
5. Standards of lipid classes are required to control the good separation of lipid classes extracted from plant material by TLC. The lipids extracted from flax seeds are composed mainly of three neutral lipids: monoacylglycerols (MAG), diacylglycerols (DAG), triacylglycerols (TAG) and four phospholipids: phosphatidylserine (PS), phosphatidylcholine (PC), phosphatidylglycerol (PG), and phosphatidylethanolamines (PE). Plant photosynthetic tissues are characterized by high abundance of glycoglycerolipids especially in monogalactosyldiacylglycerol and digalactosyldiacylglycerol (MGDG, DGDG). Standards of MGDG and DGDG can also be added and separated using the protocol described in this chapter by TLC.
6. The regulation of vapor pressure deficit (VPD) has been demonstrated as an efficient solution to simultaneously maintain optimal ranges of temperature and relative humidity under greenhouse. Therefore, plant photosynthesis is enhanced [28], and water condensation which promotes the development of fungal pathogens on the plant organs is limited [29]. Use electrical evaporative cooling device to control temperature and relative air humidity. Outside shading system and sunroof can also be used to control the temperature in greenhouse.
7. Depending on sources, water composition varies tremendously because of its ion concentrations and contaminations with biological or chemical agents. Monitor the concentration of nutrient solution using a sensor in the drain solution to measure its EC. An automation of the irrigation using a combined time and light-integral controlled water supply is advised.
8. The biochemical composition of the plant organ may vary along the day so it is important to consider the time of harvest and to limit the duration of the collect, which also means that

the number of experimenters preparing the samples should be planned. For example, one person could collect the fruits, one could record and write down morphological data (sample size, weight, etc.) and two could cut samples into pieces and freeze them.

9. The harvesting and storage of flax seeds at all stages of development is carried out in four steps: firstly, cut the capsules off the plants very quickly; secondly, open the capsules very quickly with a scalpel to recover the seeds; thirdly, immediately drop the harvested seeds into a 2.0 mL microcentrifuge tube previously placed in a liquid nitrogen bath; fourthly, write the stage of development of the harvested seed on the sample tube; and fifthly, store at −80 °C or freeze-dry. In the case of mature seeds: firstly crush the capsules; secondly, harvest the seeds and place them in a 15 mL sterile falcon; thirdly, place the falcon containing the seeds in a craft envelope; and fourthly, place the craft envelope in a cupboard dedicated to the preservation and storage of seeds at 12 °C, without light and a minimum of humidity of 50%.

10. According to our lab workflow, the SPEX SamplePrep 2010 Geno/Grinder® was adapted for scintillation vials that are used to harvest samples. The adaptations consist of a metallic form to organize the vials in the grinder and metallic caps to close vials and to avoid plastic contaminations by the regular lid. Besides the Retsch mill gives better results for fibrous organs/tissues like stems or roots when using specific metallic bowls. However, in that configuration it can only handle two samples at once, and the bowls need to be washed after each sample. A consequence is that huge amounts of liquid nitrogen are needed to cool down the bowl after each washing. Increasing the number of beads might increase the grinding efficiency but more material will be lost. Fruits are difficult to grind because of their high water content. Thus, bigger beads may improve the grinding but may also break the HDPE vials. If the sample is divided into several vials, we recommend using bigger vials, or a grinding bowl big enough to grind all the sample material at the time as the mix of a powder is not easy to achieve.

11. Start the freeze dryer at least 1 h in advance, to let it cool down to −30 °C, put under vacuum and check that there is no leak. Then place inside the frozen samples; do it gradually in order to keep the temperature at −30 °C. For tomato fruit, our freeze-drying program lasts 5 days, as detailed: vacuum is set between 6 and 12 Pa (50–100 mT), day 1 (D1) at −30 °C, D2 increase gradually in temperature up to −20 °C, D3 increase gradually in temperature up to −10 °C, D4 increase gradually in temperature up to 0 °C, D5 increase gradually in

temperature up to 20 °C, take out the vials late in the afternoon and store them under dry conditions.

12. Sample weighing is a critical step that has to be operated under cryogenic conditions, and represents a bottleneck for high throughput. The design of Cryo-weighing robots helps removing this bottleneck.
13. The amount of material, the use of fresh matter instead of dry matter, as well as the volume of extraction solvent (solvent--sample ratio) can be modified according to the matrix studied, the analytical technology chosen, the metabolites targeted, and the detection limit of the analytical equipment.
14. Ultrasonic treatment is a way to improve the extraction efficiency; we recommend staying under cold condition to avoid artifact and metabolite degradation [2]. The use of ULTRA-TURRAX® also leads to good extraction efficiency, but it is more difficult to implement and more time-consuming, as samples are processed one by one.
15. We have compared methanolic and ethanolic extractions with satisfactory results using three types of sample matrices (mature sunflower, maize, and Arabidopsis leaves).
16. Alternatively, wash the sample with 300 μL (ethanol 50%, HEPES KOH 10 mM, pH 6), and discard supernatant. This allows for pellet recovery for starch and total protein analyses.
17. Although optional, storage of lipid extracts is very important to allow carrying out experiments several days or weeks after extraction. Lipid extracts can be stored at −80 °C in glass tubes with Teflon caps, which have previously been flushed with nitrogen or argon. Alternatively, lipid samples can be stored at room temperature for 4 weeks after pretreatment using flushing nitrogen and oxygen absorber. This method is very convenient for sending samples to distant laboratories. It is also recommended to use aluminum foil to prevent the samples from the oxidation from sunlight exposure [30]. - Dissolve the lipid extracts into 300 μL of chloroform with 50 μg/mL of BHT. Rapidly flush the glass tubes with nitrogen before closing to prevent fatty acid oxidation. Store the glass tubes at −80 °C. Make sure that the extracts are protected from air and light. To transport lipid extracts to a distant laboratory, place lipid samples in a plastic bag with oxygen absorber in a refrigerator at 2 °C. After 7 days, the residual oxygen in the plastic bag will be sufficiently eliminated by the oxygen absorber. The lipid extracts in the bag will be kept almost completely unchanged during further preservation at room temperature for at least 4 weeks. However, it should be remembered that the best results are obtained with freshly prepared material.

18. Extraction and quantification of both polar and nonpolar lipids are done using the Folch method [31]. Lipids are extracted with the extraction mix 1 whereas other cellular components are removed by the addition of the extraction mix 2. The two phases are separated by centrifugation. The lower chloroform phase includes the lipids and the top methanol–KCl phase contains the nonlipid components [31].
19. The separation of lipid classes is performed by TLC. This method has the advantage of being relatively inexpensive and fast. The separation, which is performed on a polar TLC plate silica gel stationary phase, occurs due to polarity differences of the lipid classes by two different apolar mobile phases. The first apolar mobile phase (developing solvent 1) used contains chloroform–methanol–water (60:35:8, v/v/v) and migrates to 50% of the height of the TLC plate (from the baseline, 9 cm), whereas the second mobile phase (developing solvent 2) used contains hexane–diethyl ether–acetic acid (70:30:1, v/v/v), and migrates to 90% of the height of the TLC plate. The first mobile phase separates the different polar glycerolipids (PS, PC, PG, and PE) and the second separates the neutral glycerolipid fraction (MAG, DAG, TAG) [32]. To assess the separation, the developed plate is exposed to iodine vapor, a reversible staining, which will form noncovalent brown complexes with unsaturated lipids. A typical TLC plate is shown in Fig. 3 using lipid samples extracted from mature flax seeds.
20. A deposit must not exceed a diameter of 1 cm (due to spreading effect during migration). Each plate can hold up to ten samples.
21. The iodine crystals must not be directly in contact with the TLC plate. Please note that samples that contain highly saturated lipids are hardly stained. Do not expose the plates to iodine for too long, as iodine may covalently modify polyunsaturated fatty acids.
22. Silica powder is carcinogenic, so put on gloves, mask, and protective goggles.

Acknowledgments

C. Bénard thanks Patricia Ballias and Mickael Maucourt for fruitful technical discussions and Dr. Stéphane Bernillon for reviewing the manuscript and providing helpful comments. This work was supported by the projects MetaboHUB-ANR-11-INBS-0010, PHENOME-ANR-11-INBS-0012, SUNRISE-ANR-11-BTBR-0005, the Regional Council of Picardie, and by the European Union which cofunded equipment utilized within CPER 2007-2020.

References

1. Biais B, Bernillon S, Deborde C et al (2012) Precautions for harvest, sampling, storage, and transport of crop plant metabolomics samples. In: Hardy N, Hall R (eds) Plant metabolomics, Methods in molecular biology (methods and protocols), vol 860. Humana, New York, pp 51–63
2. Kim HK, Verpoorte R (2010) Sample preparation for plant metabolomics. Phytochem Anal 2:4–13
3. Allwood JW, De Vos RC, Moing A et al (2011) Plant metabolomics and its potential for systems biology research background concepts, technology, and methodology. In: Jameson D, Verma M, Westerhoff HV (eds) Methods in enzymology, vol 500. Academic, Amsterdam, pp 299–336
4. Ernst M, Silva DB, Silva RR et al (2014) Mass spectrometry in plant metabolomics strategies: from analytical platforms to data acquisition and processing. Nat Prod Rep 31:784–806
5. Hall RD (2006) Plant metabolomics: from holistic hope, to hype, to hot topic. New Phytol 169:453–468
6. Gibon Y, Rolin D (2012) Aspects of experimental design for plant metabolomics experiments and guidelines for growth of plant material. In: Hardy N, Hall R (eds) Plant metabolomics, Methods in molecular biology (methods and protocols), vol 860. Humana, New York, pp 13–30
7. Fiehn O, Sumner LW, Rhee SY et al (2007) Minimum reporting standards for plant biology context information in metabolomic studies. Metabolomics 3:195–201
8. Fiehn O, Wohlgemuth G, Scholz M et al (2008) Quality control for plant metabolomics: reporting MSI-compliant studies. Plant J 53:691–704
9. Bernillon S, Biais B, Deborde C et al (2013) Metabolomic and elemental profiling of melon fruit quality as affected by genotype and environment. Metabolomics 9:57–77
10. Bundy JG, Matthew PD, Viant MR (2009) Environmental metabolomics: a critical review and future perspectives. Metabolomics 5:3–21
11. Cabrera-Bosquet L, Fournier C, Brichet N et al (2016) High-throughput estimation of incident light, light interception and radiation-use efficiency of thousands of plants in a phenotyping platform. New Phytol 212:269–281
12. Benard C, Gautier H, Bourgaud F et al (2009) Effects of low nitrogen supply on tomato (*Solanum lycopersicum*) fruit yield and quality with special emphasis on sugars, acids, ascorbate, carotenoids, and phenolic compounds. J Agric Food Chem 57:4112–4123
13. Hannemann J, Poorter H, Usadel B et al (2009) Xeml Lab: a tool that supports the design of experiments at a graphical interface and generates computer-readable metadata files, which capture information about genotypes, growth conditions, environmental perturbations and sampling strategy. Plant Cell Environ 32:1185–1200
14. Rocca-Serra P, Brandizi M, Maguire E et al (2010) ISA software suite: supporting standards-compliant experimental annotation and enabling curation at the community level. Bioinformatics 26:2354–2356
15. Haug K, Salek RM, Steinbeck C (2017) Global open data management in metabolomics. Curr Opin Chem Biol 36:58–63
16. Schwarz D, Thompson AJ, Kläring HP (2014) Guidelines to use tomato in experiments with a controlled environment. Front Plant Sci 5:625
17. Muir AD, Westcott ND (eds) (2003) Flax: the genus Linum. CRC Press, Boca Raton
18. Weigel D, Glazebrook J (eds) (2002) Arabidopsis: a laboratory manual. CSHL Press, New York
19. Gibon Y, Usadel B, Blaesing OE et al (2006) Integration of metabolite with transcript and enzyme activity profiling during diurnal cycles in Arabidopsis rosettes. Gen Biol 7:R76
20. Benard C, Bernillon S, Biais B et al (2015) Metabolomic profiling in tomato reveals diel compositional changes in fruit affected by source–sink relationships. J Exp Bot 66:3391–3404
21. Markert B (1995) Sample preparation (cleaning, drying, homogenization) for trace element analysis in plant matrices. Science Total Environ 176:45–61
22. Mushtaq MY, Choi YH, Verpoorte R, Wilson EG (2014) Extraction for metabolomics: access to the metabolome. Phytochem Anal 25:291–306
23. Hiue CW (2002) A review of modern sample-preparation techniques for the extraction and analysis of medicinal plants. Anal Bioanal Chem 373:23–30
24. Tikunov Y, Lommen A, Ric de Vos CH et al (2005) A novel approach for nontargeted data analysis for metabolomics. large-scale profiling of tomato fruit volatiles. Plant Physiol 139:1125–1137
25. Salem MA, Jüppner J, Krzysztof B, Giavalisco P (2016) Protocol: a fast, comprehensive and reproducible one-step extraction method for the rapid preparation of polar and semi-polar

metabolites, lipids, proteins, starch and cell wall polymers from a single sample. Plant Methods 12:45

26. Li MY, Welti R, Wang XM (2006) Quantitative profiling of Arabidopsis polar glycerolipids in response to phosphorus starvation: roles of phospholipases D zeta 1 and D zeta 2 in phosphatidylcholine hydrolysis and digalactosyldiacylglycerol accumulation in phosphorus-starved plants. Plant Physiol 1428:750–761
27. Browse J, Mccourt PJ, Somerville CR (1986) Fatty acid composition of leaf lipids determined after combined digestion and fatty acid methyl ester formation from fresh tissue. Anal Biochem 152:141–145
28. Zhang D, Du Q, Zhang Z et al (2017) Vapour pressure deficit control in relation to water transport and water productivity in greenhouse tomato production during summer. Sci Rep 7:43461
29. Albajes R, Gullino ML, van Lenteren JC, Elad Y (eds) (1999) Integrated pest and disease management in greenhouse crops. Kluwer Academic Publishers, Dordrecht
30. Hirao S, Ishida Y, Tsuge S, Ohtani H (2003) A novel method for preservation of labile lipid samples at ambient temperature with oxygen absorber. J Oleo Sci 52:583–588
31. Folch J, Lees M, Sloane Stanley GH (1957) A simple method for the isolation and purification of total lipides from animal tissues. J Biol Chem 2268:497–509
32. Touchstone JC (1995) Thin-layer chromatographic procedures for lipid separation. J Chrom B Biomed Appl 671:169–195

Chapter 9

Extraction of Plant Lipids for LC-MS-Based Untargeted Plant Lipidomics

Thusitha W. T. Rupasinghe and Ute Roessner

Abstract

Lipids are defined as hydrophobic or amphipathic small molecules which consist of a number of structurally and functionally distinct molecules that span from nonpolar to neutral to polar compounds. Lipidomics is the comprehensive analysis of all lipids in a biological system. Changes in lipid metabolism and composition, as well as of distinct lipid species have been linked with altered plant growth, development, and responses to environmental stresses including salinity. Recently, improved liquid chromatography mass spectrometry (LC-MS)-based techniques have provided the rapid expansion of lipidomics research. Sample preparation and lipid extraction are important steps in lipidomics, and this chapter describes important considerations in lipid monophasic and biphasic extractions from plant tissues prior to untargeted plant lipidomics approaches with LC-MS.

Key words Lipids, Tissue extraction, Cell extraction, Monophasic lipid extraction, Biphasic lipid extraction, LC-MS, Untargeted lipidomics

1 Introduction

Plant lipids are structurally diverse hydrophobic or amphipathic molecules that have a variety of biological functions within cells, such as serving as the main building blocks of cell membranes, energy storage, transport, cell intracellular and extracellular signaling, membrane anchoring, and protein trafficking [1–3].

According to the nomenclature scheme introduced by LIPID-MAPS, lipids are classified in eight categories based on their chemical structure and biosynthetic perspectives. The eight lipid classes include fatty acids, glycerolipids, glycerophospholipids, sphingolipids, sterol lipids, prenol lipids, saccharolipids, and polyketides [4].

Fatty acids are the most common basic structural element of lipids, having unsaturated and saturated straight hydrocarbon chains. Fatty acid-derived molecules are important signaling components in interspecies communication and in plant defense, and

Carla António (ed.), *Plant Metabolomics: Methods and Protocols*, Methods in Molecular Biology, vol. 1778,
https://doi.org/10.1007/978-1-4939-7819-9_9,

are key precursor molecules for lipid, oxylipin, and plant hormone biosynthesis [5].

Glycerolipids consist of glycerol backbones with various numbers of fatty acids such as monoacyl glycerides, diacylglycerides, and triacylglycerides.

Glycerophospholipids are the major building blocks of cell membranes that are involved in cell signaling, membrane anchoring, and substrate transporting. Glycerophospholipids are derived from the phosphatidic acid (PA) and are composed of different molecular species with varied fatty acid chain length and degree of unsaturation [6]. In plants, the major glycerophospholipids are phosphatidylcholines (PC), phosphatidylethanolamine (PE), phosphatidylglycerol (PG), phosphatidylinositol (PI), phosphatidylserine (PS), and lysophosphatidylcholine (LPC), which are important components for the formation of lipid bilayers as part of cell membranes [7].

Sphingolipids are derivatives of ceramides and consist of amide groups linked to long-chain fatty acids with sphingoid backbones. Relative to sphingolipids in animals, sphingomyelin, globosides, sulfatides, or gangliosides are absent in plants, but glycosyl inositol phosphorylceramides (GIPCs), glycosylCERamide (gluCER), and ceramide account for ca. 64%, 34%, and 2%, respectively, of total sphingolipids in Arabidopsis [8]. Because of the large biomass of plants and fungi, GIPCs are therefore the most abundant sphingolipids of the biosphere.

Improvements in the resolution and sensitivity of mass spectrometers have facilitated the identification and characterization of key compounds in biological processes, including metabolites, proteins, and the large diversity of polarity within lipid species [9–11]. Due to these recent technological advances, almost all modern lipid-based analyses utilize optimized and tailored mass spectrometry (MS)-based methodologies [12], and these have successfully been applied to plant lipid research [13–15]. Liquid chromatography-mass spectrometry (LC-MS) is a powerful analytical technique which provides online chromatographic separation with information on retention times of lipid species, adding another dimension of selectivity compared to the direct infusion of lipids to the mass spectrometer, and LC-MS has been used to analyze complex lipid classes like glycerolipids, glycerophospholipids, and glycolipids [16, 17].

In sample preparation for lipid analyses, it is critical to choose the relevant organic solvent for protein precipitation; it must be suitable for effectively extracting lipids and remove insoluble particles by centrifuging rather than filtering. Filtering of biological samples can result in the loss of compounds of interest; therefore, it is advisable to avoid sample filtering at all times. In addition, stability and solubility of lipid species in extraction solvents play a crucial role in method development. Lipid extraction in different types of sample matrices has been well documented [17, 18]. Lipid extraction methods from biological samples benefit from the high solubility of hydrocarbon chains in organic solvents. One of the most common extraction methods for lipids in biological systems is the use of chloroform–methanol in a solvent ratio of 2/1 (v/v) as previously reported by Bligh and Dyer [19] and Folch [20]. The use of chloroform–methanol–water in a solvent ratio of 8:4:3 (v/v/v) has been shown to provide the most efficient biphasic partitioning of lipids into the organic phase due to the solubility and polarity of the different lipid classes, thereby improving the extraction efficiency. In addition to using chloroform–methanol, isopropanol–hexane has also been reported for the extraction of polar lipid species and fatty acids [21]; however, it has poor extraction efficiency for galactolipids.

In general, lipid extraction must be incorporated with organic solvents in lipidomics, but a single standard protocol for extracting all lipid classes has not yet been established. The efficiency of the lipid extraction is validated by spiking known amounts of lipid standards to the sample matrix prior to LC-MS analyses.

In conclusion, LC-MS-based untargeted plant lipidomics approaches have demonstrated the potential for rapid, comprehensive, and quantitative profiling of complex lipid molecular species from biological samples. The untargeted lipidomics approach is challenging but when coupled to high-resolution mass spectrometers that provide high mass accuracy, information of product ions, and retention times of the detected lipid species, it provides a higher confidence of the putatively identified lipidomics data.

2 Materials

2.1 Plant Material and Sample Preparation for LC-MS Analysis

1. Plant tissues from any plant species for lipid analysis (leaves, roots, or other tissue).
2. Liquid nitrogen for sample harvest, quenching, grinding, and weighing (*see* **Note 1**).
3. 2.0-mL safelock microcentrifuge tubes (e.g., Eppendorf).
4. 2.0-mL cryo-mill tubes.
5. Freezer safe marker pen, resistant to long-term storage at −80 °C.
6. Metal spatula for handling plant tissues.
7. Protective gloves for handling frozen objects.
8. Ceramic pestle and mortar or grinder with ball mill.
9. Autosampler vials for LC-MS analysis (e.g., 2.0-mL glass vials with 200 μL glass inserts).

2.2 Equipment, Instrumentation, and Software

1. −80 °C freezer to store plant tissues.
2. Analytical balance for accurate weighing of plant tissues and chemicals (at least four decimal places).
3. Vortex.
4. Thermoshaker with a 2.0-mL thermoblock.
5. Centrifuge suitable for 2.0-mL microcentrifuge tubes.
6. Automatic grinder if handling fresh weight (FW) plant tissues (e.g., ball mills, ULTRA-TURRAX, or cryo-mills).
7. Freeze dryer if handling dry weight (DW) plant tissues.
8. Nitrogen evaporator (e.g., TurboVap) or centrifugal concentrator.
9. LC column: C8 Ascentis Express, 2.1 mm internal diameter, 150 mm length, 2.7 μm particle size (Sigma-Aldrich).
10. LC-MS system: Agilent 6520 Series Quadrupole Time-of-Flight (Q-TOF) or equivalent equipped with an electrospray ionization (ESI) source.
11. MS data acquisition software: Agilent MassHunter Profile Finder or equivalent.

2.3 Solutions

Prepare all solutions in nitric acid-cleaned glassware and use within 5 days (*see* **Note 2**).

1. Internal standard solution: deuterium labeled lipid standard 14:0 PE-d54 (Avanti Polar lipids), prepared in 100% HPLC grade methanol.
2. Extraction solution A: isopropanol–hexane–water (55:20:25, v/v/v).
3. Extraction solution B: HPLC grade chloroform–methanol (2:1, v/v).
4. Extraction solution C: 0.1 M HCl.
5. Extraction solution D: 33% methylamine solution in ethanol-–water (7/3, v/v).
6. Reconstituting solution E: butanol–methanol (1:1, v/v) with 10 mM ammonium formate.
7. Antioxidant solution: 0.1% (w/v) butylatedhydroxytoluene (BHT) (*see* **Note 3**).
8. LC-MS eluent A: HPLC grade water–acetonitrile (40:60, v/v) with 10 mM ammonium formate.
9. LC-MS eluent B: HPLC grade isopropanol (IPA)–acetonitrile (90:10, v/v) with 10 mM ammonium formate.

3 Methods

Despite most lipid species are stable at room temperature, it is important to quench lipid metabolism as soon as possible prior to the extraction process. Lipid species could undergo enzymatic reactions which may result in the oxidation and formation of saturated lipids. During homogenizing, the usage of excess force by the beads in cryo-mills or fast mortar action can result in the generation of heat and can release acyl fatty acids from lipid species leaving lyso-lipid species. It is important to preserve lipid species and avoid any metabolic changes during the lipid extraction process. In a lipidomics approach, monophasic extraction facilitates the extraction of both apolar and polar lipid species. In this section, we will describe protocols to extract phospholipids and sphingolipids from plant tissues for subsequent LC-MS lipidomics analysis.

3.1 Lipid Extraction from Plant Tissues for LC-MS Analysis

3.1.1 Monophasic Phospholipid Extraction

1. Place ~10 mg of freeze-dried (DW) or ~50 mg of fresh frozen (FW) plant material into a precooled sample tube (e.g., 2.0-mL cryo-mill tube with ceramic beads) (*see* **Note 4**).
2. Add 1000 μL of extraction solution A to each 2.0-mL cryo-mill tube and immediately freeze it in liquid nitrogen.
3. Homogenize tissue using Program 2 of the cryo-mill (automatic grinder): set temperature to −10 °C, speed to 3510 rcf, 3 cycles of 45 s, and 30 s pause in between (*see* **Note 5**).
4. Shake at 60 °C for 15 min (*see* **Note 6**).
5. Centrifuge samples at 11,000 × *g* at room temperature for 10 min.
6. Transfer supernatant into a new 2.0-mL safelock microcentrifuge tube (e.g., Eppendorf).
7. Add 1000 μL of extraction solution A to the pellet of the 2.0-mL cryo-mill tube, and vortex.
8. Repeat **steps 4** and **5** (shake and centrifuge) and add the second supernatant to the previous one (**step 6**).
9. Dry under nitrogen, if nitrogen evaporator is available or in a centrifugal concentrator just until dry (set to a minimum vacuum of 200 bars to avoid chloroform bubbling); can lead to the loss of some short chain FAs if overdried for a longer period of time.
10. Add to the dried lipid extracts 200 μL of reconstituting solution E, vortex, and centrifuge as before (**step 5**).
11. Transfer the supernatant to a suitable glass vial and seal it tightly for LC-MS analysis (e.g., 2.0-mL glass vials with 200 μL glass inserts).

12. Take 50 μL aliquot from each sample and pool it together into a glass vial to make a pooled biological quality control (PBQC) sample (*see* **Note 7**).
13. If long-term storage is required, store dried lipid extracts at −20 °C prior to reconstituting and LC-MS analysis.

3.1.2 Biphasic Phospholipid Extraction

1. Place ~10 mg of freeze-dried (DW) or ~50 mg of fresh frozen (FW) plant material into a precooled sample tube (e.g., 2.0-mL cryo-mill tube with ceramic beads) (*see* **Note 4**).
2. Add 200 μL methanol to each 2.0-mL cryo-mill tube and immediately freeze it in liquid nitrogen.
3. Homogenize tissue using Program 2 of the cryo-mill (automatic grinder): set temperature to −10 °C, speed to 6100 rpm, 3 cycles of 45 s, and 30 s pause in between (*see* **Note 5**).
4. Add 400 μL chloroform.
5. Shake at room temperature for 15 min (*see* **Note 6**).
6. Centrifuge samples at 11,000 × *g* at room temperature for 5 min.
7. Transfer supernatant into a new 2.0-mL safelock microcentrifuge tube (e.g., Eppendorf).
8. Add 300 μL of extraction solution B to the pellet of the 2.0-mL cryo-mill tube, and vortex.
9. Repeat **steps 5** and **6** (shake and centrifuge) and add the second supernatant to the previous one (**step 7**).
10. Add 300 μL of extraction solution C to the pooled supernatant, vortex, and centrifuge as before (**step 6**).
11. Collect lower phase to a new 2.0-mL safelock microcentrifuge tube (e.g., Eppendorf).
12. Wash remaining upper phase with 200 μL chloroform, vortex, and centrifuge as before (**step 6**).
13. Collect the second lower lipid phase, and pool it with the previous one (**step 11**).
14. Dry under nitrogen, if nitrogen evaporator is available or in a centrifugal concentrator just until dry (set to a minimum vacuum of 200 bars to avoid chloroform bubbling); can lead to the loss of some short chain FAs if over-dried for a longer period of time.
15. Add to the dried lipid extracts 200 μL of reconstituting solution E, vortex, and centrifuge as before (**step 6**).
16. Transfer the supernatant to a suitable glass vial and seal it tightly for LC-MS analysis (e.g., 2.0-mL glass vials with 200 μL glass inserts).

17. Take 50 µL aliquot from each sample and pool it together into a glass vial to make a pooled biological quality control (PBQC) sample (*see* **Note 7**).
18. If long-term storage is required, store dried lipid extracts at −20 °C prior to reconstituting and LC-MS analysis.

3.1.3 Sphingolipid Extraction

1. Place ~10 mg of freeze-dried (DW) plant material into a 2.0-mL cryo-mill tube (*see* **Note 4**).
2. Add 1000 µL of extraction solution A to each 2.0-mL cryo-mill tube and immediately freeze it in liquid nitrogen.
3. Homogenize tissue using Program 2 of the cryo-mill (automatic grinder): set temperature to −10 °C, speed to 6100 rpm, 3 cycles of 45 s, and 30 s pause in between (*see* **Note 5**).
4. Shake at 60 °C for 15 min (*see* **Note 6**).
5. Centrifuge samples at 11,000 × *g* at room temperature for 10 min.
6. Transfer supernatant into a clean 2.0-mL safelock microcentrifuge tube (e.g., Eppendorf).
7. Add 1000 µL of extraction solution A to the pellet of the 2.0-mL cryo-mill tube, and vortex.
8. Repeat **steps 4** and **5** (shake and centrifuge) and add the second supernatant to the previous one (**step 6**).
9. Dry under nitrogen for de-esterification (*see* **Note 8**).
10. Add 1000 µL of extraction solution D to the pooled dried supernatant.
11. Incubate at 50 °C for 1 h (*see* **Note 9**).
12. After hydrolysis, dry under nitrogen.
13. Add to the dried lipid extracts 200 µL of reconstituting solution E, vortex, and centrifuge as before (**step 5**).
14. Transfer the supernatant to a suitable glass vial and seal it tightly for LC-MS analysis (e.g., 2.0-mL glass vials with 200 µL glass inserts).
15. Take 50 µL aliquot from each sample and pool it together into a glass vial to make a pooled biological quality control (PBQC) sample (*see* **Note 7**).
16. If long-term storage is required, store dried lipid extracts at −20 °C prior to reconstituting and LC-MS analysis.
17. If larger extraction volumes are required during the extraction protocol, please adjust the size of tubes (*see* **Note 10**).

3.2 Untargeted LC-MS Conditions and Parameters

1. Set solvent flow rate to 260 µL/min.
2. Set autosampler temperature to 12 °C.
3. Set LC column temperature to 55 °C.

Table 1
LC gradient for lipid analysis using binary solvent system

Time (min)	% Solvent A	% Solvent B	Flow rate (μL/min)
0	68	32	260
1.5	68	32	260
4.0	55	45	260
5	48	52	260
8	42	58	260
11	34	66	260
14	30	70	260
18	25	75	260
21	3	97	260
25	3	97	260
25.1	68	32	260
30	68	32	260

Table 2
Mass spectrometric conditions for detecting lipid species using negative and positive electrospray ionization (ESI) modes

Parameter	Negative ESI ion mode	Positive ESI ion mode
LC nebulizer pressure	40 psi	40 psi
Gas flow-rate	7 L/min	7 L/min
Gas temperature	325 °C	325 °C
Capillary voltage	4000 V	4500 V
Fragmentor	150 V	100 V
Skimmer	75 V	60 V

4. Set the injection sample volume to 5 μL.
5. Program de LC method for lipid analysis according to the elution gradient described in Table 1 (Total run time: 30 min).
6. Program de mass spectrometric parameters for lipid analysis according to the parameters described in Table 2.

3.3 Untargeted Lipid Data Processing

Untargeted mass spectrometry profiling data processing is significantly challenging due to the complexity of the raw data, limited spectral, reference biochemical databases, and incomplete knowledge of plant lipid metabolism and its regulation [22]. Most of the available metabolomics software often has a limited applicability in lipidomics data, i.e., proper isotopic correction or ability to use multiple internal standards for different lipid classes. Therefore, in addition to general metabolomics software that can be used for the analysis of lipidomics data, specialized lipidomics software packages have been specifically designed for lipidomics analysis and these are often combined with extensive mass spectral databases [23].

Many commercial vendors developed software packages dedicated to lipidomics applications. The most common commercial software include LipidSearch (ThermoFisher), LipidView (SCIEX) [23], Lipidyzer Platform (SCIEX), and SimLipid (PREMIER Biosoft). Although commercial lipidomics software provides enhanced functionality, they are often based on specific instrument features or can only work with particular vendor-specific raw files.

Many open-source programs have thus been developed for raw LC-MS data processing, such as MZmine [24], MSDIAL [25], and XCMS [26]; such software provide raw data conversion of many vendor-specific raw data files. Although lipid identification is challenging, some open-source software packages are available for lipids, such as LIPIDMAPS [27], Lipidblast [28], LipidXplorer [29], Lipid data analyzer [30], and Lipid-Pro [31], with additional information on product ions and predicted retention time for lipid species which provides most of the valuable information in untargeted plant lipidomics.

4 Notes

1. Prepare liquid nitrogen in appropriate storage containers (e.g., stainless steel) and use it when quenching metabolism to avoid metabolite degradation.
2. Measure required solvents using volumetric flasks where possible or the smallest possible graduated measuring cylinders. Measure all chemicals in separate glassware then pool them in the solvent bottle.
3. The use of an antioxidant, such as butylated hydroxyltoluene (BHT) is recommended in lipid extractions to avoid lipid oxidation. We recommend a concentration of BHT in the extraction solvent of 0.01% (w/v).
4. Harvest the plant tissue, rapidly wash it using iced-cold water to remove contaminants such as dust and soil, and place it into a sample tube (e.g., 2.0-mL cryo-mill tube). For frozen tissues, cut plant tissue into smaller pieces and place them into an Eppendorf tube and/or cryo-mill tube. If working with freeze-dried (DW) tissue, sample procedure is applied without the need for precooling or keeping tissue frozen.

5. Homogenization of plant tissue samples using a cryo-mill (automatic grinder) must operate using ceramic beads under cold temperature (as low as −18 °C). Operate the cryo-mill according to the manufacturer's instructions.
6. Rapidly freezing samples in liquid nitrogen or quenching samples with cold organic solvents are the most common methods of metabolic arrest. However, due to the poor solubility of lipids in cold temperatures, lipid extractions are carried out at room temperature (to avoid lipids sticking on the sample tubes) or higher than room temperature depending on the lipid class of interest.
7. Prepare a pooled biological quality control (PBQC) sample by combining the same volume (50 μL) from all samples in a glass vial for LC-MS analysis. The PBQC sample represents the matrix under study, and can be used to monitor and correct any variations in the instrument response due to technical issues. The PBQC sample should be run in every five samples.
8. Remove extraction solvents by evaporation, either using a nitrogen evaporator or a centrifugal concentrator, before adding deesterifying solutions to achieve maximum efficiency.
9. Add de-esterifying solutions to lipid extract to remove phospholipids, which contains ester bonds to de-esterified phospholipids, while sphingolipids, which contains amide linkage, will not be affected.
10. If an ultracentrifuge is available to centrifuge 15-mL or larger tubes, then do so at increased speeds for minimal time. Make sure that the same process is used for all conditions and at all-time points.

References

1. Watson AD (2006) Lipidomics: a global approach to lipid analysis in biological system. J Lipid Res 14:2101–2111
2. Welti R, Wang X (2004) Lipid species profiling: a high-throughput approach to identify lipid compositional changes and determine the function of genes involved in lipid metabolism and signaling. Curr Opin Plant Biol 7:337–344
3. Wenk MR (2005) The emerging field of lipidomics. Nat Rev Drug Discov 4:594–610
4. Fahy E, Subramaniam S, Murphy R et al (2009) Update of the LIPID MAPS comprehensive classification system for lipids. J Lipid Res 50:S9–S14
5. Weber H (2002) Fatty acid-derived signals in plants. Trends Plant Sci 7:217–224
6. Narasimhan R, Wang G, Li M et al (2013) Differential changes in galactolipid and phospholipid species in soybean leaves and roots under nitrogen deficiency and after nodulation. Phytochemistry 96:81–91
7. Van Meer G (2005) Cellular lipidomics. EMBO J 24:3159–3165
8. Markham JE, Jaworski JG (2007) Rapid measurement of sphingolipids from Arabidopsis thaliana by reversed-phase high-performance liquid chromatography coupled to electrospray ionization tandem mass spectrometry. Rapid Comm Mass Spectrom 21:1304–1314
9. Lee YJ, Perdian DC, Song Z et al (2012) Use of mass spectrometry for imaging metabolites in plants. Plant J 70:81–95
10. Okazaki Y, Kamide Y, Hirai MY, Saito K (2013) Plant lipidomics based on hydrophilic interaction chromatography coupled to ion trap time-offlight mass spectrometry. Metabolomics 9:121–131

11. Hill CB, Bacic A, Roessner U (2014) LC-MS profiling to link metabolic and phenotypic diversity in plant mapping populations. In: Raftery D (ed) Mass spectrometry in metabolomics, Methods in molecular biology (methods and protocols), vol 1198. Humana, New York, pp 29–41
12. Horn PJ, Chapman KD (2012) High-resolution measurements in plant biology: lipidomics in tissues, cells and subcellular compartments. Plant J 70:69–80
13. Samarakoon T, Shiva S, Lowe K et al (2012) Arabidopsis thaliana membrane lipid molecular species and their mass spectral analysis. In: Normanly J (ed) High-throughput phenotyping in plants, Methods in molecular biology (methods and protocols), vol 918. Humana, Totowa, pp 179–268
14. Shiva S, Vu HS, Roth MR, Zhou Z, Marepally SR, Nune DS, Lushington GH, Visvanathan M, Welti R (2013) Lipidomic analysis of plant membrane lipids by direct infusion tandem mass spectrometry. Plant Lipid Signal Protoc 1009:79–91
15. Chalbi N, Martínez-Ballesta MC, Youssef NB, Carvajal M (2015) Intrinsic stability of Brassicaceae plasma membrane in relation to changes in proteins and lipids as a response to salinity. J Plant Physiol 175:148–156
16. Hu C, Heijden R, Wang M et al (2009) Analytical strategies in lipidomics and application in disease biomarker discovery. J Chromatogr B 87:2836–2846
17. Mazumdar J, Striepen B (2007) Make it or take it: fatty acid metabolisum of apicomplexan parasites. Eukaryot Cell 6:1727–1735
18. Kofeler HC, Fauland A, Rechberger GN, Trotzmuler M (2012) Mass spectrometry based lipidomics: An overview of technological platforms. Meta 2:19–38
19. Bligh EG, Dyer WJ (1959) A rapid method of total lipid extraction and purification. Can J Biochem 37:911–917
20. Folch J, Lees M, Sloane-Stanley GH (1957) A simple method for the isolation and purification of total lipids from animal tissues. J Biol Chem 226:497–509
21. Milne S, Ivanova P, Forrester J, Brown HAH (2006) Lipidomics: an analysis of cellular lipids by ESI-MS. Methods 39:92–103
22. Hartler J, Tharakan R, Kofeler HC et al (2013) Bioinformatics tools and challenges in structural analysis of lipidomics MS/MS data. Brief Bioinform 14:375–390
23. Ejsing S, Duchoslav E, Sampaio J et al (2006) Automated identification and quantification of glycerophospholipid molecular species by multiple precursor ion scanning. Anal Chem 78:6202–6214
24. Katajamaa M, Miettinen J, Oresic M (2006) MZmine: toolbox for processing and visualization of mass spectrometry based molecular profile data. Bioinformatics 22:634–636
25. Tsugawa H, Cajka T, Kind T et al (2015) MS-DIAL: data-independent MS/MS deconvolution for comprehensive metabolome analysis. Nat Methods 12:523–526
26. Smith CA, Want EJ, O'Maille G et al (2006) XCMS: processing mass spectrometry data for metabolite profiling using nonlinear peak alignment, matching and identification. Anal Chem 78:779–787
27. Cotter D, Maer A, Guda C et al (2006) LMPD: LIPID MAPS proteome database. Nucleic Acids Res 34:507–510
28. Kind T, Okazaki Y, Saito K, Fiehn O (2014) LipidBlast templates as flexible tools for creating new in-silico tandem mass spectral libraries. Anal Chem 86:11024–11027
29. Herzog R, Schuhmann K, Schwudke D et al (2012) LipidXplorer: a software for consensual cross-platform lipidomics. PLoS One 7: e29851
30. Hartler J, Trötzmüller M, Chitraju C et al (2012) Lipid data analyzer: unattended identification and quantitation of lipids in LC-MS data. Bioinformatics 27:572–577
31. Ahmed Z, Mayr M, Zeeshan S et al (2015) Lipid-pro: a computational lipid identification solution for untargeted lipidomics on data-independent acquisition tandem mass spectrometry platforms. Bioinforma Oxf Engl 31:1150–1153

Chapter 10

Semi-targeted Lipidomics of Plant Acyl Lipids Using UPLC-HR-MS in Combination with a Data-Independent Acquisition Mode

Mohamed A. Salem and Patrick Giavalisco

Abstract

In recent years, multiple mass-spectrometric methods have been developed to tackle fundamental analytical questions in the field of biology and biochemistry. One essential approach relies on the use of liquid chromatography (LC), for efficient compound separation, coupled to high-resolution mass spectrometry (HR-MS). Even though these techniques are highly sensitive allowing for the reliable measurement of several thousand mass features, the major bottleneck is to convert the measured masses into annotated lipid species. To overcome this problem, we present a simple, example-based workflow, which provides an introduction to basic strategies for the manual validation of LC-MS-based lipidomic data. The whole strategy makes use of a data-independent acquisition (DIA) method, where alternating MS measurement cycles using high and low-energy scans are used. This measurement strategy allows to reliably annotate lipids, based on the exact mass measurements of intact, but also fragmented lipids from continuously recorded spectra.

Key words Plant lipids, Acyl lipids, Glycerolipids, Galactolipids, Glycerophospholipids, Targeted lipidomics, Ultra Performance Liquid Chromatography, High-resolution mass spectrometry, Data-independent acquisition, All-ion fragmentation

1 Introduction

Even though the complexity and number of individual lipid species in a cell are in the range of hundreds to thousands [1], the basic structures constituting the possible lipid species are highly similar. This is based on the fact that the majority of lipids is being synthesized either by carbanion- or carbcation-based condensations of thioesters or isoprene subunits, respectively [2]. Due to this reduced synthesis principles, the classification of the lipids can be restricted to eight main groups, which themselves contain different number of subgroups [2, 3].

Interestingly, it is not only possible to subdivide the entirety of lipids into the abovementioned classes, but it is also possible to

Carla António (ed.), *Plant Metabolomics: Methods and Protocols*, Methods in Molecular Biology, vol. 1778,
https://doi.org/10.1007/978-1-4939-7819-9_10,

describe the resulting lipid structures by a limited number of chemical building blocks [4]. This chemical modularity provides one of the main advantages for the systematic analysis of lipids, since it provides homogenous populations of hydrophobic compounds, which are clearly distinguishable between, but also within the different lipid classes [5].

The analytical strategies to analyze lipids are widespread and include practical low-tech methods like thin-layer chromatography (TLC) as well as more sophisticated approaches, including nuclear magnetic resonance (NMR) or mass spectrometric (MS) measurements [6]. Clearly, the advent of the soft electrospray ionization (ESI) method [7] paved the way to multiple analytical strategies, enabling the detection and quantification of complex lipid samples using mass spectrometry-based analysis methods [8, 9]. Accordingly, ESI-MS methods started to dominate the lipidomics field. The two dominant strategies are the chromatography-free, shotgun approach [4, 10] and the liquid chromatography (LC)-based method [11].

For the shotgun approach, crude lipid extracts are infused directly into the mass spectrometer and analyzed by measuring the intact masses of the lipids followed by their selected fragmentation. This approach, which is compatible with low- [10] as well as high-resolution mass spectrometers [12], is preferentially quantifying well-ionizing lipid species, since the injection of all lipids at the same time comes along with strong ion suppression effects [13].

The LC-MS method relies on lipid separation prior to the mass-spectrometric measurement. Using LC prior to MS scatters the injection of the lipids over time and therefore minimizes the ion suppression experienced by a single lipid species [13]. Therefore, even weak ionizing lipid species, like the neutral triacylglycerols (TAGs), can be frequently detected using LC-MS approaches [14]. The two most popular chromatographic systems for lipidomic studies are either based on reversed phase (RP) columns [5] or hydrophilic interaction chromatography (HILIC) systems [15]. While the HILIC approach separates lipids according to the chemical class [16], RP columns separate lipids down to the species level, even allowing for separation of structural isomers [17]. Both chromatographic systems can be combined with high- and low-resolution mass spectrometers, but high resolution is preferred, where compound annotation can be obtained from exact intact masses in combination with or without fragmentation data [5].

In the current chapter, we introduce a workflow, including a robust lipid extraction protocol next to an ultra performance (UP) LC-MS method, which is using a data-independent acquisition (DIA) MS-method [18, 19]. Based on an exemplary data set derived from *Arabidopsis thaliana* rosette leaf samples, we illustrate here the basic strategies for the manual validation and annotation of selected lipid species. By doing so, routine annotation of more than 150 lipid species can be obtained from 25 mg or less of fresh plant tissue [5, 18, 19].

2 Materials

Use solvents of high purity, at least LC grade. Perform all the extraction steps in a fume hood. Wear protective glasses and clothing during extraction.

2.1 Plant Material Harvesting and Storage

1. Harvesting bottles or tubes.
2. Scissors and tweezers.
3. Mortar and pestle, or tissue grinder.
4. Liquid nitrogen.
5. −80 °C freezer.
6. Analytical balance.

2.2 Lipid Extraction

1. Extraction solvent mixture: methyl *tert*-butyl ether (75 mL) and methanol (25 mL) (*see* **Note 1**).
2. Solvent mixture to induce phase separation: water (75 mL) and methanol (25 mL).
3. Solvent mixture for lipid resuspension for UPLC-MS analysis: acetonitrile (70 mL) and isopropanol (30 mL).
4. Microcentrifuge tubes or preferably glass vials.
5. Vortex mixer.
6. Ultrasonic bath.
7. Orbital shaker.
8. Benchtop centrifuge.
9. Nitrogen evaporator or centrifugal concentrator.

2.3 Liquid Chromatography

1. UPLC system (Waters ACQUITY UPLC system, Waters, Manchester, UK) with autosampler, column oven, and binary pumps.
2. Bridged Ethyl Hybrid (BEH) C_8 column, 2.1 mm internal diameter, 100 mm length, 1.7 μm particle size (Waters, Manchester, UK) fitted with a 2.1 internal diameter × 5 mm length guard column (ACQUITY UPLC BEH C_8 VanGuard Pre-column, Waters, Manchester, UK) (*see* **Note 2**).
3. UPLC eluent A: Mix 10 mL of 1 M ammonium acetate solution in water with 989 mL of LC-grade water (final concentration: 10 mM ammonium acetate). Add 1 mL of LC-grade acetic acid (final concentration: 0.1% acetic acid).

4. UPLC eluent B: Mix 692 mL acetonitrile and 297 mL of isopropanol. Add 10 mL of 1 M ammonium acetate in water (final concentration: 10 mM ammonium acetate) and 1 mL of acetic acid (final concentration: 0.1% acetic acid).
5. 2.0-mL autosampler glass vials with 200 μL glass inserts.

2.4 Mass Spectrometry

1. High-resolution MS instrument (Orbitrap Exactive, Thermo Scientific, Bremen, Germany), equipped with a heated electrospray ionization (HESI) source attached to an ACQUITY UPLC system (*see* **Note 2**).
2. Manual data inspection is performed in the Qual Browser of the Xcalibur software (Version 2.2 SP 1, Thermo Scientific, Bremen, Germany).

3 Methods

3.1 Tissue Harvesting, Quenching, and Homogenization

1. Harvest the plant tissue and snap-freeze the samples in liquid nitrogen (*see* **Note 3**).
2. Homogenize the plant tissue using a mortar and pestle, or tissue homogenizer (*see* **Note 4**).
3. Store samples in liquid nitrogen, dry ice, or −80 °C freezer until further extraction (*see* **Note 5**).

3.2 Extraction of Lipid Samples

1. Weigh 25 mg of the fresh-frozen (FW) ground plant tissue into 2.0 mL microcentrifuge tubes or preferably glass vials. Record the sample weight (*see* **Note 4**).
2. Add 1.0 mL of pre-cooled (−20 °C) extraction solvent mixture to each sample.
3. Vortex samples until all tissue is properly suspended.
4. Keep samples on an orbital shaker for 45 min at 4 °C.
5. Put the samples on an ice-cooled ultrasonic bath for 15 min.
6. Add 0.65 mL of solvent mixture to induce phase separation.
7. Vortex the samples for 1 min.
8. Centrifuge the samples at 10,000 × *g* for 10 min at 4 °C.
9. Transfer precisely 0.5 mL of the upper (lipid-containing) phase into new microcentrifuge tubes or preferably glass vials.
10. Dry samples under N_2 gas or in a centrifugal concentrator without heating.
11. Add 0.2 mL of solvent mixture for lipid resuspension, vortex the samples, and centrifuge at 5000 × *g* for 3 min at 4 °C.
12. Transfer the samples to 2.0 mL autosampler glass vials with 200 μL glass insert and cap tightly for UPLC-HR-MS analysis.

3.3 UPLC-HR-MS Analysis

1. Set up the UPLC separation method. Set the column temperature to 60 °C. Set the flow rate at 0.4 mL/min. Inject 2 μL per sample. Start the UPLC gradient with 45% eluent A for 1 min, use linear gradient from 45% A to 25% A for 3 min, use a linear gradient from 25% A to 11% A for 8 min, and finally use a linear gradient from 11% A to 0% A for 3 min. Wash the column for 4.5 min with 0% A before setting back to 45% A which is held for 4.5 min.
2. Acquire mass spectra from two UPLC runs per sample in positive and negative ionization modes using a mass range *m/z* 100–1500.
3. Record the spectra using alternating full scan (low-energy) and all-ion fragmentation (AIF, high-energy) scan mode (*see* **Note 2**).
4. Electronic settings for the HR-MS. Set the resolution to 10,000 with 10 scans per second. Restrict the loading time of the Orbitrap Exactive cell to a maximum of 100 ms, using a target value of 1 × E^6 ions. Set the capillary voltage to 3 kV with a sheath gas flow of 60 and an auxiliary gas flow of 35 (values are given in arbitrary units). Set the capillary temperature to 150 °C and the drying gas in the heated electrospray source to 350 °C. Set the skimmer voltage to 25 V and the tube lens to 130 V.

3.4 Lipid Identification and Data Processing

1. To manually inspect and validate the raw spectra from the UPLC-HR-MS runs, software provided by the instrument vendor is usually used. For the all-ion fragmentation (AIF) data, acquired on the Orbitrap Exactive, the vendor software (Xcalibur) was used. Total ion chromatograms (TIC), Base Peak Chromatograms (BPC) and extracted ion chromatograms (XIC) can be obtained, providing information on the general structure of the data, but also on RT and *m/z* of the measured lipid features. An example of a TIC in positive ionization mode is shown in Fig. 1.
2. To analyze selected lipid species from the acquired LC-MS measurements, it is required to determine the measured mass of the lipid(s) of interest (*see* **Note 6**). Specific lipid classes have a preference for specific adducts. This preference can be used, next to the accurate mass of the lipid, as an additional criterion for the identification (Table 1). Information on chemical formulas and masses of lipid species can be obtained from public lipid databases or the supplementary data of plant lipidomic studies (*see* **Note 7**).
3. To start the manual validation process of lipids of interest from LC-MS runs, extracted ion chromatograms (XIC), using the calculated mass of the adduct ion of the selected compounds has to be extracted. The XIC of UPLC-HR-MS measured lipids usually provides a symmetric, bell shaped peak of 3–6 s RT width. As an example the ammonium adduct of

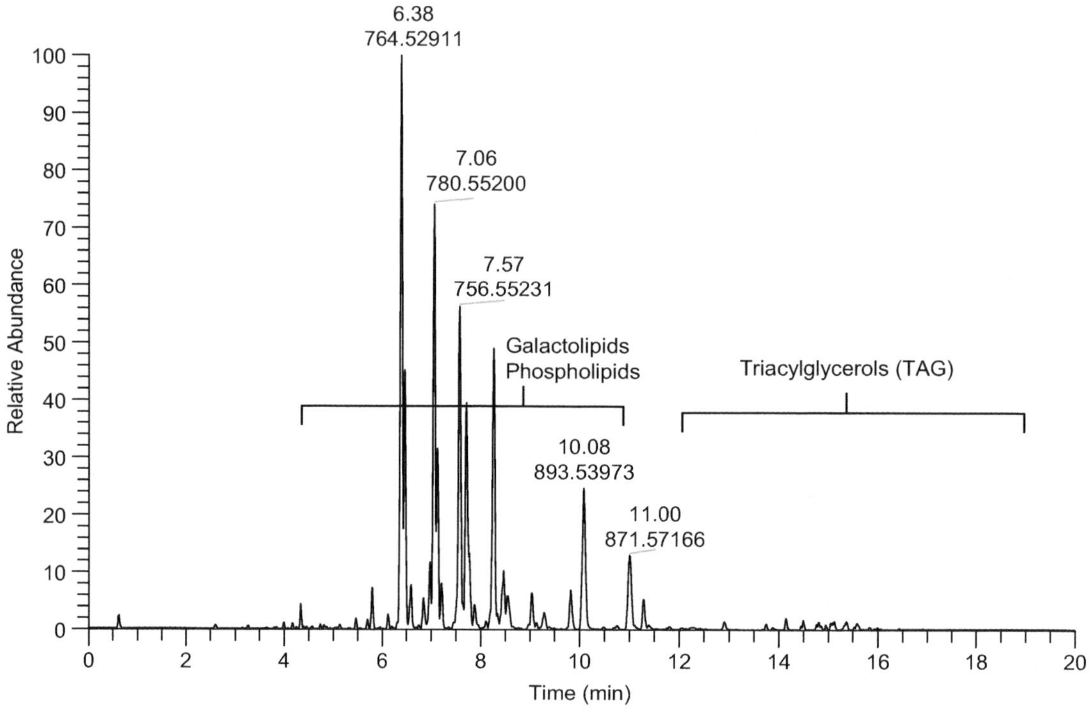

Fig. 1 Total ion chromatogram of an *Arabidopsis thaliana* lipid extract. Mass spectra are recorded in positive ionization mode and relative abundances of eluted peaks (*y*-axis) *versus* retention time in minutes (*x*-axis) are shown. The indicated regions correspond to elution of polar lipids (galacto and phosphoglycerol-lipids) and more neutral lipids (triacylglycerols)

monogalactosyldiacylglycerol (MGDG) 34:6 (*see* **Note 8**) was extracted, which results in a peak at RT 6.38 min (Fig. 2a). The observed zig-zag pattern of the chromatogram is obtained due to the employed measurement mode, where alternating low- and high-energy spectra were recorded (*see* **Note 9**). Figure 2b shows the obtained mass spectrum of the low-energy scan, where intact molecules are measured, while Fig. 2c provide the MS/MS (high-energy) spectrum, where the fragments of the measured lipids are detected.

4. To illustrate the stepwise annotation process of a plant lipid, phosphatidylcholine (PC 34:2) was selected as an example. This compound has the molecular formula $C_{42}H_{80}NO_8P$, which results in the monoisotopic mass of 757.5616063 amu. As shown in Fig. 3a, the peak is not yet visible in the TIC, while a 10 s peak, derived from *m/z* 758.56835 ($[M+H^+]^+$), is eluting at RT 8.25 min (Fig. 3b).
5. To validate the accuracy of the detected $[M+H^+]^+$ adduct, a second adduct ion has to be found, which functions as a reference (*see* **Note 10**). As can be seen in Fig. 3c a peak with *m/z* 780.54995, representing the PC 34:2 as a sodium ion ($[M+Na^+]^+$), is additionally detected. Next to the numeric

Table 1
Lipid classes and the detected adducts

Lipid class	Subclass	Adducts in positive ionization mode	Adducts in negative ionization mode	Most abundant adduct
Glycerophospholipids	Phosphatidylcholine (PC)/ Lysophosphatidylcholine (LPC)	$[M+H^+]^+$, $[M+Na^+]^+$	$[M-H^+]^-$, $[M+Ac-H^+]^-$	$[M+H^+]^+$
	Phosphatidylethanolamine (PE)/ Lysophosphatidylethanolamine (LPE)	$[M+H^+]^+$, $[M+Na^+]^+$	$[M-H^+]^-$	$[M+H^+]^+$
	Phosphatidylglycerol (PG)	$[M+H^+]^+$, $[M+NH_4{}^+]^+$, $[M+Na^+]^+$	$[M-H^+]^-$	$[M-H^+]^-$
	Phosphatidylinositol (PI)	$[M+H^+]^+$, $[M+NH_4{}^+]^+$	$[M-H^+]^-$	$[M-H^+]^-$
	Phosphatidylserine (PS)	$[M+H^+]^+$	$[M-H^+]^-$	$[M+H^+]^+$
Glycerolipids	Monogalactosyldiacylglycerol (MGDG)	$[M+NH_4{}^+]^+$, $[M+Na^+]^+$	$[M-H^+]^-$	$[M+NH_4{}^+]^+$
	Digalactosyldiacylglycerol (DGDG)	$[M+NH_4{}^+]^+$, $[M+Na^+]^+$	$[M-H^+]^-$	$[M+NH_4{}^+]^+$
	Sulfoquinovosyldiacylglycerol (SQDG)	$[M+NH_4{}^+]^+$, $[M+Na^+]^+$	$[M-H^+]^-$	$[M-H^+]^-$
	Diacylglyceride (DAG)	$[M+NH_4{}^+]^+$, $[M+Na^+]^+$	–	$[M+Na^+]^+$
	Triacylglyceride (TAG)	$[M+NH_4{}^+]^+$, $[M+Na^+]^+$	–	$[M+NH_4{}^+]^+$
Sphingolipids	Ceramide (Cer)	$[M+H^+]^+$, $[M+NH_4{}^+]^+$, $[M+Na^+]^+$	$[M-H^+]^-$	$[M+H^+]^+$
	Glucosylceramide (GlcCer)	$[M+H^+]^+$, $[M+NH_4{}^+]^+$, $[M+Na^+]^+$	$[[M-H^+]^-$	$[M+H^+]^+$
	Glycosylinositolphosphoceramide (GIPC)	$[M+H^+]^+$, $[M+NH_4{}^+]^+$, $[M+Na^+]^+$	$[M-H^+]^-$	$[M+H^+]^+$
Fatty acyls	Fatty acid (FA)	–	$[M-H^+]^-$	$[M-H^+]^-$

assignment of the two adducts it is required that these two adducts, if they are really derived from the same precursor molecule, show the same elution profile (coelute). As can be seen in the small inset in Fig. 3c, this is the case for the two observed adducts of PC 34:2 and confirms the results of expected adducts for PCs (Table 1).

6. To obtain more structural information of the measured PC 34:2, the fragmentation pattern, obtained from the AIF scans

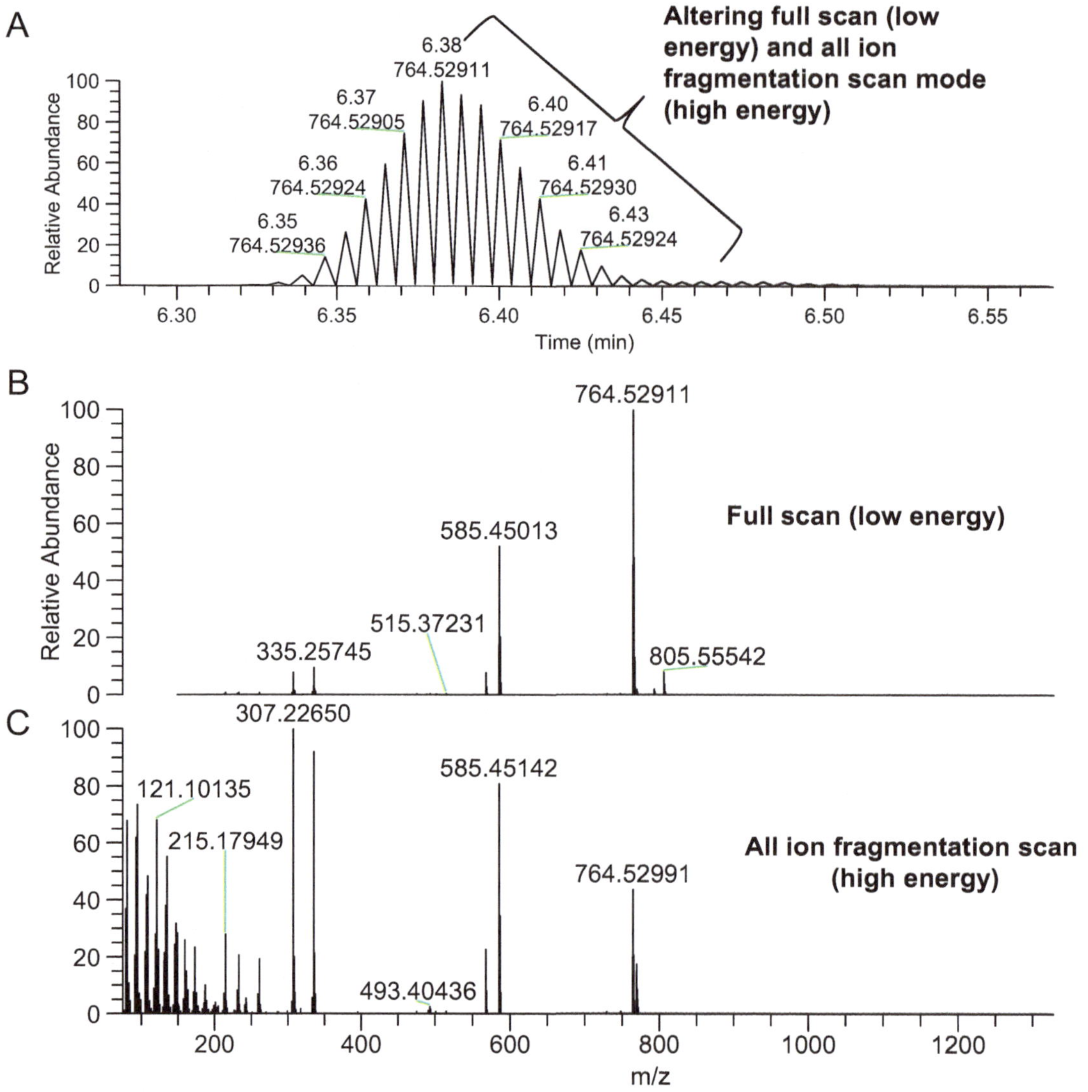

Fig. 2 High- and low-energy chromatograms and mass spectra of MGDG 34:6. (**a**) Extracted ion chromatogram of the $[M+NH_4^+]^+$ ion of MGDG 34:6. Numbers above the peaks indicate RT (min) and *m/z*. The zig-zag in the XIC is obtained due to the different intensities from the high- and low-energy spectra from the all-ion fragmentation mode. (**b**) Low-energy mass spectrum of MGDG 34:6 showing intact molecular masses (**c**) High-energy mass spectrum of MGDG 34:6 showing fragment masses

(*see* **Notes 2** and **9**) can now be utilized. As shown in Fig. 3d, the high-energy scan at RT 8.25 min contains a fragment of *m/z* 184.07234. Extracting the ion chromatogram of the fragment mass shows that this fragment clearly coelutes with the precursor of PC 34:2 (*inset in* Fig. 3d). This coelution indicates, as shown for the two adducts of PC 34:2 (*inset in* Fig. 3c), that both peaks result from the same precursor compound. The *m/z* 184.07234 can be associated to the formula

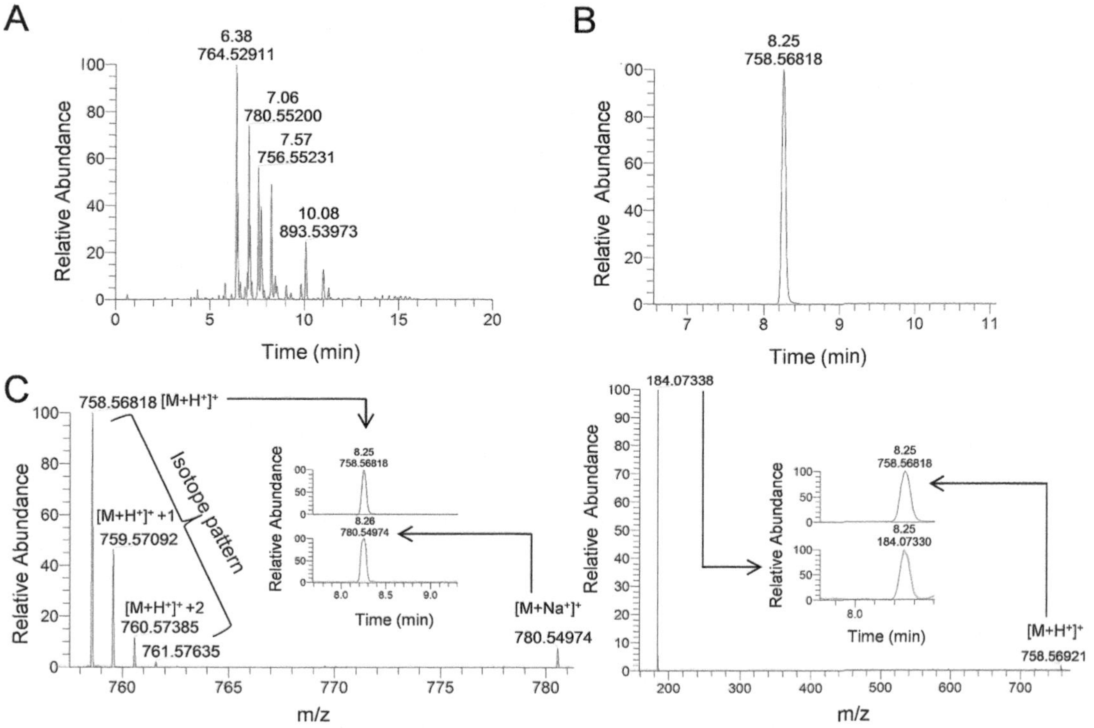

Fig. 3 Adduct and head group detection of PC 34:2. (**a**) Total ion chromatogram of lipid extract. (**b**) Extracted ion chromatogram from of phosphatidylcholine (PC) 34:2. Numbers above the peaks are RT and *m/z*. (**c**) Mass spectrum from RT 8.25 min of the low-energy mass scan. The peak at *m/z* 758.56835 represents the $[M+H^+]^+$ adduct of the lipid. The peak at *m/z* 780.54995 represents the $[M+Na^+]^+$ adduct. The insets are showing XICs of the two different adducts. (**d**) Mass spectrum from RT 8.25 min of the high-energy mass scan. The peak *m/z* 184.07324 represents the fragment of the PC head group (phosphocholine) as $[M+H^+]^+$. The insets are showing the XICs of the head group fragment and its precursor

$[C_5H_{15}O_4NP]^+$, which represents phosphocholine, a common head group fragment of PCs.

7. The use of the positive and negative ionization modes can additionally aid in the identification of compounds at multiple levels. So the combination of both polarities can be used for the proper assignment of the correct adduct. This is especially useful if the compound of interest has at least one detected adduct in each polarity. For lipids that ionize in both modes (Table 1), the independent mass spectra can be aligned and the delta of the detected *m/z* can be checked for adduct confirmation (*see* **Note 10**). As shown in Fig. 3c, PC 34:2 can be ionized in both polarities, and all the expected adducts ($[M+H^+]^+$, $[M+Na^+]^+$, and $[M+Ac\text{-}H^+]^-$) can be detected and show coelution at the RT 8.25 min (Fig. 4a).
8. To determine the acyl-chain composition of the lipid of interest, it is preferential to measure the samples in negative

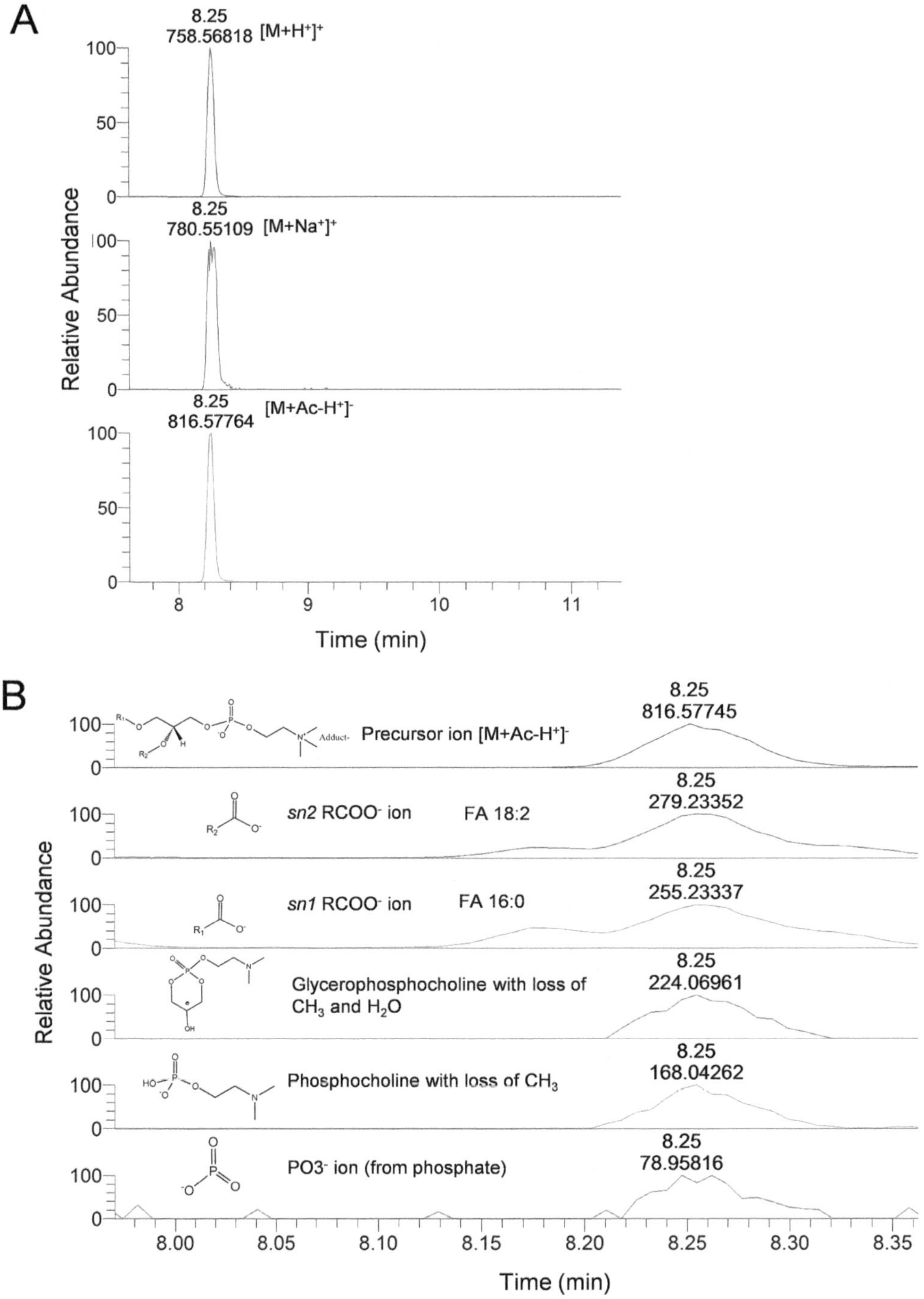

Fig. 4 The use of retention time, accurate mass and fragmentation pattern for compound validation. (**a**) XICs of PC34:2 using the masses of the $[M+H^+]^+$, $[M+Na^+]^+$ and the $[M+Ac\text{-}H^+]^-$ adducts. (**b**) Extracted ion chromatograms from negative ionization measurements for PC 34:2. The precursor ion and fragments with their structures are shown on the left-hand side of the figure

ionization mode (*see* **Note 11**). The fragments of acyl chains should, similar to the fragments of the head group or the previously shown adducts, show a coelution with their parent ion (Fig. 4b).

9. In the case where more than a single peak can be detected from an XIC for a selected *m/z*, it is possible that stereochemical isomers (*see* **Note 12**) are observed. As shown in Fig. 5a, MGDG 34:2 clearly shows such a double peak with RTs at 9.00 and 9.12 min, respectively. After having confirmed the

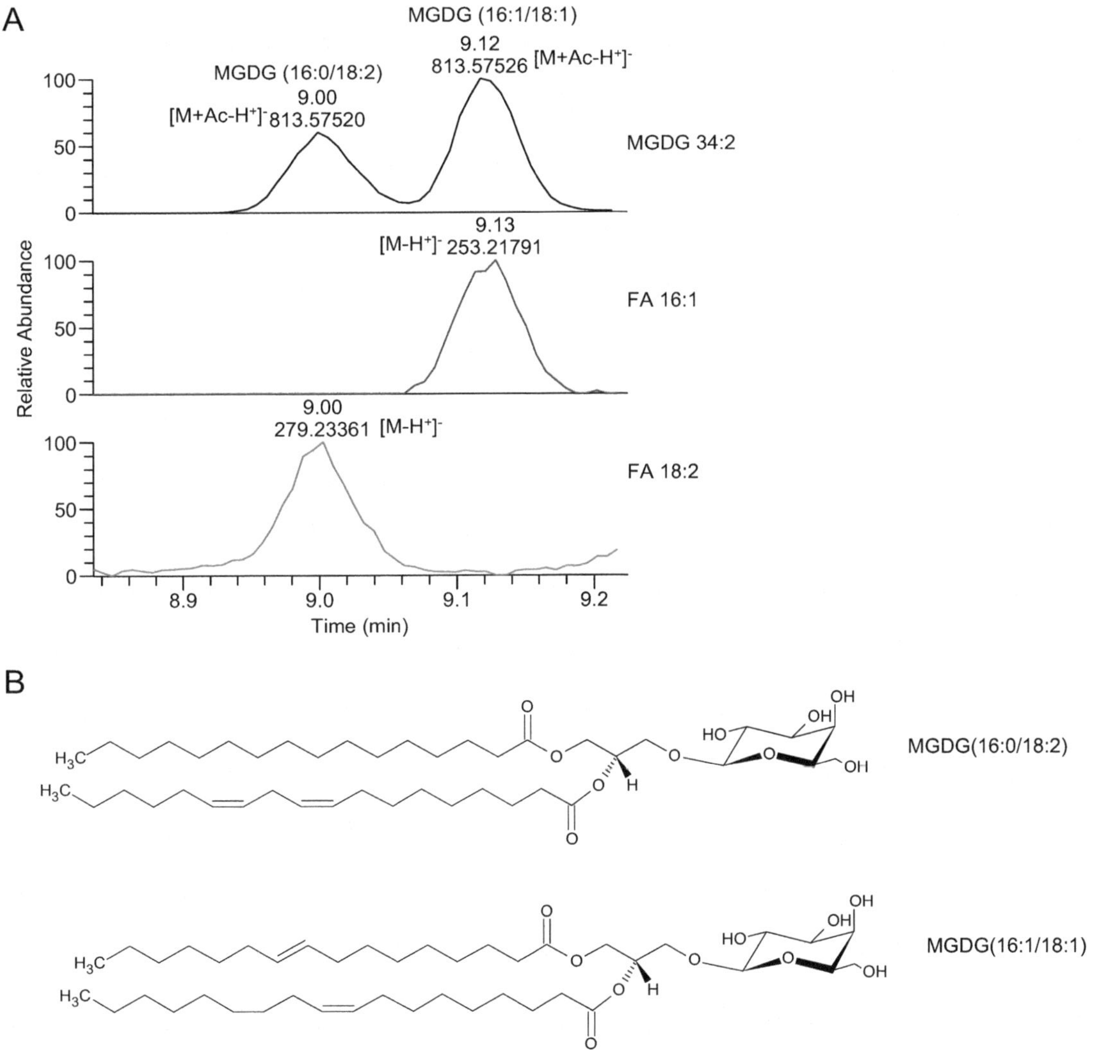

Fig. 5 Identification of acids acyl chains from stereochemically different lipid molecules. (**a**) Negative ion mode XIC for MGDG 34: 2 $[M+Ac-H^+]^-$ (upper trace). Two peaks at RT 9 and 9.12 min were detected. XIC for acids acyl chains obtained from the high-energy mass spectra (middle and lower trace) show a peak at RT 9 min that represents FA 18:2 and a second peak at 9.13 min that represents FA 16:1. (**b**) The chemical structures of two stereochemically different MGDG are shown

adduct identity of both peaks, the MS/MS analysis, using the high-energy scans, shows that one of the two MGDG isomers is containing acyl chains 16:0/18:2, while the second contains the acyl chains 16:1/18:1 (Fig. 5a, b).

10. An example, to illustrate the annotation of acyl-chain composition, if the analyzed lipid is not ionizing in negative mode (*see* **Note 11**), is given for the TAG 54:8 (Fig. 6). Here the combination of MS/MS and neutral loss scans allow for confirming that the detected TAG has a acyl-chain composition containing 18:2/18:3/18:3 (*see* **Note 13**).

11. Next to the use of adduct and fragmentation in combination with co-elution, the elution profiles of lipids from the same

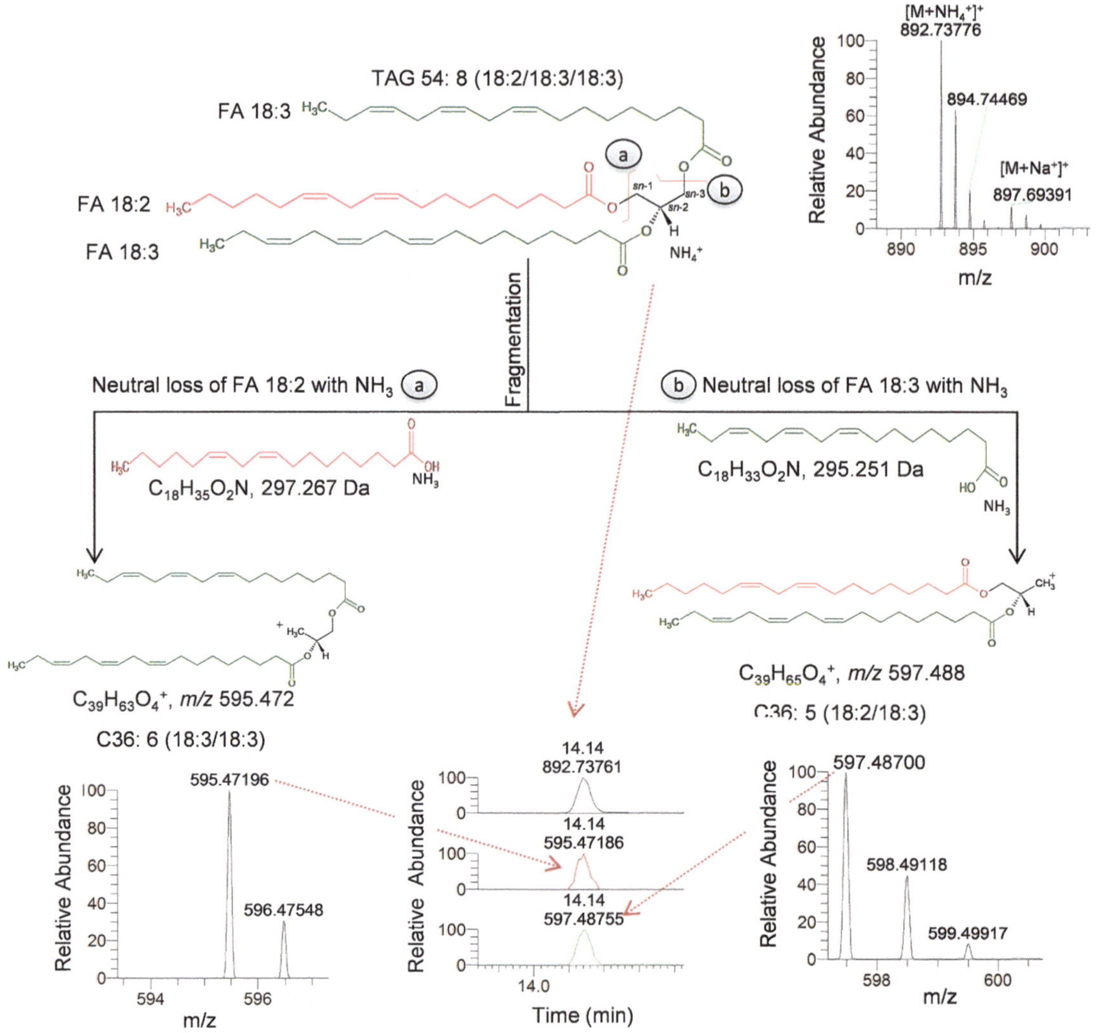

Fig. 6 High-energy analysis of neutrals losses to determine the acyl-chain composition of TAG 54:8

classes can be used for their collective identification. Lipids show a systematic RT shift, according to the acyl-chain length (Fig. 7a) or the degree of unsaturation (Fig. 7b) (*see* **Note 14**). If the lipids are plotted as scatter plots, with monoisotopic mass values on the *x*- and RT on the *y*-axis, it is easy to discriminate correctly annotated lipid species from falsely annotated ones-within the analyzed class (Fig. 7c).

12. Similar to the systematic intraclass variability, which is dominated by the acyl-chain composition, the interclass variability is mostly determined by the head group differences (*see* **Note 15**) (Fig. 8). Figure 8 shows the interclass RT differences between different plastidic galactolipids (MGDG; digalactosyldiacylglycerol, DGDG) and sulfolipids (sulfoquinovosyldiacylglycerol, SQDG) containing the same acyl-chain composition.
13. Once the manual annotation of all lipids of interest, using a combination of the abovementioned strategies, is obtained, the global data processing including peak picking, peak alignment, noise reduction, and normalization can be pursued. For this purpose dedicated vendor or open source software can be used [20–23].

4 Notes

1. If required, internal standards can be added according to the employed tissue and species, and of course, in accordance to the analytical question. Non-physiological but also stable isotope-labeled lipid standards are available from Avanti Polar lipids (Alabaster, Alabama).
2. In this chapter, lipids are identified using the chromatographic and spectral data obtained using alternating full scan (low-energy) and all-ion fragmentation (AIF) scan mode (high-energy) [5]. This provides comprehensive data for the identification of lipids by combining the precursor ion (low-energy) and the corresponding fragmentation data (high-energy) from a single chromatographic analysis. Other reversed-phase (RP) columns, UPLC systems, or mass spectrometers can be used with the described protocol, employing minor adaptations to column temperature, flow rate, scan window, and resolution.
3. The harvesting and freezing step has to be performed quickly to avoid lipid changes due to enzymatic activities.
4. Thawing of the plant tissues should be avoided during grinding or weighing steps to minimize enzymatic degradation.
5. Samples should be extracted directly after grinding, and long-term storage should be avoided.

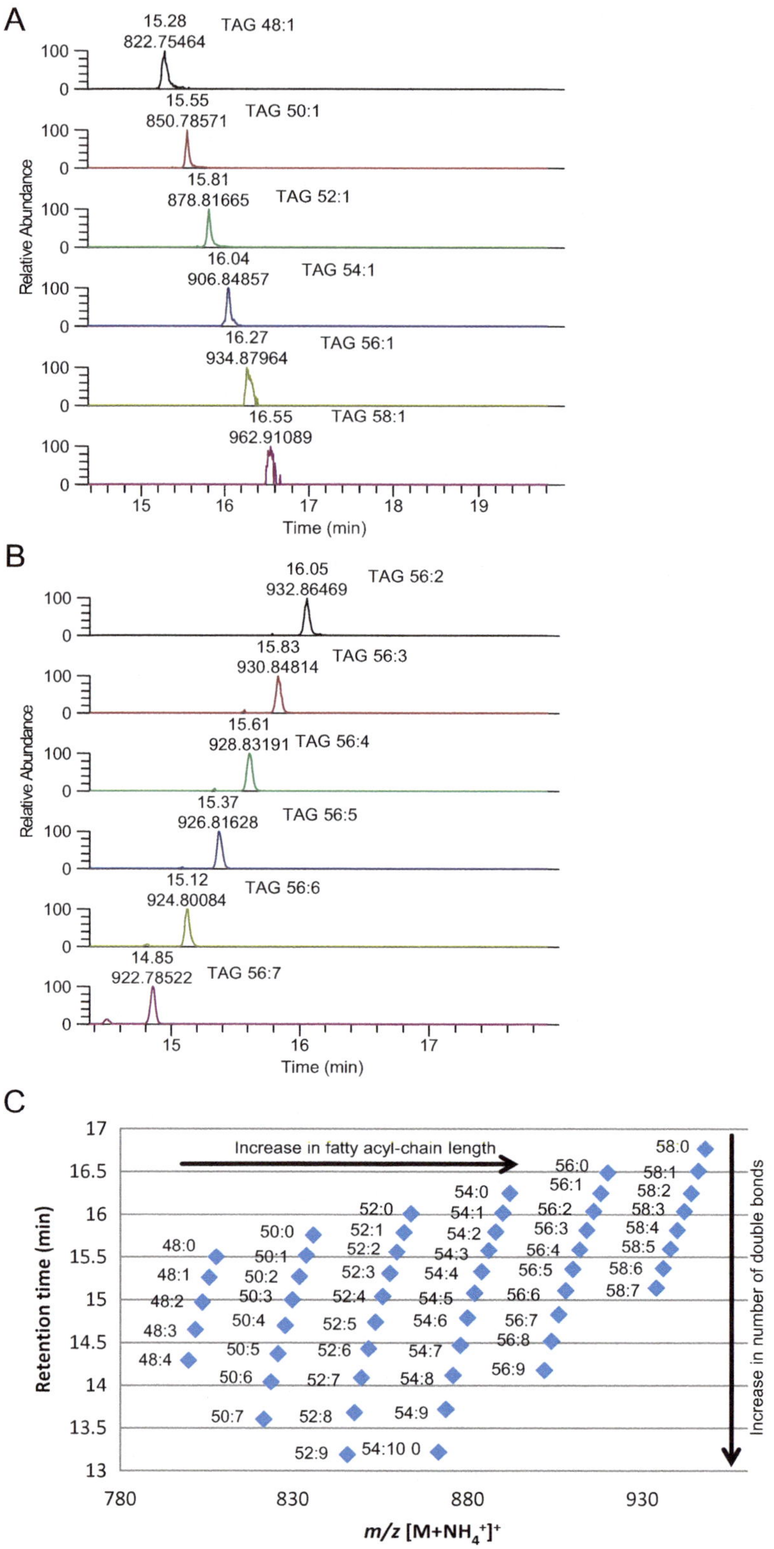

Fig. 7 Chromatographic separation for the identification of lipids within classes. (**a**) Different lipid species within the same class show a retention time pattern according to their acyl-chain length. Intraclass variability

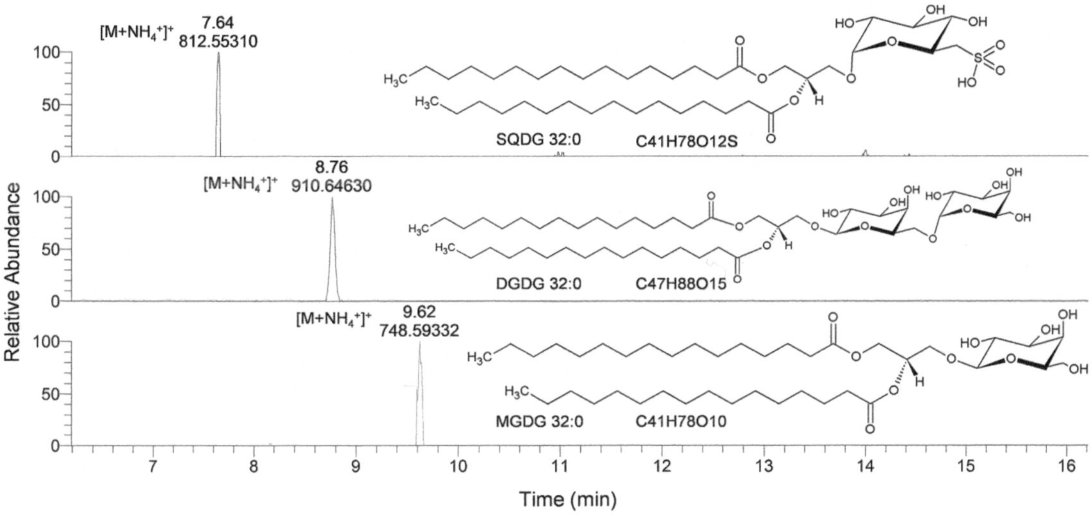

Fig. 8 The importance of chromatographic data for identification of lipids between classes. Different lipids classes show retention time pattern according to the chemical composition of the contained head groups. Interclass variability is shown by extracted ion chromatogram in positive ionization for the galactolipids (MGDG and DGDG) and the sulfur-lipid (SQDG). All lipids are 32:0

6. To measure a compound with MS, it has first to be ionized. This means that the molecule has to obtain a charge in the ionization source. Accordingly, the measured mass of a compound detected in the mass spectrometer is derived from its monoisotopic mass [M], which represents the sum of the individual elements present in the elemental formula (e.g., glucose with the elemental formula $C_6H_{12}O_6$ has a monoisotopic mass of 180.0628396 amu), plus or minus a charged residue. These charged residues are called adducts, and the ones commonly observed for the positive ionization mode are the addition of a proton ($[M+H^+]^+$), a sodium ion ($[M+Na^+]^+$), or an ammonium ion ($[M+NH4^+]^+$). The commonly observed adducts in the negative ionization mode are the loss of a proton ($[M-H^+]^-$) or the loss of a proton in combination with the association of an acetate molecule ($[M+Ac\text{-}H^+]^-$). A table with the observed adducts of the detected lipid classes is given in Table 1, while a more complete overview of possible adducts can be found at http://fiehnlab.ucdavis.edu/staff/kind/metabolomics/ms-adduct-calculator.

Fig. 7 (continued) is shown by XIC in positive ionization mode for the TAG 48:1 to 58:1. (**b**) Different lipid species within the same class show a retention time pattern according to the number of double bonds. Intraclass variability is shown by XIC in positive ionization mode for the TAG 56:2–7. (**c**) Scatter plot representation of 52 TAGs annotated from Arabidopsis leaves. The *m/z* of the protonated adducts $[M+NH_4]^+$ is given on the *x*-axis and the observed RT (min) is given in the *y*-axis. The plot illustrates the correlation between elution (RT) and number of carbons on the acyl chains or number of double bonds (degree of unsaturation)

7. A comprehensive repository of lipid data, including a large lipid database with the classification of the lipids, their chemical formulas, structures, and their monoisotopic but also ionized masses, can be found on the website of the LIPID MAPS consortium (http://www.lipidmaps.org/). Results of several hundreds of measured plant lipid species can be found in the supplementary material from several published papers [5, 24–26].
8. Lipids are named here as an acronym of the lipid class (Table 1), followed by two numbers separated by a colon. The first number denotes the sum of carbon molecules in the acyl chains bound to the glycerol backbone, while the second number denotes the sum of double bonds in the acyl chains.
9. The measurement mode of alternating low- and high-energy mode switches leads to the consecutive production of spectra containing either the intact lipid masses (low-energy, Fig. 2b) or fragmentation of all ions (high-energy, Fig. 2c). The advantage of this method is that it provides information of the elution behavior of intact lipids next to the elution behavior of the obtained fragmentation spectra [5].
10. To determine the adduct identity of the measured *m/z*, the difference of the measured *m/z* to its neighboring *m/z* has to be inspected. Specific differences of +17.0265 amu or +21.9819 amu can be detected for the ammonium $[M+NH_4^+]^+$ and the sodium adduct $[M+Na^+]^+$, compared to the protonated M $[M+H^+]^+$. Similarly, a delta of 4.9554 amu can be used to associate the ammonium and the sodium adduct. For the negative ionization mode, a delta of 60.0211 amu denotes the difference between the deprotonated $[M-H^+]^-$ ion and the deprotonated acetate adduct $[M+Ac\text{-}H^+]^-$.
11. Even though the positive ionization mode can cover most types of lipid molecules (Table 1), fatty acids (FA) strictly ionize only in negative ionization mode. Accordingly, if a comprehensive picture of all lipid species should be recorded, the measurement in both positive and negative ionization modes is advised. This is especially true to properly assign the correct adduct (*see* **Note 10**), but also due to the fact that some lipid classes ionize better in one or the other ionization mode [5]. Next to the pure adduct composition, the adduct intensity ratios can also be used for the annotation (Table 1). Next to the improved ionization specificity, the different polarities provide unique fragmentation patterns with unique features. Accordingly, the composition of the acyl chains can be directly obtained from tandem MS/MS spectra in negative ionization mode, where the FA fragments can be detected as $[M-H^+]^-$

ions (Fig. 4b). To obtain FA composition in the positive mode one has to utilize neutral loss (NL) analysis, which detects the loss of a neutral fatty acid fragment (FA) from the precursor molecule (Fig. 6).

12. Stereochemistry such as the position of double bonds on the acyl chains, stereoisomers (*E* or *Z*) or the *sn*-positions (stereospecific numbering to designate the configuration of glycerol derivatives, *sn*1, *sn*2, *sn*3) provides a source of structural heterogeneity, while the monoisotopic mass is maintained [8]. Accordingly, these compounds cannot be discriminated by the mass spectrometer, unless they are either separated by the chromatography (Fig. 5) or highly-specialized techniques such as ozone-induced dissociation, collision-induced dissociation (CID) in combination with ion mobility mass spectrometry (IM-MS) or addition of lithium in combination with fragmentation [27–29].

13. The neutral triacylglycerol (TAG) 54:8 can have different combinations of acyl chains like 18:2/18:3/18:3, 16:1/18:3/20:4 or 16:1/18:2/20:5, among others. To determine the correct acyl-chain composition of this TAG, a NL analysis has to be performed (*see* **Note 11**). TAG 54:8 is detected at RT 14.14 min in Arabidopsis rosette leaf samples, with a predominant $[M+NH_4^+]^+$ ion (Fig. 6). This ammonium ion peak can be fragmented, which results in the identification of a NL of 297.267 ($[M(FA\ 18{:}2)+NH_3]$), derived from the delta between the $[M+NH_4]^+$ ion of the TAG 54:8 precursor and the product (fragment) ion at *m/z* 595.472. Additionally, a NL of 295.251 ($[M(FA\ 18{:}3)+NH_3]$), derived from the product ion at *m/z* 597.488, was determined. Therefore, since two of the three acyl chains have been detected by NL scans, the third one has to be another 18:3 acyl chain, leading to an assigned composition of 18:2/18:3/18:3 (the order is arbitrary).

14. Lipids in this study are separated using reversed phase (RP) columns, which separate lipids according to their hydrophobic properties. Hydrophobicity is obtained from the carbon-chain length of the fatty acids, the unsaturation degree (number of double bonds) in the acyl chains of the lipids, and the chemical composition of the head group. Thus, a lipid that contains a shorter carbon chain is eluted earlier from the UPLC column than a lipid with a longer acyl chain. Additionally, a lipid with more double bonds in the fatty acid chains elutes earlier than a lipid with a fully saturated acyl chain. The same is true for the polarity of the head group. Accordingly, the lipid intraclass RT distribution can be used, in combination with the proper adduct detection and the accurate mass, to reliably annotate lipids. This is an advantage of separation-based

lipidomics compared to shotgun-based lipidomics, using direct infusion of lipid samples.

15. While the differences between lipid species within a lipid class are determined by the acyl-chain composition, the inter-class discrimination is determined by the polarity of the head group. In Fig. 8 three XICs, of plastidic lipids (MGDG, DGDG and SQDG) containing the same acyl-chain length are displayed. SQDG, which contains a glucose modified with a sulfonic acid residue as its head group, elutes earlier than DGDG, which contains a galactose dimer moiety. MGDG, which contains only a single galactose moiety, elutes the latest of the three.

Acknowledgments

We would like to thank Andrea Leisse and Dr. Vinzenz Hofferek for technical assistance. The Max Planck Society and the DAAD are kindly acknowledged for the generous funding.

References

1. Muro E, Atilla-Gokcumen GE, Eggert US (2014) Lipids in cell biology: how can we understand them better? Mol Biol Cell 25:1819–1823
2. Fahy E, Subramaniam S, Brown HA et al (2005) A comprehensive classification system for lipids. J Lipid Res 46:839–861
3. Fahy E, Subramaniam S, Murphy RC et al (2009) Update of the LIPID MAPS comprehensive classification system for lipids. J Lipid Res 50(Suppl):S9–14
4. Han X, Gross RW (2005) Shotgun lipidomics: electrospray ionization mass spectrometric analysis and quantitation of cellular lipidomes directly from crude extracts of biological samples. Mass Spectrom Rev 24:367–412
5. Hummel J, Segu S, Li Y et al (2011) Ultra performance liquid chromatography and high resolution mass spectrometry for the analysis of plant lipids. Front Plant Sci 2:54
6. Wenk MR (2005) The emerging field of lipidomics. Nat Rev Drug Discov 4:594–610
7. Fenn JB, Mann M, Meng CK et al (1989) Electrospray ionization for mass spectrometry of large biomolecules. Science 246:64–71
8. Blanksby SJ, Mitchell TW (2010) Advances in mass spectrometry for lipidomics. Ann Rev Anal Chem 3:433–465
9. Harkewicz R, Dennis EA (2011) Applications of mass spectrometry to lipids and membranes. Ann Rev Biochem 80:301–325
10. Shiva S, Vu HS, Roth MR et al (2013) Lipidomic analysis of plant membrane lipids by direct infusion tandem mass spectrometry. In: Munnik T, Heilmann I (eds) Plant lipid signaling protocols, Methods in molecular biology (methods and protocols), vol 1009. Humana, Totowa, pp 79–91
11. Cajka T, Fiehn O (2014) Comprehensive analysis of lipids in biological systems by liquid chromatography-mass spectrometry. Trends Analyt Chem 61:192–206
12. Almeida R, Pauling JK, Sokol E et al (2015) Comprehensive lipidome analysis by shotgun lipidomics on a hybrid quadrupole-orbitrap-linear ion trap mass spectrometer. J Am Soc Mass Spectrom 26:133–148
13. Annesley TM (2003) Ion suppression in mass spectrometry. Clin Chem 49:1041–1044
14. Fauland A, Kofeler H, Trotzmuller M et al (2011) A comprehensive method for lipid profiling by liquid chromatography-ion cyclotron resonance mass spectrometry. J Lipid Res 52:2314–2322
15. Okazaki Y, Kamide Y, Hirai MY, Saito K (2013) Plant lipidomics based on hydrophilic interaction chromatography coupled to ion trap time-of-flight mass spectrometry. Metabolomics 9:121–131
16. Cifkova E, Holcapek M, Lisa M et al (2012) Nontargeted quantitation of lipid classes using hydrophilic interaction liquid

chromatography-electrospray ionization mass spectrometry with single internal standard and response factor approach. Anal Chem 84:10064–10070

17. Degenkolbe T, Giavalisco P, Zuther E et al (2012) Differential remodeling of the lipidome during cold acclimation in natural accessions of Arabidopsis thaliana. Plant J 72:972–982
18. Salem MA, Juppner J, Bajdzienko K, Giavalisco P (2016) Protocol: a fast, comprehensive and reproducible one-step extraction method for the rapid preparation of polar and semi-polar metabolites, lipids, proteins, starch and cell wall polymers from a single sample. Plant Methods 12:45
19. Salem M, Bernach M, Bajdzienko K, Giavalisco P (2017) A simple fractionated extraction method for the comprehensive analysis of metabolites, lipids, and proteins from a single sample. J Vis Exp 124:e55802
20. Lommen A (2009) MetAlign: interface-driven, versatile metabolomics tool for hyphenated full-scan mass spectrometry data preprocessing. Anal Chem 81:3079–3086
21. Katajamaa M, Miettinen J, Oresic M (2006) MZmine: toolbox for processing and visualization of mass spectrometry based molecular profile data. Bioinformatics 22:634–636
22. Tautenhahn R, Patti GJ, Rinehart D, Siuzdak G (2012) XCMS online: a web-based platform to process untargeted metabolomic data. Anal Chem 84:5035–5039
23. Gowda H, Ivanisevic J, Johnson CH et al (2014) Interactive XCMS online: simplifying advanced metabolomic data processing and subsequent statistical analyses. Anal Chem 86:6931–6939
24. Devaiah SP, Roth MR, Baughman E et al (2006) Quantitative profiling of polar glycerolipid species from organs of wild-type Arabidopsis and a phospholipase Dalpha1 knockout mutant. Phytochemistry 67:1907–1924
25. Tarazona P, Feussner K, Feussner I (2015) An enhanced plant lipidomics method based on multiplexed liquid chromatography-mass spectrometry reveals additional insights into cold- and drought-induced membrane remodeling. Plant J 84:621–633
26. Vu HS, Shiva S, Roth MR et al (2014) Lipid changes after leaf wounding in Arabidopsis thaliana: expanded lipidomic data form the basis for lipid co-occurrence analysis. Plant J 80:728–743
27. Castro-Perez J, Roddy TP, Nibbering NM et al (2011) Localization of fatty acyl and double bond positions in phosphatidylcholines using a dual stage CID fragmentation coupled with ion mobility mass spectrometry. J Am Soc Mass Spectrom 22:1552–1567
28. Hsu FF, Turk J (2008) Elucidation of the double-bond position of long-chain unsaturated fatty acids by multiple-stage linear ion-trap mass spectrometry with electrospray ionization. J Am Soc Mass Spectrom 19:1673–1680
29. Poad BL, Pham HT, Thomas MC et al (2010) Ozone-induced dissociation on a modified tandem linear ion-trap: observations of different reactivity for isomeric lipids. J Am Soc Mass Spectrom 21:1989–1999

Chapter 11

Plant Lipidomics Using UPLC-QTOF-MS

Yozo Okazaki and Kazuki Saito

Abstract

Lipids are primary metabolites used for energy storage, signal transduction, and pigmentation, besides being common membrane components. Considering the importance of comprehensive lipid analysis (lipidomics) for a detailed understanding of cellular metabolic states, this chapter introduces liquid chromatography–quadrupole time-of-flight mass spectrometry as a powerful untargeted plant lipidomics technique.

Key words Lipidomics, Liquid chromatography, Quadrupole time-of-flight mass spectrometry, Data-dependent acquisition

1 Introduction

Lipids are a class of structurally diverse chemicals with a broad range of functions, being typically analyzed by gas chromatography and liquid chromatography coupled to mass spectrometry (GC-MS, LC-MS). GC-MS has been often used to analyze simple lipids and sterols, but usually not used for complex lipids such as phospholipids and glycoglycerolipids because these lipids are not easily vaporized. Conversely, LC-MS (utilizing atmospheric pressure ionization) does not require analyte vaporization for chromatographic separation and ionization, and has thus become the main technique for lipidomics, being well suited for analyzing heat-labile metabolites. In our group, LC-MS-based untargeted lipidomics has been utilized in plant research [1–5] including functional genomics [6] and discovery of new functional lipid classes [7–9].

In this chapter, we provide a detailed methodology of sample preparation by raising the error-prone process and describe lipid analysis based on liquid chromatography-quadrupole time-of-flight mass spectrometry (LC-QTOF-MS) in data-dependent MS/MS acquisition mode. In addition, we introduce data analysis using MS-DIAL software [10, 11], providing a description of the

Carla António (ed.), *Plant Metabolomics: Methods and Protocols*, Methods in Molecular Biology, vol. 1778,
https://doi.org/10.1007/978-1-4939-7819-9_11,

corresponding peak annotation and peak alignment functions. Since the above library contains MS/MS spectra of most known glycerolipids and sphingolipids, MS-DIAL provides a good starting point for plant lipidomics.

2 Materials

2.1 Reagents and Equipment Used for Plant Lipidomics

1. 1,2-didecanoyl-*sn*-glycero-3-phosphocholine (*see* **Note 1**).
2. Extraction solvent A: methanol–methyl *tert*-butyl ether (3:1, v/v). Measure the required volumes of methanol (750 mL) and methyl *tert*-butyl ether (250 mL) using volumetric cylinders and mix them in a glass bottle [8, 12, 13] (*see* **Note 2**).
3. Internal standard (IS) solution: 250 μM 1,2-didecanoyl-*sn*-glycero-3-phosphocholine. Weigh 1,2-didecanoyl-*sn*-glycero-3-phosphocholine (1.41 mg) in a 10-mL glass vial and dissolve it in extraction solvent A. Transfer the solution to a 10-mL volumetric flask, fill the flask with extraction solvent A, and store the solution at −30 °C.
4. Extraction solvent B: extraction solvent A containing 1 μM 1,2-didecanoyl-*sn*-glycero-3-phosphocholine. Fill ~500 mL of extraction solvent A into a 1000-mL volumetric flask, add 4 mL of IS solution, and fill the flask with extraction solvent A. Store the obtained solution in a refrigerator under dark conditions.
5. Sample disruption device: Shake Master Neo (BMS, Japan).
6. Solvent C: acetonitrile–water–1 M ammonium acetate–formic acid (200:800:10:1, v/v/v/v). Mix acetonitrile (200 mL), water (800 mL), 1 M ammonium acetate (10 mL), and formic acid (1 mL) in a glass bottle.
7. Solvent D: acetonitrile–2-propanol–1 M ammonium acetate–-formic acid (100:900:10:1, v/v/v/v). Mix acetonitrile (100 mL), 2-propanol (900 mL), 1 M ammonium acetate (10 mL), and formic acid (1 mL) in a glass bottle.
8. LC-MS system: ACQUITY UPLC system coupled with a Xevo G2 QTof MS (Waters, United States).
9. LC column: ACQUITY UPLC HSS T3, 1.0 mm internal diameter, 50 mm length, 1.8 μm pore size (Waters, United States).
10. 2.0-mL microcentrifuge tubes (e.g., Eppendorf).
11. Zirconia beads.
12. Sharp cutting tool for plant tissue harvest.
13. Liquid nitrogen for plant sample preparation.
14. Protective gloves for handling liquid nitrogen.
15. Micropipettes and tips suitable for handling organic solvents.
16. Vortex.

17. Orbital shaker.
18. Centrifuge suitable for 2.0-mL microcentrifuge tubes.
19. Centrifugal concentrator.
20. 2.0-mL autosampler glass vials with 200 μL glass inserts.

3 Methods

3.1 Sample Preparation

1. Prepare a 2.0-mL microcentrifuge tube with a zirconia bead, and write the sample number on the cap (Fig. 1a).
2. Cut a plant sample using a sharp cutting tool (Fig. 1b) and weigh it (*see* **Note 3**).
3. Transfer the sample into the microcentrifuge tube and freeze it after each weighing procedure by immersion into liquid nitrogen to about half of the sample tube height (Fig. 1c; *see* **Note 4**).

3.2 Lipid Extraction

1. Chill the sample holder of the Shake Master by immersing it into liquid nitrogen (Fig. 1d). Quickly place the sample tubes into the chilled sample holder (Fig. 1e) and homogenize tissues by shaking at 900 rpm for 2 min (*see* **Note 5**).
2. Transfer the samples to a chilled rack by partial immersion into liquid nitrogen. Make sure that the samples are finely powdered and not thawed (Fig. 1f).
3. Prerinse the micropipette tip (e.g., PIPETMAN of Gilson) by repeated aspiration and dispensing of extraction solvent B. Hereafter, all liquid volumes for lipid extraction are assumed to be measured using micropipettes. Prerinse the tips as needed for accurate measuring.
4. Open the sample tube. Make sure that no sample powder is attached beneath the sample cap.
5. Add the required extraction solvent B (16 μL solvent/1 mg fresh weight) and agitate the obtained mixture using a vortex mixer. Place this sample in a rack held at room temperature.
6. Open the sample tube cap, add water to each homogenate (4 μL water/1 mg fresh weight), and agitate the mixture using a vortex mixer.
7. Shake the samples well for 5 min at room temperature using an orbital shaker.
8. Keep the samples on ice for 15 min.
9. Centrifuge samples at 3000 × *g* at 4 °C for 10 min.
10. Collect 160 μL of the upper layer (Fig. 1g), and transfer the aliquot to a new microcentrifuge tube.
11. Evaporate the solvent using a centrifugal concentrator or under nitrogen stream.

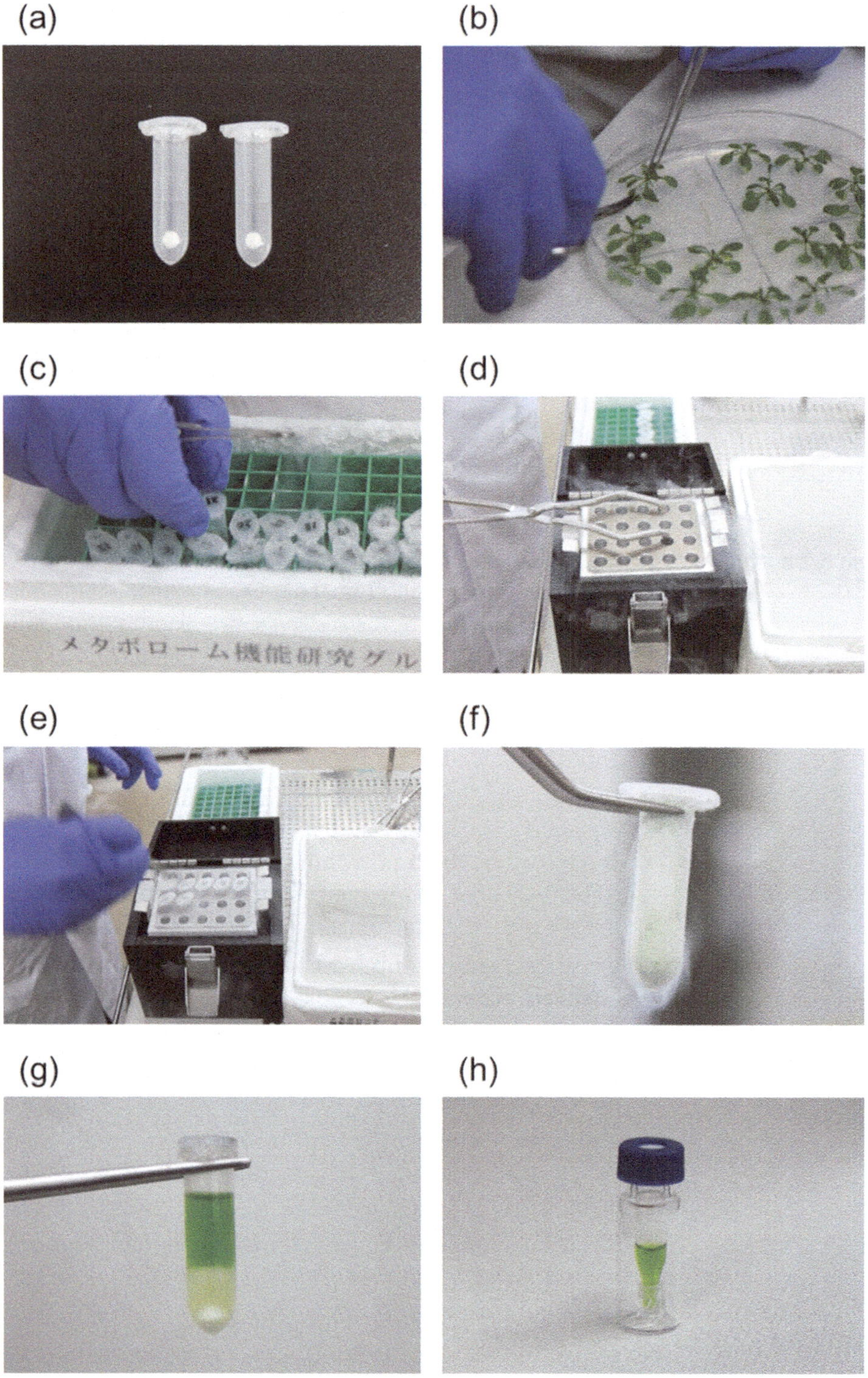

Fig. 1 Sample preparation process. (**a**) Microcentrifuge tubes with zirconia beads used for sampling. (**b**) Sampling of *Arabidopsis* plants using scissors and tweezers. (**c**) Freezing of plant samples by immersion into liquid nitrogen. (**d**) Sample holder (chilled with liquid nitrogen) used for crushing plant samples. (**e**) Sample arrangement in the chilled holder. (**f**) Crushed plant sample. (**g**) Lipid extract (upper layer) washed with water. (**h**) Sample for LC-MS analysis

12. Add 200 μL of ethanol to the above tube and shake it for 5 min at room temperature using an orbital shaker.
13. Centrifuge samples at 10,000 × *g* at 4 °C for 15 min.
14. Transfer the supernatant to a vial with a 200 μL glass insert and seal it tightly (Fig. 1h). In general, ~160 μL of the supernatant can be collected.

3.3 Data Acquisition

1. Set LC parameters as follows: flow rate = 0.15 mL/min; gradient generation = binary pumping system using solvents C and D; gradient program (linear) = 35% D at 0 min, 70% D at 3 min, 85% D at 7 min, 90% D at 10 min, 90% D at 12 min, 35% D at 12.5 min (maintained up to 15 min for reequilibration); column temperature = 55 °C (*see* **Note 6**); injection volume = 1 μL.
2. Set MS parameters as follows (*see* **Note 7**): ionization mode = ESI; correction by lock mass function = yes; MS1 (survey of precursor ions for MS/MS) scan range = *m/z* 100–1200 and scan time = 0.2 s; peak detection mode to trigger MS/MS = intensity-based detection.
3. Set MS/MS parameters as follows: scan range = *m/z* 100–1200; scan time = 0.1 s; scan repeat = 3; event number = 6; collision energy = 20–50 eV (ramp mode); deisotope peak selection = yes.
4. Activate the column oven and start column equilibration at a flow rate of 0.05/mL min.
5. Gradually increase the flow rate to 0.15 mL/min while making sure that the column temperature has reached the designated value. Moreover, pay attention to the system pressure.
6. Program the batch queue and start the data acquisition sequence after column equilibration has finished (*see* **Note 8**).

3.4 Peak Annotation and Alignment

1. Transform the raw data format using Reifycs Abf Converter (*see* **Note 9**).
2. Start the MS-DIAL software and first import the transformed data acquired in positive-ion mode according to the tutorial (*see* **Note 10**).
3. Start a new project and select "lipidomics" as target omics.
4. In analysis parameter settings, select "MSP file" (*see* **Note 11**) and modulate "retention time tolerance" (*see* **Note 12**). For modulating other settings, refer to the tutorial.
5. Press "finish" to proceed with peak picking, alignment, and annotation.
6. Check the lipids identified by positive MS/MS analysis shown in the 2D LC-MS data map (Fig. 2a) (*see* **Note 13**). Spectra matching can be confirmed by clicking on individual spots (Fig. 2b).

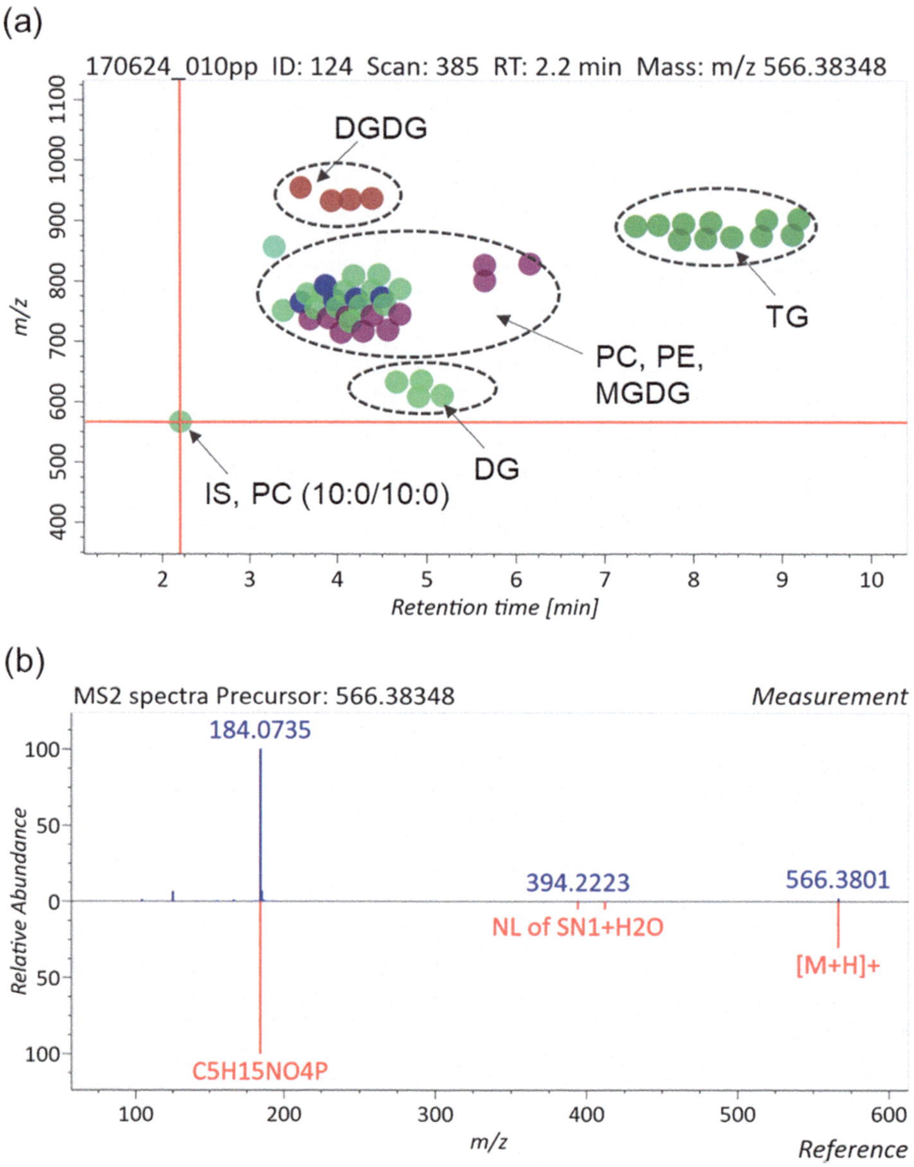

Fig. 2 Major lipids identified by data-dependent MS/MS analysis in positive-ion mode. (**a**) 2D map of detected lipids, with precursor ions of annotated species shown as spots. The following lipid classes were shown: monogalactosyldiacylglycerol (MGDG), digalactosyldiacylglycerol (DGDG), phosphatidylethanolamine (PE), phosphatidylcholine (PC), diacylglycerol (DG), and triacylglycerol (TG). Spots of PC (green), PE (purple), and MGDG (blue) showed extensive overlap. (**b**) Comparison of MS/MS spectra of detected ion, PC (10:0/10:0), and reference

7. Export the peak intensity matrix of signal alignment and use it for further analyses (e.g., *t*-test and multivariate analysis). The major glycerolipids detected in the leaves of *Arabidopsis thaliana* using the described method are listed in Table 1.

Table 1
Major glycerolipids detected in the extract of *Arabidopsis thaliana* leaves in positive-ion mode

Average retention time (min)	Average *m/z* of precursor ion	Metabolite name (compound name and precursor ion type)
5.17	610.5406	DG 34:2; DG(16:0/18:2); $[M + NH_4]^+$
4.89	608.5252	DG 34:3; DG(16:0/18:3); $[M + NH_4]^+$
5.68	638.5751	DG 36:2; DG(18:0/18:2); $[M + NH_4]^+$
4.94	634.5414	DG 36:4; DG(18:2/18:2); $[M + NH_4]^+$
4.67	632.5262	DG 36:5; DG(18:2/18:3); $[M + NH_4]^+$
4.39	936.662	DGDG 34:1; DGDG(16:0/18:1); $[M + NH_4]^+$
4.15	934.6448	DGDG 34:2; DGDG(16:0/18:2); $[M + NH_4]^+$
3.94	932.6273	DGDG 34:3; DGDG(16:0/18:3); $[M + NH_4]^+$
3.33	926.5829	DGDG 34:6; DGDG(16:3/18:3); $[M + NH_4]^+$
3.60	954.6108	DGDG 36:6; DGDG(18:3/18:3); $[M + NH_4]^+$
1.91	496.339	LysoPC 16:0; $[M + H]^+$
1.65	520.3405	LysoPC 18:2; $[M + H]^+$
4.77	774.6074	MGDG 34:1; MGDG(16:0/18:1); $[M + NH_4]^+$
4.48	772.5917	MGDG 34:2; MGDG(16:0/18:2); $[M + NH_4]^+$
4.22	770.5756	MGDG 34:3; MGDG(16:1/18:2); $[M + NH_4]^+$
3.98	768.5616	MGDG 34:4; MGDG(16:1/18:3); $[M + NH_4]^+$
3.60	764.5282	MGDG 34:6; MGDG(16:3/18:3); $[M + NH_4]^+$
3.87	792.5599	MGDG 36:6; MGDG(18:3/18:3); $[M + NH_4]^+$
4.15	732.5532	PC 32:1; PC(16:0/16:1); $[M + H]^+$
4.53	760.5849	PC 34:1; PC(16:0/18:1); $[M + H]^+$
4.26	758.5696	PC 34:2; PC(16:0/18:2); $[M + H]^+$
4.01	756.5522	PC 34:3; PC(16:0/18:3); $[M + H]^+$
3.78	754.5363	PC 34:4; PC(16:1/18:3); $[M + H]^+$
3.39	750.509	PC 34:6; PC(16:3/18:3); $[M + H]^+$
4.71	786.5997	PC 36:2; PC(18:0/18:2); $[M + H]^+$
4.39	784.5848	PC 36:3; PC(18:0/18:3); $[M + H]^+$
4.08	782.5686	PC 36:4; PC(18:2/18:2); $[M + H]^+$
3.86	780.5518	PC 36:5; PC(18:2/18:3); $[M + H]^+$
3.67	778.535	PC 36:6; PC(18:3/18:3); $[M + H]^+$
4.44	810.6003	PC 38:4; PC(18:2/20:2); $[M + H]^+$

(continued)

Table 1 (continued)

Average retention time (min)	Average *m/z* of precursor ion	Metabolite name (compound name and precursor ion type)
4.19	808.5822	PC 38:5; PC(20:2/18:3); $[M + H]^+$
4.57	718.5348	PE 34:1; PE(16:0/18:1); $[M + H]^+$
4.29	716.5235	PE 34:2; PE(16:0/18:2); $[M + H]^+$
4.05	714.5061	PE 34:3; PE(16:0/18:3); [M + H]+
4.71	744.5529	PE 36:2; PE(18:0/18:2); $[M + H]^+$
4.39	742.5392	PE 36:3; PE(18:0/18:3); $[M + H]^+$
4.11	740.5213	PE 36:4; PE(18:2/18:2); $[M + H]^+$
3.89	738.5046	PE 36:5; PE(18:2/18:3); $[M + H]^+$
3.70	736.4901	PE 36:6; PE(18:3/18:3); $[M + H]^+$
5.65	800.6165	PE 40:2; PE(22:0/18:2); $[M + H]^+$
6.16	828.6487	PE 42:2; PE(24:0/18:2); $[M + H]^+$
5.86	826.6294	PE 42:3; PE(24:0/18:3); $[M + H]^+$
5.65	826.6328	PE 42:3; PE(24:1/18:2); $[M + H]^+$
3.76	836.551	SQDG 34:2; SQDG(16:0/18:2); $[M + NH_4]^+$
3.60	834.535	SQDG 34:3; SQDG(16:0/18:3); $[M + NH_4]^+$
3.46	858.5406	SQDG 36:5; SQDG(18:2/18:3); $[M + NH_4]^+$
3.28	856.5225	SQDG 36:6; SQDG(18:3/18:3); $[M + NH_4]^+$
9.05	850.7924	TG 50:1; TG(16:0/16:0/18:1); $[M + NH_4]^+$
8.70	848.7671	TG 50:2; TG(16:0/16:0/18:2); $[M + NH_4]^+$
8.38	846.7534	TG 50:3; TG(16:0/16:0/18:3); $[M + NH_4]^+$
9.12	876.8008	TG 52:2; TG(16:0/18:1/18:1); $[M + NH_4]^+$
8.76	874.7844	TG 52:3; TG(16:0/18:1/18:2); $[M + NH_4]^+$
8.43	872.7697	TG 52:4; TG(16:0/18:2/18:2); $[M + NH_4]^+$
8.13	870.7534	TG 52:5; TG(16:0/18:2/18:3); $[M + NH_4]^+$
7.84	868.7375	TG 52:6; TG(16:0/18:3/18:3); $[M + NH_4]^+$
7.24	864.7045	TG 52:8; TG(16:3/18:2/18:3); $[M + NH_4]^+$
6.97	862.6907	TG 52:9; TG(16:3/18:3/18:3); $[M + NH_4]^+$
9.18	902.8149	TG 54:3; TG(18:1/18:1/18:1); $[M + NH_4]^+$
8.83	900.8001	TG 54:4; TG(18:1/18:1/18:2); $[M + NH_4]^+$
8.50	898.7865	TG 54:5; TG(18:1/18:2/18:2); $[M + NH_4]^+$
8.18	896.7692	TG 54:6; TG(18:1/18:2/18:3); $[M + NH_4]^+$

(continued)

Table 1 (continued)

Average retention time (min)	Average *m/z* of precursor ion	Metabolite name (compound name and precursor ion type)
7.88	894.7542	TG 54:7; TG(18:2/18:2/18:3); $[M + NH_4]^+$
7.60	892.7369	TG 54:8; TG(18:2/18:3/18:3); $[M + NH_4]^+$
7.35	890.7207	TG 54:9; TG(18:3/18:3/18:3); $[M + NH_4]^+$

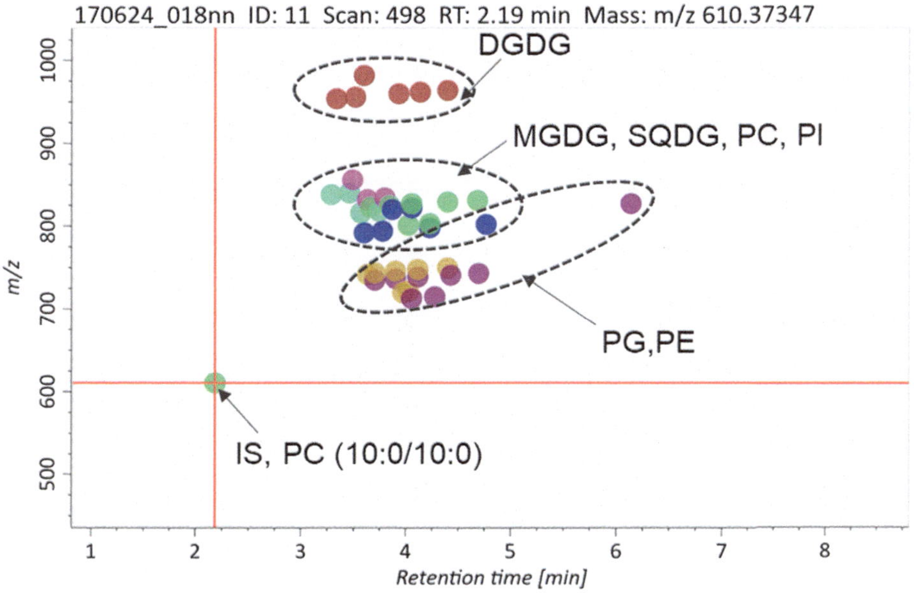

Fig. 3 Major lipid species identified by data-dependent MS/MS analysis in negative-ion mode. 2D image of detected lipid species, with precursor ions of annotated species shown as spots. The following lipid classes were shown: phosphatidylinositol (PI), phosphatidylglycerol (PG), sulfoquinovosyldiacylglycerol (SQDG), MGDG, DGDG, PE, and PC. Spots of PI (*pink*), SQDG (*bluish green*), MGDG (*blue*), PC (*green*), PE (*purple*), and PG (*yellow*) showed extensive overlap

8. Repeat **steps 1–7** using MS/MS data recorded in negative-ion mode and obtain an annotated dataset (Fig. 3 and Table 2).

4 Notes

1. Other lipids can be used as an internal standard. The internal standard described herein was selected due to being relatively inexpensive and because there is only subtle 10:0/10:0-PC in extracts from tested plant materials (e.g., *Arabidopsis* and rice).

Table 2
Major glycerolipids detected in the extract of *Arabidopsis thaliana* leaves in negative-ion mode

Average retention time (min)	Average *m/z* of precursor ion	Metabolite name (compound name and precursor ion type)
4.39	963.6245	DGDG 34:1; DGDG(16:0/18:1); $[M + FA - H]^{-a}$
4.15	961.6094	DGDG 34:2; DGDG(16:0/18:2); $[M + FA - H]^-$
3.94	959.5912	DGDG 34:3; DGDG(16:0/18:3); $[M + FA - H]^-$
3.52	955.5623	DGDG 34:5; DGDG(16:2/18:3); $[M + FA - H]^-$
3.34	953.5454	DGDG 34:6; DGDG(16:3/18:3); $[M + FA - H]^-$
3.61	981.5762	DGDG 36:6; DGDG(18:3/18:3); $[M + FA - H]^-$
2.95	255.2322	Fatty acid 16:0; $[M - H]^-$
2.73	279.2338	Fatty acid 18:2; $[M - H]^-$
2.42	277.216	Fatty acid 18:3; $[M - H]^-$
4.77	801.5718	MGDG 34:1; MGDG(16:0/18:1); $[M + FA - H]^-$
4.49	799.5559	MGDG 34:2; MGDG(16:0/18:2); $[M + FA - H]^-$
4.22	797.5425	MGDG 34:3; MGDG(16:0/18:3); $[M + FA - H]^-$
3.99	795.5243	MGDG 34:4; MGDG(16:2/18:2); $[M + FA - H]^-$
3.81	793.5098	MGDG 34:5; MGDG(16:2/18:3); $[M + FA - H]^-$
3.58	791.4935	MGDG 34:6; MGDG(16:3/18:3); $[M + FA - H]^-$
4.11	821.5399	MGDG 36:5; MGDG(18:2/18:3); $[M + FA - H]^-$
3.87	819.5242	MGDG 36:6; MGDG(18:3/18:3); $[M + FA - H]^-$
4.24	802.5593	PC 34:2; PC(16:0/18:2); $[M + FA - H]^-$
4.03	800.5432	PC 34:3; PC(16:0/18:3); $[M + FA - H]^-$
4.69	830.5895	PC 36:2; PC(18:0/18:2); $[M + FA - H]^-$
4.40	828.5746	PC 36:3; PC(18:1/18:2); $[M + FA - H]^-$
4.08	826.5563	PC 36:4; PC(18:2/18:2); $[M + FA - H]^-$

(continued)

Table 2
(continued)

Average retention time (min)	Average *m/z* of precursor ion	Metabolite name (compound name and precursor ion type)
3.85	824.5421	PC 36:5; PC(18:2/18:3); $[M + FA - H]^-$
3.68	822.5283	PC 36:6; PC(18:3/18:3); $[M + FA - H]^-$
4.29	714.5078	PE 34:2; PE(16:0/18:2); $[M - H]^-$
4.06	712.4913	PE 34:3; PE(16:0/18:3); $[M - H]^-$
4.70	742.5387	PE 36:2; PE(18:0/18:2); $[M - H]^-$
4.41	740.5236	PE 36:3; PE(18:0/18:3); $[M - H]^-$
4.12	738.505	PE 36:4; PE(18:2/18:2); $[M - H]^-$
3.55	738.5062	PE 36:4; PE(18:2/18:2); $[M - H]^-$
3.39	736.4903	PE 36:5; PE(18:2/18:3); $[M - H]^-$
3.89	736.4907	PE 36:5; PE(18:2/18:3); $[M - H]^-$
3.70	734.4748	PE 36:6; PE(18:3/18:3); $[M - H]^-$
6.15	826.6323	PE 42:2; PE(24:0/18:2); $[M - H]^-$
4.03	721.5013	PG 32:0; PG(16:0/16:0); $[M - H]^-$
3.96	719.4866	PG 32:1; PG(16:0/16:1); $[M - H]^-$
4.40	749.5332	PG 34:0; PG(16:0/18:0); $[M - H]^-$
4.12	747.5155	PG 34:1; PG(16:0/18:1); $[M - H]^-$
3.89	745.5005	PG 34:2; PG(16:0/18:2); $[M - H]^-$
3.71	743.4848	PG 34:3; PG(16:0/18:3); $[M - H]^-$
3.64	741.4694	PG 34:4; PG(16:1/18:3); $[M - H]^-$
3.80	833.516	PI 34:2; PI(16:0/18:2); $[M - H]^-$
3.64	831.4998	PI 34:3; PI(16:0/18:3); $[M - H]^-$
3.50	855.4987	PI 36:5; PI(18:2/18:3); $[M - H]^-$
3.75	817.5114	SQDG 34:2; SQDG(16:0/18:2); $[M - H]^-$
3.58	815.4965	SQDG 34:3; SQDG(16:0/18:3); $[M - H]^-$
3.46	839.4955	SQDG 36:5; SQDG(18:2/18:3); $[M - H]^-$
3.29	837.48	SQDG 36:6; SQDG(18:3/18:3); $[M - H]^-$

[a]FA denotes formic acid

2. Although the required amounts of solvents can be measured using cylinders, an analytical balance generally provides superior accuracy and was utilized in this work.

3. In general, 25–80 mg of fresh weight is sufficient for analysis of vegetative tissues.
4. Samples should not be thawed until the completion of homogenization in the extraction solvent B. Plant tissues contain substantial amounts of lipid catabolism enzymes such lipases, lipoxygenases, and acyltransferases. Sample thawing enhances lipid decomposition by these enzymes, leading to a significant decrease of phospholipid content and an increase of oxylipin and acyl-MGDG contents [14, 15].
5. Conventional mortar and pestle can be used instead of an electric sample disruption device.
6. Although 2-propanol in solvent D is required to elute highly hydrophobic lipids such as triacylglycerols from C18 columns, its viscosity increases the pressure in the LC system, which can be mitigated by using a high column temperature.
7. Since most parameters are highly dependent on the mass spectrometer used, no excessive details are provided.
8. Samples should be analyzed in random order. To stabilize the retention time of each metabolite, several repeated analyses of blank samples (e.g., absolute ethanol or extracts of a microcentrifuge tube) or QC samples (e.g., mixtures of whole sets of analytes) should be scheduled at the beginning of the batch queue.
9. Abf converter is provided by Reifycs (http://www.reifycs.com/AbfConverter/). Despite some MS vendor-specific requirements, details are also available at the provider's website.
10. MS-DIAL is provided by Dr. Tsugawa and available at the RIKEN PRIMe website (http://prime.psc.riken.jp) [16].
11. Three different libraries (MSDIAL-LipidDBs-VS23-FiehnO.lbm, MSDIAL-LipidDBs-VS23-AritaM.lbm, and MSDIAL-LipidDBs-VS23-SaitoK.lbm) can be selected in the "identification" tab of "analysis parameter setting." The MSP of SaitoK contains retention times of polar plant glycerolipids detected in our laboratory. If you utilize our chromatographic conditions, the MSP file name should be changed manually (default library is MSDIAL-LipidDBs-VS23-FiehnO.lbm).
12. Not all retention times originate from our dataset. Thus, the "retention time tolerance" should be increased to assign various lipids in your samples, and the retention time difference between experimental and reference data can be checked after processing.
13. In general, positive ionization is more suited for lipid detection than negative ionization, but the preferred mode for every class of lipids needs to be selected considering their physicochemical properties, e.g., glycerolipids without polar head groups (triacylglycerols, diacylglycerols, and monoacylglycerols) and sterol

derivatives are hardly detectable in the negative-ion mode. In contrast, negative ionization is suited for the structural analysis of sulfoquinovosyldiacylglycerol (SQDG), phosphatidylglycerol (PG), phosphatidylinositol (PI), and free fatty acids.

Acknowledgments

We acknowledge technical assistance from Mr. Kouji Takano (RIKEN Center for Sustainable Resource Science). This work has been partly supported by grants from the Strategic International Collaborative Research Program of Japan Science and Technology Agency (Metabolomics for a Low Carbon Society, JST-NSF) and the Integrated Lipidology Program of RIKEN. Moreover, support from the Japan Advanced Plant Science Network is gratefully acknowledged.

References

1. Guan X, Okazaki Y, Lithio A et al (2017) Discovery and characterization of the 3-hydroxyacyl-ACP dehydratase component of the plant mitochondrial fatty acid synthase system. Plant Physiol 173:2010–2028
2. Higashi Y, Okazaki Y, Myouga F, Shinozaki K, Saito K (2015) Landscape of the lipidome and transcriptome under heat stress in Arabidopsis thaliana. Sci Rep 5:10533
3. Okazaki Y, Kamide Y, Hirai MY, Saito K (2013) Plant lipidomics based on hydrophilic interaction chromatography coupled to ion trap time-of-flight mass spectrometry. Metabolomics 9(Suppl 1):121–131
4. Okazaki Y, Saito K (2014) Roles of lipids as signaling molecules and mitigators during stress response in plants. Plant J 79:584–596
5. Shimada TL, Takano Y, Shimada T et al (2014) Leaf oil body functions as a subcellular factory for the production of a phytoalexin in Arabidopsis. Plant Physiol 164:105–118
6. Okazaki Y, Shimojima M, Sawada Y et al (2009) A chloroplastic UDP-glucose pyrophosphorylase from Arabidopsis is the committed enzyme for the first step of sulfolipid biosynthesis. Plant Cell 21:892–909
7. Okazaki Y, Takano K, Saito K (2017) Lipidomic analysis of soybean leaves revealed tissue-dependent difference in lipid remodeling under phosphorus-limited growth conditions. Plant Biotechnol 34:57–63
8. Okazaki Y, Nishizawa T, Takano K et al (2015) Induced accumulation of glucuronosyldiacylglycerol in tomato and soybean under phosphorus deprivation. Physiol Plant 155:33–42
9. Okazaki Y, Otsuki H, Narisawa T et al (2013) A new class of plant lipid is essential for protection against phosphorus depletion. Nat Commun 4:1510
10. Tsugawa H, Ikeda K, Tanaka W et al (2017) Comprehensive identification of sphingolipid species by in silico retention time and tandem mass spectral library. J Cheminform 9:19
11. Tsugawa H, Cajka T, Kind T et al (2015) MS-DIAL: data-independent MS/MS deconvolution for comprehensive metabolome analysis. Nat Methods 12:523–526
12. Giavalisco P, Li Y, Matthes A et al (2011) Elemental formula annotation of polar and lipophilic metabolites using (13) C, (15) N and (34) S isotope labelling, in combination with high-resolution mass spectrometry. Plant J 68:364–376
13. Matyash V, Liebisch G, Kurzchalia TV et al (2008) Lipid extraction by methyl-tert-butyl ether for high-throughput lipidomics. J Lipid Res 49:1137–1146
14. Nilsson AK, Johansson ON, Fahlberg P et al (2015) Acylated monogalactosyl diacylglycerol: prevalence in the plant kingdom and identification of an enzyme catalyzing galactolipid head group acylation in Arabidopsis thaliana. Plant J 84:1152–1166
15. Heinz E (1967) Acylgalactosyl diglyceride from leaf homogenates. Biochim Biophys Acta, Lipids Lipid Metab 144:321–332
16. Akiyama K, Chikayama E, Yuasa H et al (2008) PRIMe: a Web site that assembles tools for metabolomics and transcriptomics. In Silico Biol 8:339–345

Chapter 12

Targeted LC-MS Analysis for Plant Secondary Metabolites

Takafumi Shimizu, Mutsumi Watanabe, Alisdair R. Fernie, and Takayuki Tohge

Abstract

Recent technological developments and methodological advances of both liquid chromatography (LC) and mass spectrometry (MS) have allowed LC-MS-based plant metabolomics to become a common tool for investigating quantity, quality, and chemical diversity of plant metabolites. Targeted LC-MS metabolite analysis focuses on the detection and quantitation of the researcher's target metabolites. Whilst the word "target analysis" has been used for the analytical measurement to obtain the absolute concentrations evaluated by authentic and/or stable-isotope-labeled standards, over time the phrase came to be also used in a broad sense for the measurement of annotatable metabolites by structural information obtained from the combination of different strategies such as MS/MS analysis, reference extracts, mutant analysis and database search. Here, we describe a general protocol for targeted LC-MS metabolite profiling of plant secondary metabolites. Additionally, we introduce some examples of peak annotation using the combination approach.

Key words Liquid chromatography–tandem mass spectrometry, LC-MS/MS, Plant secondary metabolites, Targeted analysis, Peak identification, Peak annotation

1 Introduction

Land plants have evolved unique metabolic abilities, producing highly divergent specialized metabolites (the so-called secondary metabolites), in order to adapt to various environmental conditions. In plant science, it is highly important to understand the types and amounts of secondary metabolites produced in the plant kingdom, since such productivity indicates the adaptive evolutions of plants in nature. Furthermore, better understanding of plant secondary metabolism can provide hints for metabolic engineering of those compounds that are also known as health beneficial or medical bioactive compounds [1].

Since metabolomics does not directly depend on the genome sequence, its scope was rapidly expanded to various species without completion of the genome sequencing. This point is an advantage

Carla António (ed.), *Plant Metabolomics: Methods and Protocols*, Methods in Molecular Biology, vol. 1778,
https://doi.org/10.1007/978-1-4939-7819-9_12, © Springer Science+Business Media, LLC, part of Springer Nature 2018

of metabolomics over other omics technologies such as transcriptomics and proteomics [2, 3], but becomes a disadvantage because of the lack of linear relationship between the genome and the metabolome resulting in the difficulty of identification of unknown peaks [4, 5]. Due to the large variety of chemical classes and properties as well as their huge dynamic range of metabolite concentrations in plant, we need to employ an extensive array of different analytical techniques to obtain better coverage of metabolites for plant metabolomics. Liquid chromatography–mass spectrometry (LC-MS)-based plant metabolomics, which is a contrast to other analytical technologies such as gas chromatography–mass spectrometry (GC-MS)- or nuclear magnetic resonance (NMR)-based plant metabolomics, finds a wide application measuring a large number of compound families using a combination of different extraction methods, separation columns and mass detection.

To date, several different LC-MS-based metabolomics platforms have been developed for the target analysis of primary metabolites [6], photosynthetic intermediates [7], lipids and fatty acids [8, 9], phytohormones [10] and secondary metabolites [11–16] as well as for untargeted metabolome analysis [17–19]. Methodologically, targeted LC-MS and untargeted LC-MS approaches for secondary metabolites are often very similar, or even the same, but they are significantly distinct in the data analysis. Whilst untargeted metabolomics is the comprehensive analysis focusing on all detected metabolite signals to provide a metabolic overview, targeted metabolomics is the profiling of defined groups of chemically characterized and biochemically annotated metabolites. Nowadays, "target analysis" refers not only to the quantification of the absolute concentrations of metabolites by authentic and/or stable-isotope-labeled standards, but also to the profiles of annotatable metabolites by structural information from tandem mass spectrometric (MS/MS) analysis, reference extracts, mutant analysis, and web resources.

In this chapter, we describe a general methodology for LC-MS-based targeted metabolomics for plant secondary metabolites, and several strategies for peak identification and annotation of detected metabolites.

2 Materials

Use ultrapure water and high quality chemicals for the experiments.

2.1 Plant Material Harvest, Sampling, and Aliquoting

1. Plant material of model plant close to the plant species of your interest (*see* Subheading 3.6, **step 4**).
2. Clean scissors or razor.
3. 1.5-mL and 2.0-mL round bottom-shaped microcentrifuge tubes (e.g., Eppendorf).

4. Zirconia bead or metal ball. The ball size depends on tube volume (e.g., 3 mm ϕ ball for 2.0-mL microcentrifuge tube).
5. Mixer Mill.
6. Liquid nitrogen.
7. –80 °C freezer.
8. Electric analytical balance.
9. Micro spatula.

2.2 Extraction of Plant Secondary Metabolites

1. Methanol (MeOH) ULC/MS grade. <CAUTION> Methanol is toxic and should be handled under the fume hood.
2. Distilled water ULC/MS grade.
3. Extraction solution of methanol–distilled water (8:2, v/v) containing appropriate concentration of internal standard (IS) (e.g., 5 μg/mL) (*see* **Note 1**).

2.3 LC-MS Analysis

1. LC-MS system with the operation software provided by MS manufacturer (e.g., Xcalibur by Thermofisher, Analyst by SCIEX, MassHunter by Agilent)
2. Suitable glass vial and screw cap for LC-MS analysis.
3. Distilled water ULC/MS grade.
4. Acetonitrile (ACN) ULC/MS grade. <CAUTION> Acetonitrile is toxic and should be handled under the fume hood.
5. Formic acid (HCOOH) ULC/MS grade. <CAUTION> Formic acid is corrosive and should be handled under the fume hood.
6. Mobile Phase: eluent (A) 0.1% formic acid in water, eluent (B) 0.1% formic acid in ACN) (*see* **Note 2**).
7. Computer and software for mass spectral data analysis (e.g., Xcalibur, Analyst, MassHunter, MSFact, XCMS2, BINBASE, MZmine2).

3 Methods

This chapter introduces an optimization procedure for the measurement of plant leaves of common model plants such as Arabidopsis, rice, and tomato. Plants produce metabolites with a large chemical diversity and dynamic range. Additionally, such metabolites are strongly influenced by growth conditions and largely different between tissues/organ types. Therefore, due to the difference of metabolite concentrations in plant species and plant tissues, the volume of extraction buffer may have to be optimized.

3.1 Plant Cultivation And Sampling

1. Cultivate plants of your interest under optimal growth conditions.
2. Harvest plant leaves using clean scissors or razor.
3. Place collected plant material into 2.0-mL round bottom-shaped microcentrifuge tubes.
4. Put a zirconia bead or metal ball into each sample tube, close the lid properly, and freeze it immediately in liquid nitrogen.
5. Grind the frozen plant tissue with a Mixer Mill for 2 min at 25 Hz.
6. Store frozen ground samples at −80 °C until further use.

3.2 Sample Aliquoting

1. Keep all the samples in liquid nitrogen; make sure that liquid nitrogen does not enter to the sample tubes.
2. Dip a new 2.0-mL round bottom-shaped microcentrifuge tube into liquid nitrogen for 5 s, put it on steady position of the electric analytical balance and switch "automatic zero correction" on quickly.
3. Aliquot and transfer the frozen powder (e.g., 50 mg ± 5 mg) into the sample tube as quickly as possible to prevent sample from thawing.
4. Check the sample weight for each sample tube.
5. Close the lid of the sample tube and put the sample back into liquid nitrogen.
6. Write down the sample weight.
7. Store all sample aliquots at −80 °C until extraction.

3.3 Extraction of Plant Metabolites

1. Open the lid of the sample tube containing frozen ground sample (*see* **Note 3**).
2. Put a precooled zirconia bead into the sample tube.
3. Add 5 μL of extraction solution per 1 mg of frozen ground sample (e.g., 256 μL extraction solution for 51.2 mg of frozen ground sample).
4. Close the lid of the sample tube and put the sample back into liquid nitrogen.
5. Homogenize the sample with a Mixer Mill for 4 min at 25 Hz.
6. Centrifuge the sample tube for 10 min at 20,000 × *g* at 4 °C.
7. Transfer the supernatant to a new 1.5-mL round bottom-shaped microcentrifuge tube.
8. Centrifuge the new sample tube for 5 min at 20,000 × *g* at 4 °C.
9. Transfer the supernatant to a new 1.5-mL round bottom-shaped microcentrifuge tube.
10. Store extracts at 4 °C until LC-MS analysis (*see* **Note 4**).

3.4 LC-MS Analysis

1. Set up an LC-MS system with autosampler for plant secondary metabolite analysis (*see* **Note 5**).
2. Set HPLC bottles of both elution buffers.
3. Transfer the sample extract to a glass vial suitable for LC-MS analysis and put it in the autosampler tray.
4. Analyze your samples after the measurement of extraction buffer blanks to equilibrate the LC-MS system (*see* **Note 6**).

3.5 Data Analysis

1. Analyze the obtained data with software provided by MS manufacturer (e.g., Xcalibur by Thermofisher, Analyst by SCIEX, MassHunter by Agilent) or freely available software (e.g., MSFact, XCMS2, BINBASE, MZmine2, and TagFinder).
2. Evaluate the data by peak area or peak height considering peak shape and peak shift.
3. Export the signal intensity data of all the detected peaks and make a whole data matrix.
4. Normalize the whole data using sample weight and peak intensity of IS.

3.6 Peak Prediction, Annotation, and Identification

In general, peak identification and annotation of detected plant secondary metabolites are very difficult because of their structural complexity. To perform "peak identification," best practices are (1) identification by NMR studies of the target peak purified from plant extracts, or (2) identification by coelution profiling with authentic standard compound. But in case of a metabolomics approach for plant secondary metabolites, the availability of authentic standard compounds is limited because of their structural complexity and diversity. Therefore, under the circumstances "peak annotation" is the greatest challenge in plant metabolomics. Unlike "peak identification" for the complete characterization, "peak annotation" is a way to provide chemical and structural specifications to the peak using structural information from MS/MS analysis, reference extracts, mutant analysis and web-resources.

In this section, we introduce some examples of peak identification and annotation using these approaches.

1. *Chromatographic coelution profiling of standard compounds for peak identification:* To identify detected peaks, compare the *m/z* value and the retention time of detected peaks with authentic standard compounds. Confirmation of MS/MS spectrum patterns is also a reliable way to identify the peak. Some chemical companies offer secondary metabolites as standard compounds (for example, EXTRASYNTHESE: http://www.extrasynthese.com/; AnalytiCon discovery: http://ac-discovery.com/).
2. *MS/MS analysis for peak annotation:* Perform MS/MS analysis with different intensity of collision energy to obtain different

fragmentation patterns. Importantly fragmentation pattern of some secondary metabolites in LC-MS analysis provides some key fragments such as aglycones and moiety detached fragments (e.g., fragments without sugar moieties). Such information provides a great hint of which metabolite classes the peak is. Additionally, the MS/MS fragment pattern can be used for peak prediction in comparison with the information in public MS/MS database such as MASSBANK (http://www.massbank.jp/index.html?lang=en) [20].

3. *Retention time index of reference compounds for peak annotation:* Reference compounds here mean similar compounds detected at different elution time with different *m/z* value but having same modification judging from the fragmentation pattern. If the compound class and tentative structure of your metabolites is known, the retention time index of reference compounds provides you some hints about a part of chemical structure of your compounds (e.g., types of chemical moieties such as glycosylation forms, methoxylation, and hydroxylation).
4. *Chromatographic coelution profiling of extracts from reference plants for peak annotation:* Annotation of detected peaks is possible by comparing *m/z* value and retention time index according to reference compounds, literature data with reference extracts (e.g., leaf extracts of Arabidopsis grown under the optimal condition). In Table 1, we present an example of a LC-MS peak table of plant secondary metabolites, which can be detected in both the leaf and flower of Arabidopsis previously published by Tohge et al. [11, 16]. When extracts of Arabidopsis leaf or flower are profiled, the detected peaks can be evaluated and annotated with those previously published in the literature.
5. *Reference plant extracts of biosynthetic gene mutants for peak annotation:* The effect of mutation in an enzymatic gene on the metabolic profile sometimes results in no-production of all downstream metabolites. The knockout mutants of biosynthetic genes, which have been previously characterized in order to identify chromatographic features associated to a given metabolite structure can be used for peak annotation. Any mutants, including natural mutants, can be also employed for this approach. Several mutants having specific metabolic phenotypes are publically available (e.g., anthocyanin over accumulator, flavonoid-less mutant, glucosinolate over accumulator, phenylpropanoid-less mutant, and lines having a metabolic phenotype by natural polymorphism or mutation) in several bioresource centers for major model plants such as Arabidopsis, rice and tomato (Fig. 1). In Fig. 1, an example

Table 1
Major secondary metabolites found in both leaf and flower extracts of Arabidopsis using LC-MS analysis, adapted from [13, 16]

Rt (min)	*m/z*	Compound class	Compound name
3.14	422(neg)	Aliphatic glucosinolate	3-methylsulfinylpropyl glucosinolate
4.24	436(neg)	Aliphatic glucosinolate	4-methylsulfinylbutyl glucosinolate
7.57	450(neg)	Aliphatic glucosinolate	5-methylsulfinylpentyl glucosinolate
9.37	464(neg)	Aliphatic glucosinolate	6-methylsulfinyl hexyl glucosinolate
11.26	1134(posi)	Anthocyanin derivative	A11 [11]
11.54	478(neg)	Aliphatic glucosinolate	7-methylsulfinylheptyl glucosinolate
12.19	385(neg)	Hydroxycinnamate derivative	Sinapoyl glucoside
12.81	420(neg)	Aliphatic glucosinolate	4-methylthiobutyl glucosinolate
14.04	492(neg)	Aliphatic glucosinolate	8-methylsulfinyloctyl glucosinolate
14.11	447(neg)	Indole glucosinolate	Indole-3-methylglucosinolate
14.77	739(neg)	Flavonol glycoside	Kaempferol 3-O-[2''-*O*-(Rha)-Glc] 7-*O*-Rha
15.49	591(neg)	Hydroxycinnamate derivative	1,2-*bis*-*O*-sinapoyl-*β*-D-glucoside
16.67	477(neg)	Indole glucosinolate	4-Methoxy-indol-3-ylmethyl-glucosinolate
16.76	593(neg)	Flavonol glycoside	Kaempferol 3-*O*-Glc 7-*O*-Rha
18.30	577(neg)	Flavonol glycoside	Kaempferol 3-*O*-Rha 7-*O*-Rha
18.67	339(neg)	Hydroxycinnamate derivative	Sinapoyl (*S*)-malate
19.63	477(neg)	Indole glucosinolate	1-methoxyindol-3-ylmethylglucosinolate
24.00	462(neg)	Aliphatic glucosinolate	7-methylthioheptyl glucosinolate
28.13	476(neg)	Aliphatic glucosinolate	8-methylthiooctyl glucosinolate

Rt retention time; *Glc* glucosyl or glucoside; *Rha* rhanosyl or rhamnoside; *A11* cyanidin 3-*O*-[2''-*O*-(6'''-*O*-(sinapoyl) xylosyl) 6''-*O*-(*p*-*O*-(glucosyl)-*p*-coumaroyl) glucoside] 5-*O*-(6''''-*O*-malonyl) glucoside

of this "peak annotation by mutant analysis" is presented [16, 21].

6. *Refinement of biosynthetic framework by predicting a link between enzymatic steps and biosynthetic intermediates for peak prediction:* For example, if nicotiflorin (kaempferol-3-*O*-[6''-*O*-(rhamnosyl)-glucoside]) was detected in tomato leaves and could characterize it by chromatographic coelution of standard compound, you can describe the biosynthetic framework for the production of nicotiflorin. Using the knowledge of this biosynthetic pathway characterized in other plant species, you can predict that more than seven enzymatic steps are necessary for the production of nicotiflorin from *p*-coumaroyl-CoA (the

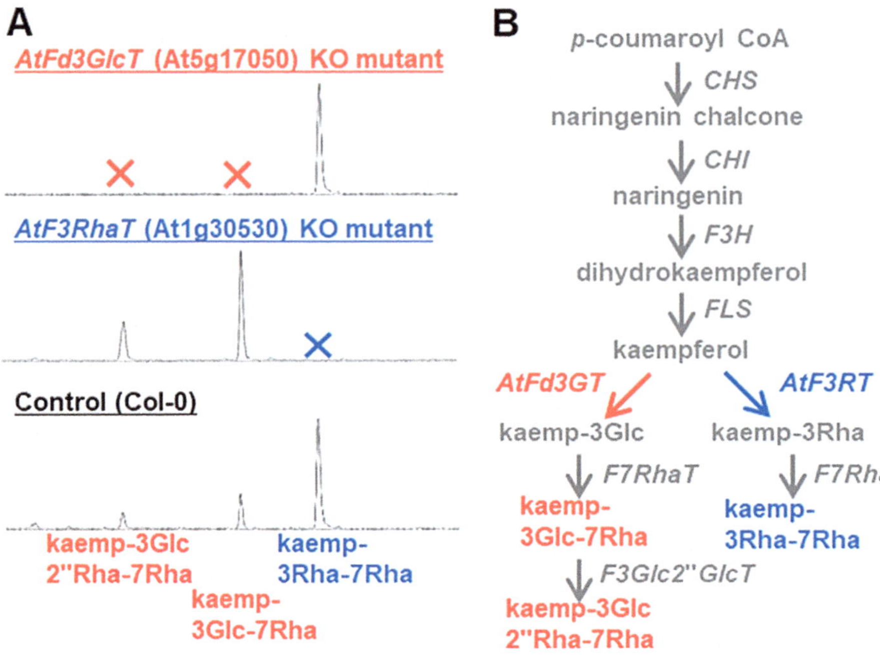

Fig. 1 An example of mutant analysis for peak annotation of secondary metabolites. (**a**) Ion extracted LC-MS chromatograms (*m/z* 287 in positive ion detection) of the extracts from wild type and knockout mutants of genes encoding flavonoid glycosyltransferase (At5g17050 and At1g30530). Peaks at *m/z* 287 in positive ion detection indicate kaempferol or cyanidin related compounds in Arabidopsis thaliana. The peaks, which are disappeared in each mutant, correspond to the downstream metabolites in the biosynthetic pathway. (**b**) Flavonol biosynthetic pathway in Arabidopsis. Abbreviations: Glc, glucosyl or glucoside; Rha, rhamnosyl or rhamnoside; KO, knockout, AtFd3GlcT, flavonoid-3-*O*-glucosyltransferase; AtF3RhaT, flavonoid-3-*O*-rhamosyltransferase; CHS, chalcone synthase; CHI, Chalcone isomerase; F3H, flavanone 3-hydroxylase; FLS, flavonol synthase; F7RhaT, flavonol-7-*O*-rhamnosyltranferase; F3Glc2″GlcT, Flavonol-3-*O*-glucoside: 2″-*O*-glucosyltrasneferase

whole pathway comprise CHS, chalcone synthase; CHI, chalcone isomerase; F3H, flavanone 3-hydroxylase; F3′H, flavonoid 3′-hydroxylase; FLS, flavonol synthase; F3GT, flavonol 3-*O*-glucosyltransferase; F3G6″RT, flavonol 3-*O*-glucoside 6″-*O*-rhamnyltransferase). Then if a peak having fragmentation of kaempferol-hexoside after the peak of nicotiflorin in the same chromatogram was detected, this peak can be predicted as a "putative kaempferol-3-*O*-glucoside." Further confirmation is required, but such analysis is useful for peak prediction. An example of this approach, focusing on tomato glycoalkaloid biosynthesis, is described in Schwahn et al. [22].

7. *A survey of metabolite databases for peak prediction:* Database search is one of the common approaches for obtaining supporting information for peak prediction. de Souza et al. [5] summarized recent web-resource databases for peak annotation on MS based plant metabolomics. Universal Natural Products

Database (UNPD) [23] and KNApSAcK (http://kanaya.naist.jp/KNApSAcK/) [24] are the largest class of database containing information of 50,000–22,000 metabolites from over 22,000 species. Since the number of combinations of atom–atom connections is incalculably large, when you search the predicted molecular weight value of your detected metabolite, you may find too many candidate compounds in metabolite database. Database search should, therefore, be used for the acquisition of hints toward predicting the compound family with other information such as possible modification type of your metabolite predicted from several approaches described above. However metabolite database survey is very powerful and a convenient tool for peak prediction on target LC-MS plant metabolomics. It is, therefore, worth trying.

4 Notes

1. To evaluate the extraction efficiency and instrument performance, internal standards (IS) should be added to the extraction buffer. The use of a stable-isotope-labeled compound of a target metabolite as IS is one of the best approaches to determine the exact concentration of an endogenous target metabolite in the plant matrix. However, most stable-isotope-labeled compounds are not commercially available for specialized secondary metabolites. Therefore, a chemical that is not present in the biological sample is typically used as IS, e.g., isovitexin (apigenin-6-*C*-glucoside) for dicot plants (a specific glycoflavone in monocots) or hesperidin (hesperetin-7-*O*-rutinoside) for monocot plants.
2. Premixed elution buffers are also available from some chemical companies.
3. Keep the bottom of the tube in liquid nitrogen making sure that the liquid nitrogen does not enter the sample tube.
4. We recommend preparing extracts before LC-MS analysis; however, extracts can be stored for at least 1 month at 4 °C protected from light exposure.
5. For the targeted LC-MS analysis for plant secondary metabolites we recommend using the following LC-MS system: common Reverse Phase (RP)-HPLC system coupled with Electrospray Ionization (ESI)-MS. Several examples, HPLC and MS parameters can be found in the literature [13–19].
6. It is recommended to setup a suitable number of HPLC preruns to warm up the system; if this procedure is not performed long enough, a problem of retention time (RT)-shift caused by the HPLC pumps or bad reequilibration of the LC column

might occur and affect the first set of samples. Such RT shift between samples gives you problematic results and data analysis.

Acknowledgments

We gratefully acknowledge the support of the Nara Institute of Science and Technology (NAIST) and the Max Planck Society (MPG) for funding.

References

1. Tohge T, Watanabe M, Hoefgen R, Fernie AR (2013) The evolution of phenylpropanoid metabolism in the green lineage. Crit Rev Biochem Mol Biol 48:123–152
2. Roessner U, Luedemann A, Brust D et al (2001) Metabolic profiling allows comprehensive phenotyping of genetically or environmentally modified plant systems. Plant Cell 13:11–29
3. Schauer N, Fernie AR (2006) Plant metabolomics: towards biological function and mechanism. Trends Plant Sci 11:508–516
4. Fernie AR, Stitt M (2012) On the discordance of metabolomics with proteomics and transcriptomics: coping with increasing complexity in logic, chemistry, and network interactions scientific correspondence. Plant Physiol 158:1139–1145
5. de Souza LP, Naake T, Tohge T, Fernie AR (2017) From chromatogram to analyte to metabolite: how to pick horses for courses from the massive web-resources for mass spectral plant metabolomics. Gigascience. https://doi.org/10.1093/gigascience/gix037
6. Sawada Y, Kuwahara A, Nagano M et al (2009) Omics-based approaches to methionine side chain elongation in Arabidopsis: characterization of the genes encoding methylthioalkylmalate isomerase and methylthioalkylmalate dehydrogenase. Plant Cell Physiol 50:1181–1190
7. Arrivault S, Guenther M, Ivakov A et al (2009) Use of reverse-phase liquid chromatography, linked to tandem mass spectrometry, to profile the Calvin cycle and other metabolic intermediates in Arabidopsis rosettes at different carbon dioxide concentrations. Plant J 59:826–839
8. Okazaki Y, Otsuki H, Narisawa T et al (2013) A new class of plant lipid is essential for protection against phosphorus depletion. Nat Commun 4:1510
9. Bromke MA, Hochmuth A, Tohge T et al (2015) Liquid chromatography high-resolution mass spectrometry for fatty acid profiling. Plant J 81:529–536
10. Seo M, Jikumaru Y, Kamiya Y (2011) Profiling of hormones and related metabolites in seed dormancy and germination studies. Methods Mol Biol 773:99–111
11. Tohge T, Nishiyama Y, Hirai MY et al (2005) Functional genomics by integrated analysis of metabolome and transcriptome of Arabidopsis plants over-expressing an MYB transcription factor. Plant J 42:218–235
12. Moco S, Bino RJ, Vorst O et al (2006) A liquid chromatography-mass spectrometry-based metabolome database for tomato. Plant Physiol 141:1205–1218
13. Tohge T, Fernie AR (2010) Combining genetic diversity, informatics and metabolomics to facilitate annotation of plant gene function. Nat Protoc 5:1210–1227
14. Rogachev I, Aharoni A (2012) UPLC-MS-based metabolite analysis in tomato. Methods Mol Biol 860:129–144
15. Itkin M, Heinig U, Tzfadia O et al (2013) Biosynthesis of antinutritional alkaloids in solanaceous crops is mediated by clustered genes. Science 341:175–179
16. Tohge T, Wendenburg R, Ishihara H et al (2016) Characterization of a recently evolved flavonol-phenylacyltransferase gene provides signatures of natural light selection in Brassicaceae. Nat Commun 7:12399
17. De Vos RC, Moco S, Lommen A et al (2007) Untargeted large-scale plant metabolomics using liquid chromatography coupled to mass spectrometry. Nat Protoc 2:778–791
18. Morreel K, Saeys Y, Dima O et al (2014) Systematic structural characterization of metabolites in Arabidopsis via candidate substrate-product pair networks. Plant Cell 26:929–945

19. Shahaf N, Rogachev I, Heinig U et al (2016) The WEIZMASS spectral library for high-confidence metabolite identification. Nat Commun 7:12423
20. Horai H, Arita M, Kanaya S et al (2010) MassBank: a public repository for sharing mass spectral data for life sciences. J Mass Spectrom 45:703–714
21. Tohge T, Yonekura-Sakakibara K, Niida R et al (2007) Phytochemical genomics in Arabidopsis thaliana: a case study for functional identification of flavonoid biosynthesis genes. Pure Appl Chem 79:811–823
22. Schwahn K, Perez de Souza L, Fernie AR, Tohge T (2014) Metabolomics-assisted refinement of the pathways of steroidal glycoalkaloid biosynthesis in the tomato clade. J Integr Plant Biol 56:864–875
23. Gu J, Gui Y, Chen L et al (2013) Use of natural products as chemical library for drug discovery and network pharmacology. PLoS One 25: e62839
24. Afendi FM, Okada T, Yamazaki M et al (2012) KNApSAcK family databases: integrated metabolite-plant species databases for multifaceted plant research. Plant Cell Physiol 53:e1

Chapter 13

An UHPLC-MS/MS Method for Target Profiling of Stress-Related Phytohormones

Ondřej Novák and Kristýna Floková

Abstract

The methodology described here represents an improved strategy for analysis of a broad range of stress-related plant hormones including jasmonates, salicylic acid, abscisic acid, and auxin metabolites. The method conditions are optimized in order to reduce the background effect of complicated plant matrix, allow effective preconcentration and thus perform highly sensitive profiling of multiple plant hormones by ultrahigh performance liquid chromatography–tandem mass spectrometry (UHPLC-MS/MS).

Key words Phytohormones, Jasmonates, Auxins, Abscisic acid, Salicylic acid, Solid-phase extraction (SPE), Liquid chromatography–mass spectrometry, LC-MS

1 Introduction

Plants have developed unique systems to persist and survive in hostile and ever-changing environment. Adaptive plants respond to current surrounding conditions result from stress-induced modulation of the phytohormonal balance. Therefore, simultaneous determination of multiple plant hormone classes, in the same tissue, has become increasingly requisite to provide comprehensive information about hormonal distribution and action [1]. Salicylic acid (SA), abscisic acid (ABA), indole-3-acetic acid (IAA) and jasmonates (JAs) have been shown to play critical roles in signals integrating the biotic and abiotic stresses with endogenous responses [2]. Their concentration levels in fresh leaf tissue range between nmol/g and fmol/g, varying across individual metabolites and depending upon stress conditions [3]. The presence of highly abundant plant tissue substances such as phenolic compounds, proteins, pigments and other lipids complicate the selective phytohormone extraction and determination.

Sample preparation using an appropriate preconcentration protocol based on solid-phase extraction (SPE) can successfully increase sample purity and minimize the analytical challenges

Carla António (ed.), *Plant Metabolomics: Methods and Protocols*, Methods in Molecular Biology, vol. 1778,
https://doi.org/10.1007/978-1-4939-7819-9_13,

presented by plant matrices [4]. Our microextraction protocol was developed with respect to chemical diversity and stability issues of analyzed compounds [5]. The efficient solubility of all analytes can be achieved with 10% methanol that excludes the extraction of highly lipophilic interfering contaminants such as pigments. Moreover, JAs with a broad range of polarity, such as biosynthetic precursors *cis*-(+)-12-oxo-phytodienic acid (OPDA) containing long lipophilic chain, together with more polar JA-amino acid conjugates and water-soluble hydroxylated derivatives, limit the adoption of selective extraction and SPE retention mechanism. The employment of fast and one-step purification procedure is optimal solution in order to avoid JAs degradation. Therefore, the low-specific but high capacity polymer-based sorbents with both hydrophilic and lipophilic retention characteristics have been applied [5].

Recent progress in the sensitivity of analytical methods based on combination of liquid chromatography and mass spectrometry (LC-MS) allows parallel analysis of multiple compounds. A number of analytical methods for determination of specific phytohormone classes have been described [4]. However, the increasing interest in multiple hormone screenings and the availability of the advancing technology calls for method development for simultaneous analysis of phytohormones [6–12]. Overall, hyphenated techniques have become the most versatile, rapid, selective, and sensitive approach available for identifying and quantifying small molecules. The power of MS-based methods is greatly enhanced by the development of effective sample enrichment procedures, especially when studying low-abundance compounds such as certain phytohormones [13].

Here we describe an ultra-high performance liquid chromatography–tandem mass spectrometry (UHPLC-MS/MS) method, set at multiple reaction monitoring (MRM) mode for simultaneous profiling of stress-related phytohormones such as JA, its biosynthetic precursors and amino acid conjugates, SA, ABA, and IAA including its metabolites (Fig. 1). In summary, the combination of an effective SPE procedure with a sensitive UHPLC-MS/MS method enables the accurate quantification of 19 compounds in a minute plant tissue sample of less than 20 mg fresh weight (FW).

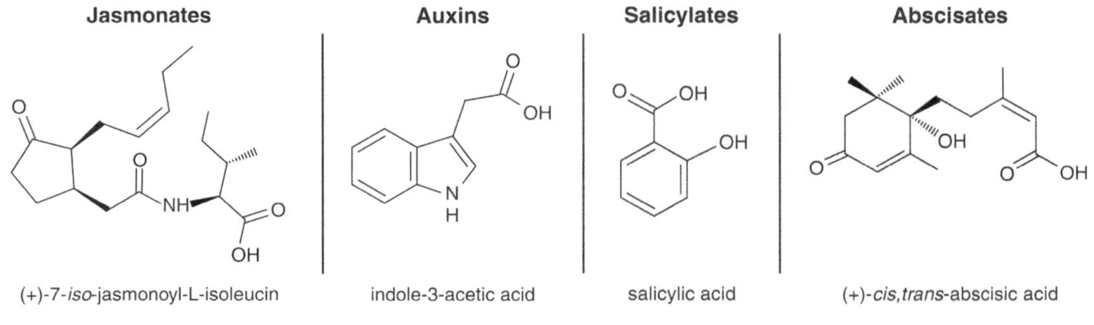

Fig. 1 Stress-related phytohormone groups

2 Materials

All aqueous solutions are prepared using deionized (Milli-Q) water with ≤18 M W/cm resistance at 25 °C (e.g., Simplicity 185 System Millipore, Bedford, MA, USA). All chemicals and solvents should be at least of analytical reagent grade (e.g., LC-MS grade).

2.1 Sampling, Extraction, and Purification

The volumes in parenthesis represent quantity per single sample.

1. Plant tissue samples (e.g., leaves, roots, and fruits) (*see* **Note 1**).
2. Liquid nitrogen.
3. Analytical balances.
4. Ceramic mortar and pestle precooled with liquid nitrogen.
5. 2.0-mL safelock microcentrifuge tubes with safe-lock lid (2×).
6. Stable-isotope-labeled internal standards (IS) (*see* **Note 2**).
7. IS mixture prepared in 30% acetonitrile–water: 5×10^{-7} M (5 pmol/10 μL) of [$^2H_6$]-(+)-*cis,trans*-abscisic acid (ABA), [$^{13}C_6$]-indole-3-acetic acid (IAA), [$^{13}C_6$]-2-oxindole-3-acetic acid (oxIAA), [$^{13}C_6$]-IAA-aspartate (IAAsp) and [$^{13}C_6$]-IAA-glutamate (IAGlu); 1×10^{-6} M (10 pmol/10 μL) of [$^2H_6$]-(±)-jasmonic acid (JA), [$^2H_5$]-*cis*-(+)-12-oxo-phytodienoic acid (*cis*-OPDA) and [$^2H_2$]-(−)-jasmonoyl-isoleucine (JA-Ile); 1.5×10^{-6} M (15 pmol/10 μL) of [$^2H_4$]-salicylic acid (SA).
8. Crushed-ice bath.
9. 100% methanol (1 mL).
10. Deionized water (1 mL).
11. Extraction solution: ice-cold 10% methanol–water (v/v) (5 mL).
12. 80% methanol–water (v/v) (3 mL).
13. Laboratory bead-mill homogenizer (e.g., Mixer Mill MM 301, Retsch GmbH).
14. 2.0-mm ceria-stabilized zirconium oxide beads (4×).
15. Ultrasonication bath (frequency 37 kHz) cooled with an ice block.
16. Benchtop laboratory rotator (designed for incubators and cold rooms down to 4 °C).
17. Refrigerated centrifuge.
18. Solid phase extraction (SPE) vacuum manifold (e.g., 12-port extraction plate) and membrane vacuum pump stand (~120 torr vacuum).
19. SPE column Oasis® HLB (1 cc/30 mg) (Waters, Milford, MA, USA).
20. Glass tubes.

21. Nitrogen evaporator (e.g., TurboVap® LV evaporation system).
22. 2.0-mL screw-top autosampler vials and 250 μL glass inserts.

2.2 UHPLC-MS/MS Method

1. Unlabeled and labeled phytohormone standards.
2. Calibration solutions (*see* **Note 2**).
3. Reversed-phase LC column: ACQUITY UPLC® CSH C18 Column, 130 Å, 1.7 μm particle size, 2.1 mm internal diameter, 100 mm length (Waters, Milford, MA, USA).
4. Mobile phase: eluent (A) acetonitrile; eluent (B) 10 mM formic acid. Add 0.377 μL of formic acid (98–100%) into 1 L of deionized water.
5. UHPLC-MS/MS system and associated equipment: ACQUITY UPLC® I-Class System and Xevo® TQ-S MS (Waters). The LC system should be equipped with at least binary high-pressure pumps, a refrigerated autosampler and a thermostatic column compartment. The MS system should include a tandem quadrupole mass spectrometer, and each technical parameter should be modified and optimized according to the apparatus used.
6. Mass spectrometry software: MassLynx™ software (version 4.1, Waters).

3 Methods

The extraction with 10% methanol–water provides sufficient solubility for compounds of interest and decreases sample contamination with lipophilic pigments or other interfering compounds [5]. Moreover, the optimization of sample purification process significantly influences the sensitivity of final UHPLC-MS/MS analysis [14]. The used methods must be also thoroughly validated and at least seven parameters (e.g., method selectivity, sensitivity, extraction recovery, matrix effects, linearity, accuracy, and precision) should be evaluated [15, 16].

3.1 Sampling for Phytohormone Profiling

1. Grind the frozen plant material into fine powder using a mortar and a pestle under the liquid nitrogen (*see* **Note 3**).
2. Transfer individually amounts of 20–25 mg of fresh weight into 2.0-mL safe-lock microcentrifuge tubes (at least three replicates per sample) (*see* **Note 4**).
3. Freeze the plant tissues immediately in liquid nitrogen and store them at −80 °C for no longer than several weeks before sample extraction (*see* **Note 5**) (Fig. 2a).

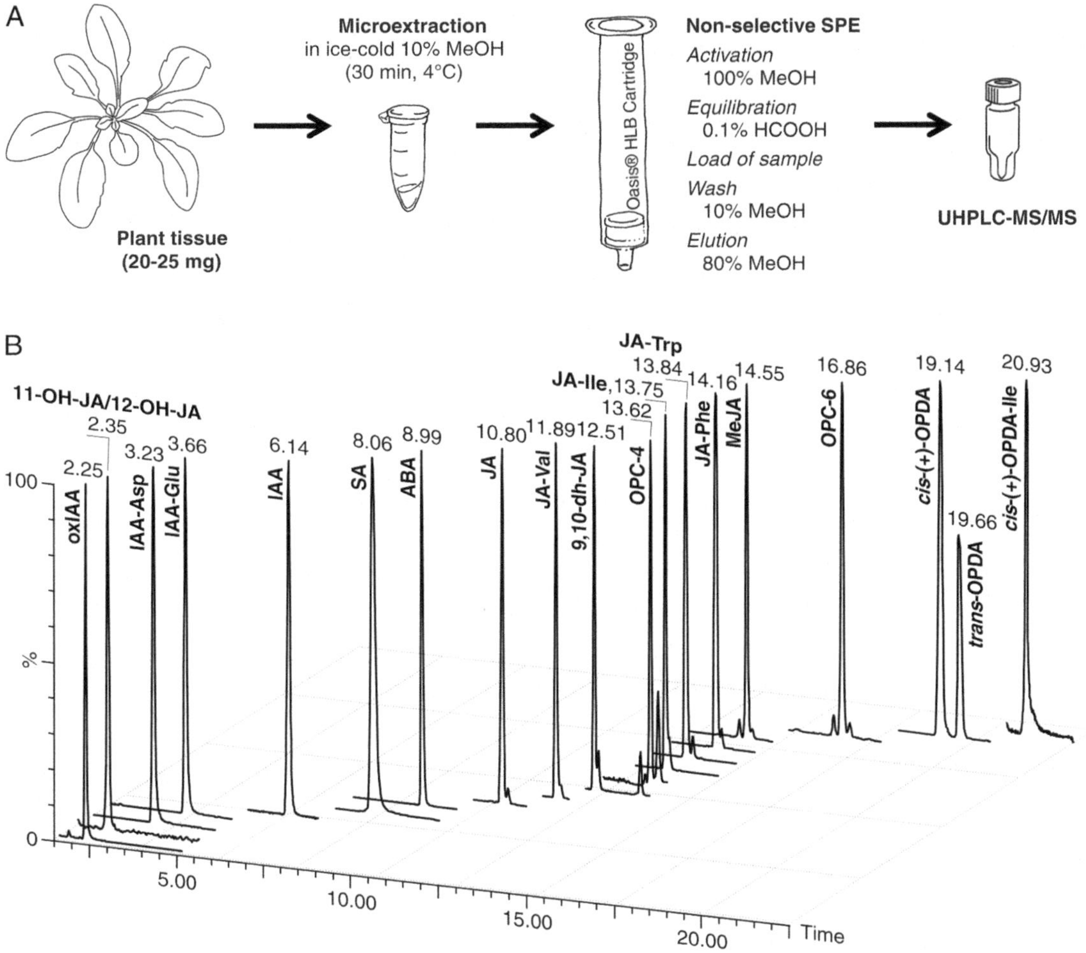

Fig. 2 Extraction and purification protocol. (**a**) Plant material (20–25 mg) is homogenized and extracted in 10% methanol containing stable isotope-labeled internal standards. Pooled supernatant is applied to a preconditioned HLB column (30 mg), which is washed and eluted with the indicated solutions. The eluate (80% methanol) is evaporated to dryness and stored at −20 °C prior to UHPLC-MS/MS analysis. (**b**) Separation of 13 jasmonates, 4 auxins, SA, and ABA standards by UHPLC-MS/MS. All compounds were injected onto a reversed-phase column, and the chromatographic analysis was split into ten scan segments to increase sensitivity (this can reduce the number of transitions monitored at the same time and maximize cycle time). The list of compounds separated: oxIAA, 2-oxindole-3-acetic acid; 11-OH-JA, (±)-11-hydroxy-jasmonic acid; 12-OH-JA, (±)-12- hydroxy-jasmonic acid; IAAsp, IAA-aspartate; IAGlu, IAA-glutamate; IAA, indole-3-acetic acid; SA, salicylic acid; ABA, abscisic acid; JA, (−)-jasmonic acid; JA-Val, (−)-jasmonoyl-L-valine; 9,10-dh-JA, (±)-9,10-dihydrojasmonic acid; OPC-4, 3-oxo-2-(2-(Z)-pentenyl) cyclopentane-1-butyric acid; JA-Ile, (−)-jasmonoyl-L-isoleucine; JA-Trp, (−)-jasmonoyl-S-tryptophan; JA-Phe, (−)-jasmonoyl-L-phenylalanine; MeJA, (−)-jasmonic acid methyl ester; OPC-6, 3-oxo-2-(2-(Z)-pentenyl)cyclopentane-1-hexanoic acid; *cis*-(+)-OPDA, *cis*-(+)-12-oxo-phytodienoic acid; *trans*-OPDA, *trans*-12-oxo-phytodienoic acid, *cis*-(+)-OPDA-Ile, *cis*-(+)-OPDA-isoleucine

3.2 Extraction of Stress-Related Plant Hormones

1. Place frozen samples from the −80 °C in a crushed-ice bath (*see* **Note 6**).
2. Add 1.0 mL of ice-cold extraction solution into each microcentrifuge tube (*see* **Note 7**).
3. Place 2-mm ceria-stabilized zirconium oxide beads (4 in total) into each tube together with 10 μL of IS mixture (*see* **Note 8**).
4. Place microcentrifuge tubes in a precooled cassette of a vibration bead mill and homogenize at a frequency of 27 Hz for 3 min.
5. Sonicate homogenized samples in a precooled ultrasonic bath at 37 kHz for 3 min.
6. Extract samples for 30 min on the benchtop laboratory rotator at 4 °C.
7. Centrifuge samples at 14,000 × *g* for 10 min at 4 °C.
8. Collect supernatants of sample extracts into new 2.0-mL microcentrifuge tubes and place them in a crushed-ice bath. Keep the remaining pellet on ice for further reextraction.
9. Reextract samples by dissolving the pellet in 1.0 mL of ice-cold extraction solution, vortex for 10 s, and centrifuge as before (step 7).
10. Pool the supernatants of one sample in a new 2.0-mL microcentrifuge tube (Fig. 2a).

3.3 Sample Purification by Nonselective SPE

1. Prior to sample loading onto SPE Oasis® HLB columns, perform sorbent conditioning with 1 mL of 100% methanol and subsequent equilibration with 1 mL of 0.1% formic acid–water (v/v).
2. Apply the sample extract (~2 mL) onto SPE column (*see* **Note 9**). Discard flow-through fraction.
3. Remove interfering compounds by washing with 1 mL of extraction solution. Discard flow-through fraction.
4. Insert clean glass tubes into the manifold rack, and then elute the sample from the column with 2 mL of 80% methanol (v/v).
5. Evaporate the collected fractions to dryness under a gentle stream of nitrogen (*see* **Note 10**).
6. Store dried sample extracts at −20 °C until LC–MS/MS analysis (Fig. 2a).

3.4 Phytohormonal Metabolite Profiling

1. Dissolve the samples in 40 μL of 10% methanol, and place the vials in an autosampler with a temperature controlled system set at 4 °C (*see* **Note 11**).
2. Separate and detect JA, its biosynthetic precursors and amino acid conjugates, SA, ABA, IAA, and its metabolites by

Table 1
Optimized instrument settings (scan modes, MRM transitions, cone voltages and collision energies)

Compound	ESI[a] mode	RT[b] (min)	MRM transition	CV[c] (V)	CE[d] (eV)	IS[e]	MRM transition
oxIAA	+	2.25	192 > 146	20	14	[$^{13}C_6$]-oxIAA	198 > 152
(±)-11-oh-JA/ (±)-12-oh-JA	−	2.35	225 > 58	20	23	[$^2H_6$]-(±)-JA	215 > 59
IAAsp	+	3.23	291 > 130	20	20	[$^{13}C_6$]-IAAsp	297 > 136
IAGlu	+	3.66	305 > 130	12	24	[$^{13}C_6$]-IAGlu	311 > 136
IAA	+	6.14	176 > 130	17	15	[$^{13}C_6$]-IAA	182 > 136
SA	−	8.06	137 > 93	25	15	[$^2H_4$]-Sa	141 > 97
(+)-*cis,trans*-ABA	−	8.99	263 > 153	23	12	[$^2H_6$]-(+)-*cis,trans*-ABA	269 > 159
(−)-JA	−	10.80	209 > 59	30	23	[$^2H_6$]-(±)-JA	215 > 59
(−)-JA-Val	+	11.89	310 > 151	30	20	[$^2H_2$]-(−)-JA-Ile	326 > 151
(±)-9,10-dh-JA	−	12.51	211 > 59	30	23	[$^2H_6$]-(±)-JA	215 > 59
(±)-OPC-4	−	13.62	237 > 59	30	22	[$^2H_6$]-(±)-JA	215 > 59
(−)-JA-Ile	+	13.75	324 > 151	30	23	[$^2H_2$]-(−)-JA-Ile	326 > 151
(−)-JA-Trp	+	13.84	397 > 351	30	17	[$^2H_2$]-(−)-JA-Ile	326 > 151
(−)-JA-Phe	+	14.16	358 > 151	30	23	[$^2H_2$]-(−)-JA-Ile	326 > 151
(±)-MeJA	+	14.55	225 > 151	30	15	[$^2H_2$]-(−)-JA-Ile	326 > 151
(±)-OPC-6	+	16.86	267 > 147	30	20	[$^2H_2$]-(−)-JA-Ile	326 > 151
cis-(+)-OPDA	+	19.14	293 > 275	30	18	[$^2H_5$]-OPDA	298 > 279
trans-OPDA	+	19.66	293 > 275	30	18	[$^2H_5$]-OPDA	298 > 279
cis-(+)-OPDA-Ile	+	20.93	406 > 360	30	14	[$^2H_5$]-OPDA	298 > 279

In our system, MS conditions are as follows: capillary voltage, 3 kV; desolvation gas flow, 1000 L/h; collision gas flow, 0.15 mL/min; nebulizer gas flow, 7 Bar; desolvation temperature, 600 °C; source block temperature, 150 °C; source offset, 50 V

[a]Electrospray ionization
[b]Retention time
[c]Cone voltage
[d]Collision energy
[e]Internal standards used for stable-isotope dilution

UHPLC-MS/MS. Retention times for each analyte and IS are given in Table 1 (Fig. 2b).

3. Inject 10 μL of each sample onto an ACQUITY UPLC® CSH C18 column (1.7 μm, 2.1 × 100 mm).
4. Elute the samples using an 28-min gradient composed of eluent (A) acetonitrile and eluent (B) 10 mM formic acid in water,

at a flow rate of 0.4 mL/min, column temperature of 36 °C, and a binary linear gradient: 0 min, 15:85 (A:B); 5.0 min, 15:85 (A:B); 15.0 min, 45:55 (A:B); 22.5 min, 47:53 (A:B).

5. At the end of the gradient, wash the column with 100% methanol for 1.5 min, and reequilibrate to initial conditions for 4.0 min.
6. Determine stress-related phytohormones by multiple reaction monitoring (MRM) of the precursor $[M + H]^+$ and the appropriate product ions according to retention times of the analytes. Use optimal collision energy and dwell time for various diagnostic transitions listed in Table 1 (*see* **Note 12**).
7. Run all calibration solutions and construct calibration curves for each analyte (*see* **Note 13**).
8. Process all data using appropriate mass spectrometry software, and determine the concentrations of each phytohormone using the stable isotope dilution method (*see* **Note 14**).

4 Notes

1. Different types of plant material can affect extraction (poor extraction efficiency), purification (low recovery of analytes), matrix effects (ion suppression and high chemical background), and chromatography (coelution of structurally related compounds). It is therefore necessary to test the method on plant materials of interest, and modify the method slightly if needed (e.g., change extraction volume and time, amount of internal standards, SPE column capacity, and washing and elution steps).
2. Dissolve all standards in 50% acetonitrile–water (v/v) to a final concentration of 1 mM (stock solutions) and store at −20 °C. Working IS mixture containing each of the stable-isotope-labeled standards can be made by diluting the stock solutions of the standards to the final concentrations. All standards can be purchased from Sigma-Aldrich (www.sigmaaldrich.com), OlChemim (www.olchemim.cz), Cambridge Isotope Laboratories (www.isotope.com), or C/D/N Isotopes (www.cdnisotopes.com).
3. Metabolic changes can happen during harvesting of the plant material. For instance, jasmonate levels are strongly induced by mechanical wounding. This effect can be reduced by short sampling times and by rapidly freezing plant material in liquid nitrogen.
4. Accurate determination of initial sample weight is crucial, especially, when minute amounts of fresh plant tissue are used.

Accurate weights successfully correct the concentration of metabolites that is usually given in the unit pmol/g FW.

5. Alternatively, samples can be lyophilized in a freeze-drier and stored in the fridge at 4 °C until extraction.
6. The powder must not thaw before the extraction solution is added.
7. The sample preparation performed at low temperature (4 °C) avoids possible metabolic conversions or enzymatic degradation of analytes.
8. It is recommended to add internal standards at the concentration level similar to that of endogenous levels of analytes.
9. The number of extracted samples in one batch is limited by the number of positions on the SPE vacuum manifold and/or in the refrigerated centrifuge.
10. Evaporation steps can significantly reduce the effectiveness of SPE methods for determining diverse JA metabolites (e.g., jasmonic acid methyl ester) due to their high volatility; therefore, the use of a nitrogen evaporator is highly recommended.
11. Use an ultrasonic bath to improve dissolution of analytes, and transfer the samples into clean LC-MS glass vials with glass inserts.
12. The dwell time should be optimized so that the cycle time provides 10–15 data points across the chromatographic peaks.
13. The analyte content of the dilutions ranged from 1.0 fmol to 500 pmol and each dilution also contained a known amount of stable-isotope-labeled internal standards. Create calibration curves by plotting known concentrations of each unlabeled analyte against analyte–IS ratios (Table 1).
14. The stable-isotope dilution assay is a powerful method, which compensates for losses or inefficiencies in the sample preparation process and ion suppression effects during MS analysis. The method is used to quantify levels of endogenous phytohormones in plant tissues, determine their ratios to the appropriate IS added (Table 1), calculate the hormone concentrations based on the established calibration curves, and then normalize to the mass of fresh tissue weighed before extraction.

Acknowledgments

We thank Karel Doležal and Aleš Pěnčík for critical reading of the protocol and Helena Mazáčová for careful revision of the manuscript. This work was funded by the Czech Science Foundation (Nr. GA17-06613S).

References

1. Müller M, Munné-Bosch S (2017) Hormone profiling in plant tissues. In: Kleine-Vehn J, Sauer M (eds) Plant hormones: methods in molecular biology, vol 1497. Humana Press, New York
2. Wasternack C, Strnad M (2016) Jasmonate signaling in plant stress responses and development – active and inactive compounds. N Biotechnol 33:604–613
3. Balcke GU, Handrick V, Bergau N et al (2012) An UPLC-MS/MS method for highly sensitive high-throughput analysis of phytohormones in plant tissues. Plant Methods 8:47
4. Tarkowská D, Novák O, Floková K et al (2014) Quo vadis plant hormone analysis? Planta 240:55–76
5. Floková K, Tarkowská D, Miersch O et al (2014) UHPLC-MS/MS based target profiling of stress-induced phytohormones. Phytochemistry 105:147–157
6. Chiwocha SD, Abrams SR, Ambrose SJ et al (2003) A method for profiling classes of plant hormones and their metabolites using liquid chromatography-electrospray ionization tandem mass spectrometry: an analysis of hormone regulation of thermodormancy of lettuce (*Lactuca sativa* L.) seeds. Plant J 35:405–417
7. Pan X, Welti R, Wang X (2008) Simultaneous quantification of major phytohormones and related compounds in crude plant extracts by liquid chromatography-electrospray tandem mass spectrometry. Phytochemistry 69:1773–1781
8. Kojima M, Kamada-Nobusada T, Komatsu H et al (2009) Highly sensitive and high-throughput analysis of plant hormones using MS-probe modification and liquid chromatography-tandem mass spectrometry: an application for hormone profiling in *Oryza sativa*. Plant Cell Physiol 50:1201–1214
9. Liu H, Li X, Xiao J, Wang S (2012) A convenient method for simultaneous quantification of multiple phytohormones and metabolites: application in study of rice-bacterium interaction. Plant Methods 8:2
10. Cai WJ, Ye TT, Wang Q et al (2016) A rapid approach to investigate spatiotemporal distribution of phytohormones in rice. Plant Methods 12:47
11. Schäfer M, Brütting C, Baldwin IT et al (2016) High throughput quantification of more than 100 primary and secondary metabolites, and phytohormones by a single solid phase extraction based sample preparation with analysis by UHPLC-HESI-MS/MS. Plant Methods 12:30
12. Luo XT, Cau BD, Chen X et al (2017) Improved methodology for analysis of multiple phytohormones using sequential magnetic solid-phase extraction coupled with liquid chromatography-tandem mass spektrometry. Anal Chim Acta 983:112–120
13. Novák O, Napier R, Ljung K (2017) Zooming in on plant hormone analysis: tissue- and cell-specific approaches. Annu Rev Plant Biol 68:323–348
14. Floková K, Feussner K, Herrfurth C et al (2016) A previously undescribed jasmonate compound in flowering *Arabidopsis thaliana*—the identification of *cis*-(+)-OPDA-Ile. Phytochemistry 122:230–237
15. Matuszewski BK, Constanzer ML, Chavez-Eng CM (2003) Strategies for the assessment of matrix effect in quantitative bioanalytical methods based on HPLC-MS/MS. Anal Chem 75:3019–3030
16. Nováková L (2013) Challenges in the development of bioanalytical liquid chromatography--mass spectrometry method with emphasis on fast analysis. J Chromatogr A 1292:25–37

Chapter 14

A Complete Pipeline for Generating a High-Resolution LC-MS-Based Reference Mass Spectra Library

Nir Shahaf, Asaph Aharoni, and Ilana Rogachev

Abstract

Databases containing mass spectrometry (MS) spectral data (i.e., MS reference libraries) are currently the most reliable and widely accepted approach to annotate unknown features in MS-based metabolomics. While for gas chromatography (GC)-MS data, a strategy for collecting, storing, and comparing to raw data has been established, this is not the case for liquid chromatography (LC)-MS data. Here, we present our approach for high-throughput data collection and automated MS reference library generation, as applied recently in the WEIZMASS library of plant metabolites. Methodologies to experimentally generate pools of chemical standards and computationally convert them into a unique source of reference data are detailed.

Key words Mass spectrometry, Metabolites, LC-MS, Reference libraries, Chemoinformatics, Metabolomics, Plant metabolism

1 Introduction

The most accurate and reliable knowledge regarding metabolic composition of biological samples is currently limited to metabolites that were either structurally characterized by NMR or those previously identified by comparison to a reference chemical standards using mass spectrometry (MS) analysis. Consequently, the foremost source of reference for metabolite annotation in a metabolomics study would be MS spectral libraries generated by similar analytical methods and biological source [1, 2]. More specifically, spectral libraries based on liquid chromatography (LC)-MS data are commonly used for identification of soluble polar and semipolar compounds in metabolomics studies [3–6] and for lipophylic compounds in lipidomics assays [7–10]. One of the biggest limitations of all available LC-MS reference libraries is their poor coverage of the chemical space related to so called "natural products" (i.e., secondary metabolites), estimated at several hundreds of thousands, most of them currently unknown [11–13]. A second limitation is the large variability in chromatographic and MS

Carla António (ed.), *Plant Metabolomics: Methods and Protocols*, Methods in Molecular Biology, vol. 1778,
https://doi.org/10.1007/978-1-4939-7819-9_14, © Springer Science+Business Media, LLC, part of Springer Nature 2018

methods and the additional variability within equivalent methods due to experimental conditions such as: temperature conditions during the time of acquisition, age of the chromatographic column, condition of the MS detector, and different ionization efficiencies of various MS technologies (resulting in highly variable spectra). These factors greatly reduce the possibilities of sharing LC-MS spectra data. One major challenge is therefore to generate large reference MS libraries that will be useful for different experimental conditions. Some of the approaches reported to date related to generating or developing open-source, community driven, MS reference libraries are: (1) the MassBank database [14] providing a set of tools to construct and share LC-MS data either internally, via a local installation, or by uploading data in a predefined format to one of the online repositories (there are current MassBank repositories in Japan, Europe and the USA). A tool to automatically generate MassBank records within the R programming environment (www.R-project.org) has additionally been developed [15]. The MassBank approach is based on manually curated LC-MS data which can be either manually or automatically be converted and uploaded to the repository; (2) the Global Natural Products Social molecular networking (GNPS) [16] was created as a universal repository for MS spectra originating from metabolomics studies, where most measurements are derived from biological matrices and a majority of the data is not annotated (i.e., MS peaks are not assigned to any chemical entity) or that many are putatively annotated without a direct comparison to a chemical standard; (3) the third approach, aiming for automatic, high throughput MS library generation, based on injections of chemical standards, is exemplified here by the WEIZMASS library of plant metabolites [6]. The idea behind the WEIZMASS approach is to provide an analytical and computational pipeline for generating LC-MS reference libraries and to provide a complementary software platform for annotating MS peaks to metabolite entries in the generated LC-MS libraries. Here, we describe this pipeline and the way in which it is coupled to the software package, freely available in the supplementary material.

2 Materials

2.1 Reagents and Equipment

1. Double deionized water, resistivity 18.2 MΩ-cm, filtered through a 0.22 μm membrane filter (*see* **Note 1**).
2. Acetonitrile, LC-MS grade.
3. Formic acid, LC-MS grade.
4. Methanol, ethanol, and DMSO for standards dilution, analytical grade.

5. Chemical standards for QC (quality control) samples, i.e., "QC-Mix-15" (*see* **Note 2**): L-tyrosine (Fluka), L-tryptophan (Sigma), D-phenylalanine (Fluka), chlorogenic acid (Fluka), caffeic acid (Sigma), *p*-Coumaric acid (Sigma), cinnamic acid (Fluka), benzoic acid (Sigma), ferulic acid (Aldrich), sinapic acid (Sigma), rutin hydrate (Sigma), quercetin dehydrate (Sigma), tomatine (ExtraSynthese; *see* **Note 3**), naringenin (Fluka), kaempferol (Fluka).
6. 96-well polypropylene (PP) plates with cover mat.
7. Ultrasonic bath.
8. Centrifuge suitable for 96-well plates.
9. Lyophilizer suitable for organic solvents (if DMSO evaporation is needed).
10. For samples injection: 96-well PP plates with cover or 2.0-mL autosampler vials and caps with a PTFE/Silicone septum. Use suitable (e.g., 250-μL glass inserts) when you have a small volume of solution for injection in vials (less than 1 mL).

2.2 Solutions for UPLC

1. Mobile phase A: 5% acetonitrile–water (v/v), containing 0.1% formic acid (v/v).
2. Mobile phase B: 100% acetonitrile, containing 0.1% formic acid (v/v).
3. Strong needle wash solution: 80% methanol (a strong organic solution that dissolves most components of the sample matrix).
4. Weak needle wash solution: 5% acetonitrile–water (v/v) (composition similar to the initial conditions of the gradient).
5. Seal wash solution: 10% methanol–water (v/v).

2.3 Instrumentation and Software

1. UPLC-PDA-qTOF system: e.g., UPLC Waters Acquity instrument connected inline to an ACQUITY PDA (photodiode array) detector and a Synapt HDMS detector (tandem quadrupole/time-of-flight mass spectrometer). The MS detector is equipped with an electrospray ion source (ESI). The Synapt HDMS system is operated in the standard qTOF mode, without using the ion mobility capabilities (*see* **Note 4**).
2. UPLC BEH C18 column (Waters Acquity), 100 mm length × 2.1 mm internal diameter, 1.7 μm particle size, with a column prefilter.
3. MassLynx 4.1 instrument software (Waters) to control the instrument and calculate accurate masses.
4. Waters databridge (www.waters.com) software or any suitable tool for conversion of raw data to an open source data format.
5. The R programing language (www.r-project.org).

6. R software packages: "xcms," "CAMERA" and "Rdisop" (https://bioconductor.org).
7. The supplementary software package (*see* postscript https://github.com/AharoniLab/MatchWeiz).

3 Methods

3.1 Preparation of Chemical Standard Pools

1. Divide the available chemicals standards into pools of a predefined size. We recommend any number between 5 and 20 compounds per pool, depending on the amount of the available chemical standards and type of the liquid chromatography used (*see* **Note 5**). For the WEIZMASS reference library, containing 3450 chemical standards and injected using a reverse-phase column of 40 min, we set the pool size to 20 chemical standards per pool. *See* Fig. 1 for an example of a pool injection.
2. Assign each chemical standard to a designated injection pool, considering: its nominal mass value, chemical structure (or class) and chromatographic retention time (if such information is available). Use the analytical information available on individual chemical standards to assign chemicals to pools in a way which will minimize the risk of overlapping chromatographic peaks (*see* **Note 6** and Table 1). Examples of potentially overlapping peaks within the same pool include: peaks of isobaric chemical standards, peaks of chemical standards eluting at similar retention times (thus potentially suppressing each other) and peaks of chemical standards with the same chemical backbone (thus potentially having equivalent distinctive mass fragments). Depending on the amount of chemical standards to inject,

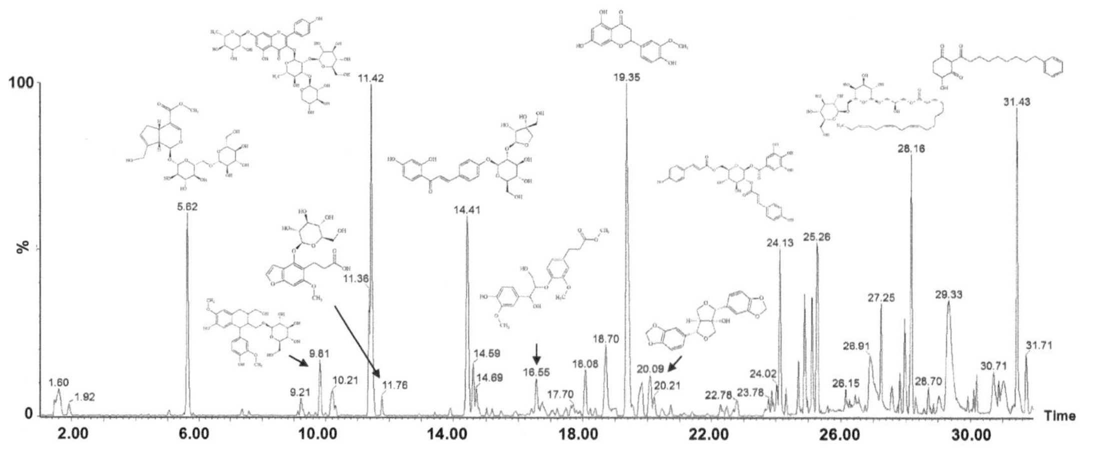

Fig. 1 Example chromatogram of a pool of chemical standards analyzed by high-resolution MS in the negative ionization mode. The chemical structures of 11 standards are given above the corresponding peaks. *See* Table 1 for description of compound parameters

Table 1
Compound data corresponding with a pool of 20 chemical standards, partially presented in Fig. 1

ID	MW	RT-Window	logD
NP-015506	330.373	Window-01	−1.71
NP-008010	287.352	Window-02	−0.81
NP-015876	550.507	Window-04	−3.99
NP-000520	522.542	Window-06	0.01
NP-001890	398.361	Window-07	−0.52
NP-005644	872.774	Window-07	−2.93
NP-014789	550.509	Window-08	0.09
NP-015480	452.495	Window-09	0.81
NP-006619	406.426	Window-09	1.89
NP-006459	684.854	Window-10	−0.05
NP-011993	624.546	Window-10	4.11
NP-001581	302.279	Window-11	2.68
NP-004447	370.353	Window-11	1.9
NP-002655	534.681	Window-12	2.98
NP-015153	354.353	Window-13	4.5
NP-015563	348.433	Window-14	3.94
NP-004813	334.45	Window-14	3.02
NP-000216	676.789	Window-15	1.34
NP-004221	344.445	Window-17	4.96
NP-015141	364.52	Window-18	7.44

The first column ("ID") corresponds with the vendor's compound ID; the second column ("MW") corresponds with the compound's molecular weight; the third column ("RT-Window") with the chromatographic retention time window (as measured by the vendor); and the fourth ("logD") with the calculated distribution coefficient, which frequently shows a good correspondence with chromatographic elution order on reverse-phase columns

assigning chemical standards to pools using the above criteria can be carried out either manually (by minimizing ambiguity within the pools) or by using the function "createStdPool" in the supplementary software package.

3. When the chemical standards are supplied as dry matter, they should first be dissolved in order to allow proper mixing. A mixture of water, formic acid, methanol, ethanol or DMSO in different proportions can be used, depending on the solubility of the compound (*see* **Note 7**). For example: 0.2 mg of dry matter is

dissolved in 500 μL of solvent for making a stock solution. From the stock solution, take an equal amount of each chemical standard to make a pool with a final concentration of 10 μL/mg per standard (*see* **Note 8**). The pools can be directly delivered to the designated well by using either a robotic system (e.g., Freedom EVO, Tecan) or through manual dispensing.

4. Store the concentrated stocks and diluted pools of chemical standards at −80 °C freezer.

3.2 Preparation of Reference Mixture "QC-Mix-15" for Quality Control

1. Prepare individual stock solutions at a concentration of 1 mg/mL in methanol of the following chemical standards: phenylalanine, chlorogenic acid, caffeic acid, coumaric acid, cinnamic acid, benzoic acid, ferulic acid, sinapic acid, rutin, quercetin, naringenin, and kaempferol. Sonicate all stock solutions for several minutes for better solubility of the compounds.
2. Prepare a α-tomatine stock solution at a concentration of 0.5 mg/mL in methanol, and sonicate for several minutes for better solubility of the compound (*see* **Note 3**).
3. Prepare tyrosine and tryptophan stock solutions at a concentration of 1 mg/mL in 80% aqueous methanol (v/v) containing 2% formic acid (v/v), and sonicate for several minutes for better solubility of the compounds.
4. Prepare a stock mixture of the chemical standards listed in **step 1** by combining equal amounts of the individual stock solutions at a final concentration 70 μg/mL of each compound. For the chemical standard α-tomatine listed in **step 2**, use a final concentration of 33 μg/mL. Aliquots of this solution can be stored at −20 °C for 3–4 months without significant changes in compounds concentration.
5. Prepare the final mixture of reference chemical standards (thereby termed "QC-Mix-15") by diluting the stock mixture tenfold with methanol. The final concentration of the chemical standards should be about 7 μg/mL (with α-tomatine at about 3 μg/mL) (*see* **Note 2**).

3.3 Injecting Chemical Standard Pools and Reference Mixture "QC-Mix-15"

1. Sonicate a plate with diluted pools after removing from the freezer in order to solubilize possible precipitates, centrifuge and transfer the supernatant to HPLC vials or to a 96-well plate for injection.
2. Use the following parameters for the chromatographic method: mobile phase (A) 5% acetonitrile containing 0.1% formic acid, mobile phase (B) 100% acetonitrile, containing 0.1% formic acid. Use a linear gradient from 100 to 72% phase (A) over 22 min, from 72 to 0% phase (A) over 14 min, then hold at 100% phase (B) for further 2 min; return to the initial conditions (100% phase A) in 0.5 min, and condition the

LC-column at 100% phase (A) for 1.5 min. It is recommended to use a divert valve to exclude minute 0–1 and minutes 38–40 from the chromatographic injection. Set the flow rate to 0.3 mL/min, the autosampler temperature to 10 °C, chromatographic column temperature to 35 °C and the injection volume between 1 and 4 μL (*see* **Note 9**). Wash the injection needle with 200 μL of a weak needle wash solution (5% aqueous ACN) and then with 600 μL of a strong needle wash solution (80% aqueous MeOH).

3. Use the following parameters for setting-up the Synapt qTOF HDMS detector coupled to an ESI source: the TOF part operating in the V-mode with mass resolution of 9000 and the MS spectra acquired in the mass range from 50 to 1500 Da, in the centroid mode. Acquisition in the centroid mode is essential for further data treatment with the preprocessing software. Use argon as a collision gas and leucine enkephalin for lock mass calibration.
4. Use the following settings during the LC-MS run: a capillary voltage of 2.4 kV, cone voltage of 28 eV, source temperature of 125 °C, desolvation temperature of 275 °C and desolvation gas flow of 650 L/h. MS data acquisition should be performed in the MS^E ramp mode (*see* **Note 10**). The MS^E ramp mode is used in order to add more specific fragmentation information for each chemical standard as compared to a fixed collision energy value. Set the collision energy to 4 eV for low-energy function and to a range of 10–30 eV ramp for the high-energy function in the positive ion mode (15–35 eV in the negative ion mode, *see* **Note 11**). Set the scan time for each function to 0.25 s. Additionally, UV/VIS spectra can be acquired by enabling the PDA detector.
5. Inject the SST (system suitability test) solution at the beginning of sequence by using the "QC-Mix-15" solution. Check that the UPLC column is properly conditioned: retention times of the reference chemical standards should correspond to the values obtained previously on a fully functioning (or a new) column, and the peaks of ferulic and sinapic acids should be resolved. Next, check that the MS system is properly calibrated: the reference calibration compound signal (i.e., Leucine Enkephalin) should be stable and the specific calibration masses should be detected by the software (*see* **Note 10**). Finally, check that the accuracy of mass determination for the mixture of chemical standards in the "QC-Mix-15" is within the expected mass measurement error of the instrument used.
6. Inject samples of the "QC-Mix-15" solution at the beginning and at the end of each sequence and within the sequence about every ten other injections.
7. Inject a blank sample (i.e., methanol solution) once at the beginning of each sequence.

3.4 Preprocessing of Chemical Standard Pools and Reference Mixture "QC-Mix-15"

1. Convert raw data of injected samples from any propriety (i.e., vendor's) format to an open source data format to allow computational preprocessing using nonvendor's software (*see* **Note 12**). Some commonly used open source formats for LC-MS data include: mzML, mzXML, and netCDF. We use the MassLynx Databridge to convert from the Waters raw format to the netCDF format, due to the fact that the converter supports acquisition signals in two MS channel (i.e., MS1 and MS2) and converts them into two separate netCDF files: the first channel (MS1) corresponding with netCDF files with the postfix "01" and the second channel (MS2) corresponding with netCDF files with the postfix "02".
2. Use the open source R package "xcms" for extracting, aligning and grouping mass and retention-time features from the raw netCDF files. Process each sample (i.e., pairs of netCDF files, corresponding with a unique pool of chemical standards) separately using the following "xcms" parameters: peak picking function "xcmsSet" ("method = matchedFilter", "fwhm = 20", "step = 0.02", "steps = 3", "snthresh = 5", "max = 100"); grouping function "group" ("method = density","bw = 20", "minsamp = 1","mzwidth = 0.1"), RT alignment function "retcor" ("method = loess"," span = 1"," "missing = 0", "extra = 0"), fill peaks ("method = chrom") (*see* **Note 13**).
3. Use the function "annotate" from the open source R package "CAMERA" to group coeluting mass features and annotate the naturally occurring isotopes, with the following parameters: "ppm = 25", "cor_eic = 0.8", "multiplier = 2", "intval = maxo".
4. Convert the object returned by the "annotate" function into a peak table using function "getPeaklist" from the CAMERA package. The resulting table should contain the following columns: "mz," "rt," "isotopes," "adduct," "pcgroup" as well as the intensities of the injected samples, corresponding with the MS1 and MS2 channels of the injected sample.
5. Repeat the procedure in steps one to four above to process data corresponding with the injected "QC-Mix-15" samples. This should result in a peak-table file for each injection (*see* **Note 14**).
6. Steps two to four can be automated by using the function "preprocessLibrary" in the supplementary software package. Step five can be automated by using the function "preprocessQCmixes" in the supplementary software package.

3.5 Converting Peak Tables to a Reference Library

To convert the generated peak-tables to a reference library, first install the supplementary software package, following the online installation instructions. Then, prepare a tab-separated table containing the following meta-data for each chemical standard in the injected pools:

1. *Peak table*: the file name corresponding with the peak table of the pool into which the chemical standard was assigned.
2. *Compound ID*: a common identifier by which the chemical standard can be identified, such as the PubChem ID (https://pubchem.ncbi.nlm.nih.gov), or the CAS number (http://www.cas.org).
3. *Compound name*: a common chemical name (optional).
4. *Chemical formula*: the molecular formula corresponding with the chemical standard.
5. *Retention time*: the expected chromatographic retention time (or retention time window) of the chemical standard in the chromatographic conditions used. The chromatographic retention time (or chromatographic retention time window) can sometimes be provided by vendors of the chemical standards, however, in case no such information exists, fill this field with "NA" (i.e., information is not available).

Use the supplementary software package function "buildMSlib" to automatically convert the peak tables into a reference library in the form of an R data object. *See the supplementary* software documentation ("?buildMSlib" in R) and Fig. 2 for further details.

Use the supplementary software package function "convertToMSP" to export the generated reference library to a text file which can be uploaded into public databases such as MassBank (*see*: www.massbank.jp or www.massbank.eu) or be processed by several software packages. *See* the software documentation ("?convertToMSP" in R) for further details.

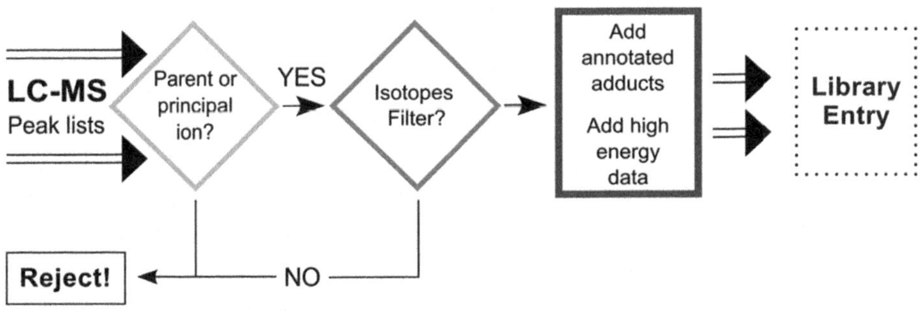

Fig. 2 An outline of the procedure to automatically create a reference library. Peak-tables generated during the preprocessing step are matched with given values related to the injected chemical standards. Next, the matched peak feature (i.e., a peak with *m/z*, retention time and intensity values), is validated by comparison to the theoretical monoisotopic mass value and by isotope decomposition, which relates the detected peak (and corresponding heavy isotope peaks) to the chemical formula of the relevant chemical standard. In case the matched feature fails one of the tests, it is being rejected, and the next matched feature is inspected. Otherwise, additional features corresponding with highly correlated peaks to the matched feature are being grouped and inserted as an item into the MS reference library. *See* also Subheading 3.5.

3.6 Metabolite Annotation Using the Reference Library

1. Inject and preprocess biological samples using the procedures previously outlined. Note that the specific analytical conditions should be kept exactly the same as in the injections of chemical standards, while the computational preprocessing parameters should be optimized per experiment. In addition, sample files originating from a biological matrix should generally be preprocessed together, in contrast to the injections of chemical standard pools, which are processed individually. Keep the output formatting exactly as described in Subheading 3.4, including the naming definitions for the MS1 and MS2 channels (i.e., with the postfix "01" and the postfix "02", respectively). Place the resulting peak table files in a designated folder.
2. Preprocess the injections of "QC-Mix-15" as described in Subheading 3.4. Place the resulting peak table files in a designated folder.
3. Use the function "runMatchWeiz" in the supplementary software package in order to search through the generated MS reference library using the experimental data. *See* the software documentation (e.g., "?runMatchWeiz" in R) for further details.

4 Notes

1. It is essential that the water used for the preparation of the UPLC solutions does not contain sodium or potassium ions, as a large amount of sodium or potassium adducts would be present during the analysis in the positive ionization mode.
2. The "QC-Mix-15" consists of a mixture of 15 standard compounds belonging to different chemical classes: amino acids (tyrosine, phenylalanine, and tryptophan); organic acids (cinnamic, benzoic, chlorogenic, caffeic, coumaric, ferulic, and synapic); flavonols (quercetin, rutin, and kaempferol); flavanone (naringenin), and a glycoalkaloid (α-tomatine). These compounds cover a wide lipophilicity range, detected in both ionization modes and their mass signals span a wide range of the m/z values, starting from 120.08 Da (the immonium ion of phenylalanine, detected in the positive ionization mode) to 1034.5 Da (the pseudomolecular ion of α-tomatine, detected in the positive ionization mode) or 1078.5 Da (the α-tomatine-formic acid adduct, detected in the negative ionization mode). Use this "QC-Mix-15" solution for QC (quality control) and SST (system suitability test) samples.
3. A commercial α-tomatine chemical standard typically contains a certain amount of dehydrotomatine impurity. Dehydrotomatine elutes approx. 1 min earlier than α-tomatine in the 40-min run, and it is therefore detectable in the chromatogram of the α-tomatine standard.

4. Apart from the UPLC-qTOF system of Waters, any other combination of High Pressure/Ultra High Pressure Liquid Chromatography with a Hybrid High Resolution Mass Spectrometry from other vendors can be used. The MS detector should produce two types of data in parallel: MS spectrum with no compound fragmentation in order to measure the molecular ion and possible adducts, as well as MS spectrum at high energy in order to measure fragments.
5. It is important to keep the number of overall injections as low as possible, in order to minimize the intrabatch and interbatch retention time drifts between injections. On the other hand, pooling too many chemical standards into one injection can lead to interpretation errors. Therefore, set the number of compounds per pool according to the available set of chemical standards: for a small set of structurally similar compounds, select a lower number of compounds (e.g., 5–10 per pool) while for a large structurally diverse set, pool a higher number of compounds (e.g., 10–20 per pool).
6. The primary consideration in assigning chemical standards to designated injection pools is to assign compounds with the same chemical formula (i.e., structural isomers) to separate pools. Next, it is essential to separate compounds with similar chromatographic retention times. This can be done either by considering preexisting knowledge regarding the compounds or by comparing their respective logD (i.e., the distribution coefficient) values, which roughly correlate with retention data. The logD is a descriptor of lipophilicity, which takes into account the pH of the aqueous phase in which the compound is solvated. Thus, use the estimated pH of the chromatographic mobile phase when calculating logD. We generally use the ChemAxon JChem software (www.chemaxon.com) to automatically calculate the logD descriptor in batch mode, with an estimated pH value of 3 as an input parameter. Structural information of the chemical standards must be available in order to calculate logD values. This information can in general be provided by the vendors of the chemical standard, in any of the common formats for storing chemical structure information. Such formats include SMILES strings, MDL (MOL or SDF), or the InChi molecular identifier. Additionally, the open-source software OpenBabel (www.openbabel.org) can be used to convert between the different chemical data formats. The third parameter to consider when deciding on the composition of pools is the chemical class of the available chemical standards. While not as important as the mass and retention time parameters, overlapping compounds belonging to the same chemical class can share fragmentation patterns and thus, potentially, some identical mass fragments. These

potential fragmentation overlaps should be avoided as they can lead to errors in peak identification during the automatic peak extraction process.

7. Use a mixture of acidic water and methanol for more hydrophilic compounds and ethanol-DMSO for the most lipophilic ones. Not more than 1–2% DMSO is allowed in the final mixture to avoid MS detector contamination.

8. Another option is to dissolve all dry standards in DMSO, take an equal amount of each stock solution to generate a pool, evaporate pool solution to dryness and resuspend in 80% methanol. DMSO is a very good solvent for both hydrophilic and hydrophobic compounds but it is highly hygroscopic. Presence of water increases its boiling point and therefore hydrated DMSO solution can melt during lyophilization. Cover the sample plate containing the DMSO solutions and make a tiny hole on top of each well (e.g. by using a small cardboard pin or a needle) in order to prevent loss of material and cross-contamination through evaporation.

9. Inject volumes of 1–4 μL, depending on the UPLC instrument used. Large injection volumes lead to broadening of the chromatographic peaks of polar compounds, which can easily be seen in the "QC-Mix-15" chromatogram.

10. The MS^E mode [17] records an exact mass precursor and fragment ion information from every detectable component in a sample. In this mode the MS rapidly alternates between two functions: the first acquiring low-energy exact mass precursor ion spectra and the second acquiring elevated-energy exact mass fragment ion spectra.

11. It is advisable to increase fragmentation collision energies to 40 eV when running in the positive ion mode and to 45 eV for the negative mode in order to break stronger molecular bonds.

12. If a commercial software, such as Progenesis QI (www.nonlinear.com) is available, the data can be maintained in the raw format and preprocessed directly, i.e., skip **steps 2–5** (Subheading 3.4). Else, if the Waters Databridge is unavailable, or the source raw data is from another vendor, then the open source data converter msConverter [18] can be used to generate either mzML or mzXML files. In this case, make sure that the MS1 and MS2 channels are converted into separate files named with the corresponding "01" and "02" postfix labels.

13. The specific function names used in R packages can change over time, due to version updates of the software packages. Therefore, verify that function names used throughout the text correspond with the software documentation of the packages currently installed on your system. This can be done from

within R, using the "?" symbol followed by the function name, or online with the relevant software package web page.

14. As the samples of "QC-Mix-15" are sequentially injected over the whole run, processing each of them individually provides more information regarding RT variability within the experiment. Therefore, use the function "preprocessQCmixes" in the supporting software package in order to generate a number of peak lists equal to the number of "QC-Mix-15" samples injected.

Acknowledgments

The authors thank the Adelis Foundation; Leona M. and Harry B. Helmsley Charitable Trust; Jeanne and Joseph Nissim Foundation for Life Sciences; Tom and Sondra Rykoff Family Foundation Research; and the Raymond Burton Plant Genome Research Fund for supporting the A. Aharoni lab activity. A. Aharoni is the incumbent of the Peter J. Cohn Professorial Chair.

References

1. Sumner LW, Amberg A, Barrett D et al (2007) Proposed minimum reporting standards for chemical analysis chemical analysis working group (CAWG) metabolomics standards initiative (MSI). Metabolomics 3:211–221
2. Schymanski EL, Jeon J, Gulde R et al (2014) Identifying small molecules via high resolution mass spectrometry: communicating confidence. Environ Sci Technol 48:2097–2098
3. Iijima Y, Nakamura Y, Ogata Y et al (2008) Metabolite annotations based on the integration of mass spectral information. Plant J 54:949–962
4. Matsuda F, Nakabayashi R, Sawada Y et al (2011) Mass spectra-based framework for automated structural elucidation of metabolome data to explore phytochemical diversity. Front Plant Sci 2:40
5. Sawada Y, Nakabayashi R, Yamada Y et al (2012) RIKEN tandem mass spectral database (ReSpect) for phytochemicals: a plant-specific MS/MS-based data resource and database. Phytochemistry 82:38–45
6. Shahaf N, Rogachev I, Heinig U et al (2016) The WEIZMASS spectral library for high-confidence metabolite identification. Nat Commun 7:12423
7. Yetukuri L, Katajamaa M, Medina-Gome G et al (2007) Bioinformatics strategies for lipidomics analysis: characterization of obesity related hepatic steatosis. BMC Syst Biol 1:12
8. Taguchi R, Ishikawa M (2010) Precise and global identification of phospholipid molecular species by an Orbitrap mass spectrometer and automated search engine lipid search. J Chromatogr A 1217:4229–4239
9. Fahy E, Sud M, Cotter D, Subramaniam S (2007) LIPID MAPS online tools for lipid research. Nucleic Acids Res 35(Web Server issue):W606–W612
10. Malitsky S, Ziv C, Rosenwasser S et al (2016) Viral infection of the marine alga Emiliania huxleyi triggers lipidome remodeling and induces the production of highly saturated triacylglycerol. New Phytol 210:88–96
11. Hartmann T (2007) From waste products to ecochemicals: fifty years research of plant secondary metabolism. Phytochemistry 68:2831–2846
12. Barupal DK, Haldiya PK, Wohlgemuth G et al (2012) MetaMapp: mapping and visualizing metabolomic data by integrating information from biochemical pathways and chemical and mass spectral similarity. BMC Bioinformatics 13:99
13. Kind T, Liu KH, Lee DY, DeFelice B, Meissen JK, Fiehn O (2013) LipidBlast in silico tandem mass spectrometry database for lipid identification. Nat Methods 10:755–758

14. Horai H, Arita M, Kanaya S et al (2010) MassBank: a public repository for sharing mass spectral data for life sciences. J Mass Spectrom 45:703–714
15. Stravs MA, Schymanski EL, Singer HP, Hollender J (2013) Automatic recalibration and processing of tandem mass spectra using formula annotation. J Mass Spectrom 48:89–99
16. Wang M, Carver JJ, Phelan VV et al (2016) Sharing and community curation of mass spectrometry data with global natural products social molecular networking. Nat Biotechnol 34:828–837
17. Bateman KP, Castro-Perez J, Wrona M et al (2007) MSE with mass defect filtering for in vitro and in vivo metabolite identification. Rapid Comm Mass Spectrom 21:1485–1496
18. Kessner D, Chambers M, Burke R et al (2008) ProteoWizard: open source software for rapid proteomics tools development. Bioinformatics 24:2534–2536

Chapter 15

Quantification of Sugars and Organic Acids in Biological Matrices Using GC-QqQ-MS

Nirupama Samanmalie Jayasinghe, Himasha Mendis, Ute Roessner, and Daniel Anthony Dias

Abstract

Gas chromatography coupled with triple quadrupole mass spectrometry (GC-QqQ-MS) can be used to accurately quantify endogenous small molecules extracted from biological samples such as plants and human fluids including sera and urine. In order to quantify primary metabolites typically from central carbon metabolism such as sugars from glycolysis and the pentose phosphate pathway; and organic acids involved in the tricarboxylic acid (TCA) cycle; polar endogenous metabolites must be extracted from the samples of interest, chemically derivatized and quantified against a linear calibration curve to a corresponding authentic standard. This chapter describes how to quantify a combination of 48 primary metabolites belonging to classes of sugars, sugar alcohols, sugar acids, sugar phosphates, and organic acids using a robust, optimized, multiple reaction monitoring (MRM)-based GC-QqQ-MS method.

Key words GC-QqQ-MS, Quantification, Sugars, Organic acids, Carbon central metabolism

1 Introduction

Gas chromatography–mass spectrometry (GC-MS) is a highly efficient, sensitive analytical platform which is widely used for the qualitative and quantitative analysis of small molecules. Although GC-MS is limited to the analysis of small molecules typically in the range from 50 to 1000 Da and requires chemical derivatization for the analysis of nonvolatile compounds, metabolite identification is reproducible due to unique fragmentation patterns of derivatized molecules acquired at standard 70 eV. Generated mass spectral fragmentation of derivatized metabolites can easily be compared to commercially available mass spectral databases allowing for comparison and has been extensively used in untargeted GC-MS metabolomics over the last two decades [1].

However, in recent years, there has been a strong emphasis within the metabolomics community that quantitative data is far more important in biological studies since it accurately describes

Carla António (ed.), *Plant Metabolomics: Methods and Protocols*, Methods in Molecular Biology, vol. 1778,
https://doi.org/10.1007/978-1-4939-7819-9_15,

the endogenous concentration of metabolites of interest providing a better and clear understanding of the biochemistry in comparison to semiquantitative data [2]. GC-QqQ-MS has recently become a routine and robust analytical tool for the quantification of a small set of known metabolites (typically primary) due to improvement in column stationary phases selectivity, instrument sensitivity, dynamic range, and minimum chromatographic interferences in MRM mode and deconvolution algorithms, therefore becoming a popular technique for quantitative metabolomics analysis.

Quantitative data of primary metabolites such as sugars, organic acids and amines provide valuable information regarding carbon metabolism and as a result can assist in potentially providing knowledge to answer a biological question, especially when other "omics" data such as genomics, transcriptomics, and proteomics data are available and integrated. However, a single analytical platform or methodology cannot be used to quantify different classes of metabolites due to their structural diversity. Therefore, a combination of complementary analytical platforms such as liquid chromatography (LC)-QqQ-MS for amine quantification [3] and GC-QqQ-MS for sugars and organic acids quantification can be used [2] to provide a comprehensive coverage of the metabolome.

Biological samples must be efficiently extracted to quantify these hydrophilic metabolites. The extraction protocol will differ based on the sample type and targeted metabolites of interest. For example, when extracting plant metabolites, one must consider that the plant metabolome is dynamic and concentrations of metabolites may range over several magnitudes [4]. Therefore, sample extraction optimization is crucial for the determination of levels of sugars and organic acids.

2 Materials

A wide variety of extraction methods documented in the literature can be used to extract polar metabolites; however, an effective and efficient extraction method must be carefully selected. In general, liquid–liquid extractions (LLE) using aqueous solvents such methanol and water are often used for this type of analysis. Once the samples are extracted, the appropriate aliquots (high and low aliquots to cover the varying concentrations of metabolites) must undergo chemical derivatization which in return enables the complex mixture of polar metabolites to be chromatographically separated via gas chromatography. All solvents and derivatization reagents must be freshly prepared and mass spectrometry grade chemicals must be used to avoid contamination and interference. High quality data can be acquired with good GC separation coupled with a MRM acquisition method. Optimum collision energies, dwell times, resolution parameters for each MRM and effective time segment separation would enhance the sensitivity and selectivity of quantitative data. Samples must be acquired together and compared to a diluted series of calibration standards mixtures to accurately measure the unknown concentration of a given metabolite. Calibration standard mixtures can be prepared in bulk, evaporated to dryness, and stored at −20 °C in the dark to save time and prevent degradation.

2.1 Chemicals

All chemicals and authentic standards used in this protocol are analytical and/or mass spectrometry grade (i.e., >95% purity; e.g., Sigma-Aldrich). All solvents are LC-MS grade (e.g., Merck Lichrosolv hypergrade). The protocol here described refers to the quantitation of a range of 48 primary metabolites in biological matrices using authentic standards.

2.2 Solutions

1. Prepare a $^{13}C_6$ Sorbitol internal standard working stock solution (5 μg/mL) in methanol. Firstly prepare 250 μg/mL $^{13}C_6$ Sorbitol internal standard stock by weighing 12.5 mg of $^{13}C_6$ Sorbitol and dissolving it in 50 mL of methanol. Then transfer 1 mL of the stock into a volumetric flask and top up to 50 mL using methanol as the diluent to make a 5 μg/mL working stock solution (*see* **Note 1**).
2. Prepare 10 mM stock solutions of each authentic standard (sugars, sugar alcohols, sugar acids, sugar phosphates and organic acids) in 50% methanol.
3. Prepare a methoxyamine hydrochloride stock solution in pyridine anhydrous (30 mg/mL pyridine) (*see* **Note 2**).
4. GC derivatization solution: *N,O-bis*-(trimethylsilyl) trifluoroacetamide with 1% trimethylchlorosilane (BSTFA +1% TMCS) (Thermo Scientific).
5. Syringe washing solution: 100% methanol followed by 100% acetonitrile or ethyl acetate/n-hexane (1/1, v/v).

2.3 Equipment, Instrumentation, and Software

1. 1.5-mL safelock microcentrifuge tubes (e.g., Eppendorf).
2. 50-mL volumetric flask.
3. 10-mL falcon tubes.
4. 2.0-mL autosampler glass vials with magnetic crimp caps (Agilent Technologies).
5. 350-μL glass vial inserts (Agilent Technologies).
6. Centrifugal concentrator (e.g., Christ Alpha RVC 2-33).
7. Nitrogen evaporator (e.g., TurboVap).
8. Thermoshaker with a 2.0-mL thermoblock.
9. GC-MS system: 7890A gas chromatograph equipped with a Gerstel 2.5.2 autosampler and electron ionization (EI) ion

source coupled with a 7000 GC-QqQ-MS (Agilent, Santa Clara, USA).

10. GC column: VF-5 MS column (30 m × 0.25 mm internal diameter × 0.25 μm film thickness) with 10 m integrated Eziguard (Agilent Technologies).
11. General purpose split/splitless liner with glass wool tapered and deactivated (Agilent Technologies).
12. MS data acquisition and quantification software: Agilent MassHunter (version B 0.08.00).

3 Methods

3.1 Preparation of the Calibration Standards Mix

3.1.1 Sugar Stock Mix and Dilution Series

1. Pool and mix aliquots of 160 μL from each 10 mM stock of sugar, sugar alcohol, sugar phosphate and sugar acid prepared in 50% methanol into a 10-mL falcon tube. The total volume will be 4.48 mL.
2. Top up the volume to 5 mL with an aliquot of 520 μL of 50% methanol.
3. Final concentration of each compound in the mix will be 320 μM.
4. Prepare a 9 point calibration series by 1:2 serial dilution of stock mix using 50% methanol as the diluent (160, 80, 40, 20, 10, 5, 2.5, 1.25, 0.625 μM).

3.1.2 Organic Acid Stock Mix and Dilution Series

1. Pool and mix aliquots of 160 μL from each 10 mM stock of each organic acid prepared in 50% methanol into a 10-mL falcon tube. The total volume will be 3.36 mL.
2. Top up the volume to 5 mL with an aliquot of 1.64 mL of 50% methanol.
3. Final concentration of each compound in the mix will be 320 μM.
4. Prepare a 9 point calibration series by 1:2 serial dilution of stock mix using 50% methanol as the diluent (160, 80, 40, 20, 10, 5, 2.5, 1.25, 0.625 μM).

3.1.3 Amalgamation of Sugars and Organic Acids Calibration Series

1. Transfer aliquots of 40 μL from each sugar calibration mix into 350-μL glass vial inserts. Then transfer the vial insert into 1.5-mL microcentrifuge tube and evaporate to dryness (Fig. 1).
2. Add 40 μL of organic acid calibration mix from each corresponding concentration into the dried glass vial inserts from **step 1** containing the sugar mix and evaporate to dryness (Fig. 1).

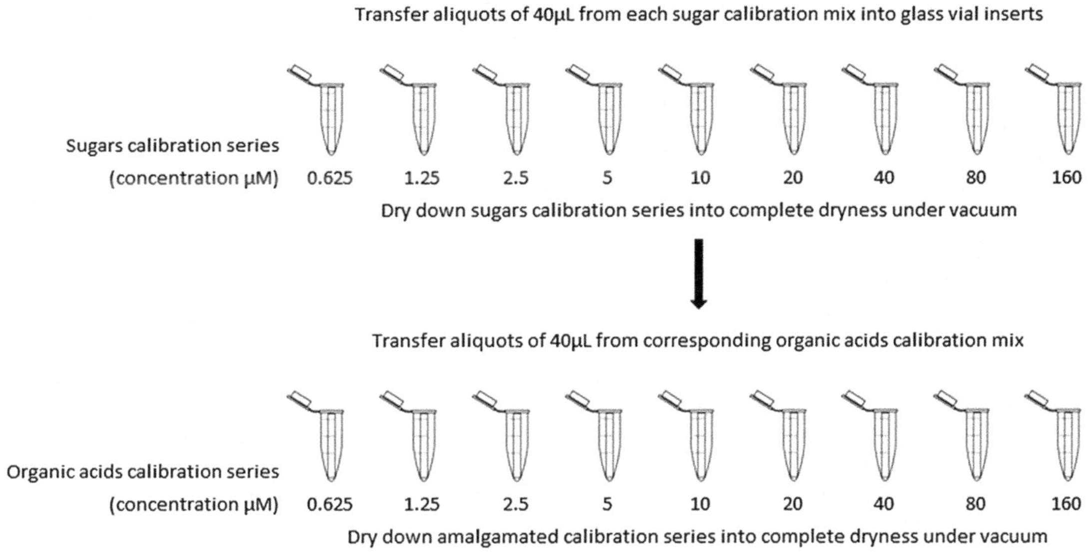

Fig. 1 Illustrations showing how sugars and organic acids are combined to prepare a quantitative calibration series

3. Finally, add an aliquot of 40 μL from the $^{13}C_6$ Sorbitol internal standard working stock solution (5 μg/mL).
4. Evaporate to dryness using centrifugal concentrator (*see* **Note 1**).
5. Transfer the dried glass inserts into 2.0-mL autosampler glass vials and seal with magnetic cap.

3.2 Sample Preparation and Metabolite Extraction

The extraction of metabolites is challenging in terms of the physical and chemical diversity of metabolites and their abundances which can vary by several orders of magnitude. Multiple extraction methods must be tested and recoveries must be carried out to obtain an optimized extraction protocol. Polar metabolites from biological samples of interest can be extracted using many different homogenizing techniques, solvents, temperature treatments and is dependent on the biological matrix of interest [1–8].

1. Optimize the amount of sample required.
2. Optimize the extraction methodology to determine the optimum extract volume and sample amount required to achieve the best concentration where metabolites of interest fall within the linear range of the calibration curve (i.e., 0.625–160 μM.
3. Then transfer 40 μL of the extract containing 5 μg/mL of $^{13}C_6$ Sorbitol into the 350-μL glass-vial inserts into a 1.5-mL microcentrifuge tube and evaporate to dryness using a centrifugal concentrator.

4. Add 10 μL of GC-MS grade 100% methanol to all dried calibration standards and samples and evaporate to complete dryness using a centrifugal concentrator (*see* **Note 3**).
5. Transfer the dried glass inserts into 2.0-mL autosampler glass vials and seal with magnetic cap.

3.3 GC-QqQ-MS Analysis

3.3.1 Online Derivatization

1. Wash syringe with the washing solution before and after reagent addition and sample injection.
2. Add 20 μL of methoxyamine hydrochloride (30 mg/mL pyridine) into 350-μL glass-vial inserts in 2.0-mL autosampler glass vials (*see* **Note 2**).
3. Incubate for 2 h in agitator (500 rpm) at 37 °C.
4. Add 20 μL of BSTFA +1% TMCS.
5. Incubate for 30 min at 37 °C in the thermomixer (500 rpm).
6. Rest 1 h at room temperature.
7. Inject 1 μL of derivatized sample into GC-QqQ-MS system.

3.3.2 GC Parameters

1. Inlet temperature 250 °C.
2. Splitless mode.
3. Constant flow mode 1 mL/min, ultrahigh purity (UHP) helium gas.
4. Oven temperature start at 50 °C, hold for 1 min, then ramp 15 °C/min up to 325 °C and hold for 4.7 min.
5. Transfer line 290 °C.

3.3.3 QqQ-MS Parameters

1. Source temperature 230 °C.
2. Collision cell gas, UHP N_2 flow rate of 1.5 mL/min.
3. Quenching gas, UHP helium flow rate of 2.25 mL/min.
4. Electron energy 70 eV.

3.3.4 QqQ-MS Acquisition Parameters

Electron ionization (EI) is a harsh ionization method where each chemically derivatized molecule is fragmented into reproducible mass spectra at 70 eV. To develop an MRM method based on EI mass spectra, a unique precursor ion must be selected followed at optimum collision energy (CE) to further fragment the precursor ion into product ions. A simplified step-by-step guide to MRM development is explained below.

1. *Select precursor ion(s)*: Acquire the mass spectrum of a given standard on MS1 scan mode, i.e., fumaric acid (Fig. 2). Select one or a few unique fragment/s, medium to highly abundant and closer to the molecular ion (if observed) (i.e., *m/z* 245 for fumaric acid).

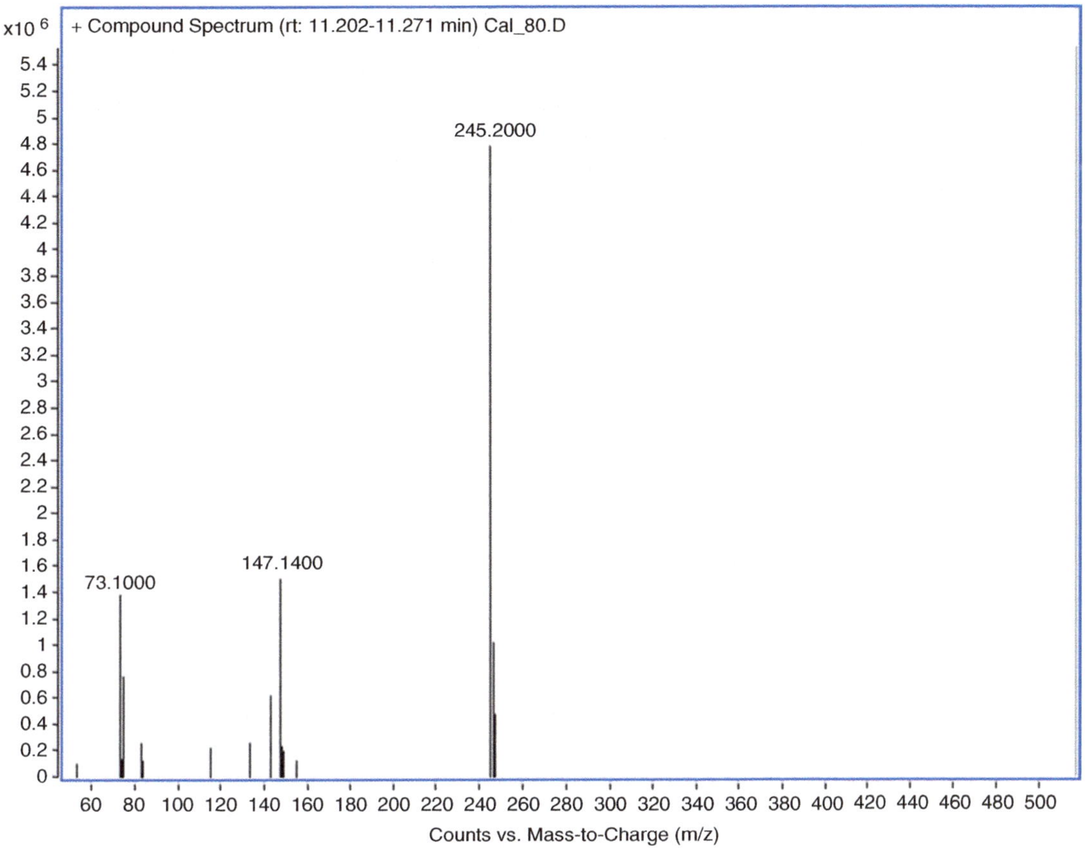

Fig. 2 MS1 mass spectrum of fumaric acid

2. *Identify product ion(s)*: Apply four different collision energies (CEs) on the selected precursor ion(s) and acquire data on product ion scan mode to identify possible product ion(s), i.e., apply CEs of 0, 5, 10, and 20 to precursor mass of *m/z* 245, select product ions *m/z* 217and *m/z* 171 as product ions based on abundance and stability.

3. *Optimize CE*: Once the MRM transitions (precursor ion -> product ion) are selected, optimize the CE in order to maximize the MRM count. Select a series of CEs with small increments and acquire the chosen MRM on "MRM acquisition mode" to report peak area. Plot the peak area vs collision energy to identify the optimum CE, i.e., peak area vs CE for fumaric acid (Fig. 3).

4. *Dwell time, scans and time segments*: The MRM method described here has over 70 MRMs that is simultaneously acquired during a 23 min run-time. In order to improve the scans, dwell time and peak shape the chromatographic run time is divided into 11 time segments, therefore only the group of

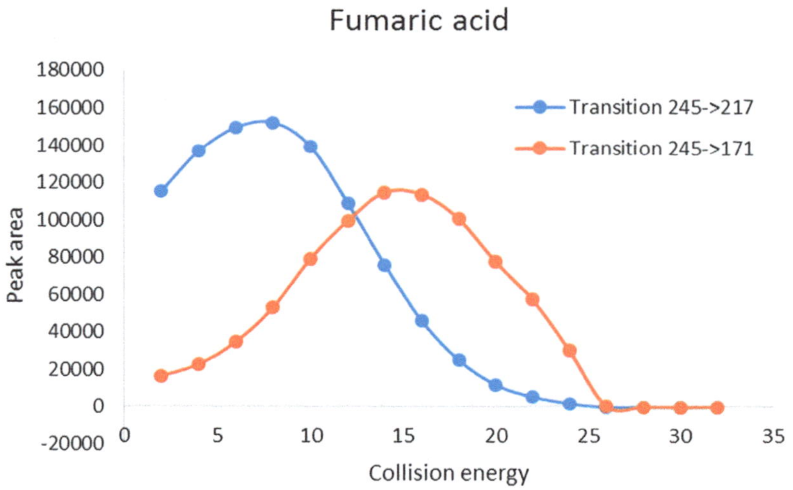

Fig. 3 Graph of peak area vs CE for fumaric acid

MRMs eluting in each time segment is acquired within that time segment. Refer to the (Table 1) below for the time segments, MRMs, CE, dwell time and peak resolution.

3.3.5 *Data Analysis*

In this section, a step-by-step guide to general processing of GC-QqQ-MS raw data is described.

1. Open Agilent MassHunter GC-QqQ-MS quantification software (version B 0.08.00).
2. Create a batch:

   File -> New Batch (Fig. 4)
3. Add samples:

   File -> Add samples -> Browse to copy samples -> Select samples -> OK (Fig. 5)
4. Select sample types from the drop-down box in "Type" column and specify concentration level (Fig. 6).
5. Create method. A "method' is the target list of compounds, containing their respective MRMs, retention times (RT) and the concentration levels of calibration series. This will be used to determine the peak area, plot calibration curves and finally the samples that have been acquired will be compared, to determine the levels of these compounds in the samples. To create a new method, go to:

   Method → New method → New method from acquired MRM data/

   This will allow the user to create a new data analysis method based on the data acquisition method. Browse and select a calibration mix data file which falls between the mid to high level concentration range (i.e., 20–80 μM range) to create a new method (Fig. 7).

Table 1
MRM acquisition parameters for all authentic sugars and organic acids standards

Time segment	Time segment start time (min)	Compound name	Precursor ion	MS1 resolution	Product ion	MS2 resolution	Dwell	Collision energy
1	6.0	Malonate	233	Unit	217	Unit	70	2
		Malonate	233	Unit	143	Unit	70	8
2	8.5	Fumaric acid	245	Unit	217	Unit	15	6
		Fumaric acid	245	Unit	171	Unit	15	12
		Itaconic acid	215	Unit	133	Unit	15	18
		Itaconic acid	259	Unit	131	Unit	15	20
		Maleic acid	245	Unit	217	Unit	15	4
		Maleic acid	245	Unit	133	Unit	15	12
		Nicotinic acid	180	Unit	136	Unit	15	14
		Nicotinic acid	180	Unit	106	Unit	15	8
		Pipecolic acid	156	Unit	128	Unit	15	6
		Pipecolic acid	156	Unit	84	Unit	15	6
		Succinic acid	172	Unit	156	Unit	15	0
		Succinic acid	172	Unit	113	Unit	15	4
3	10.3	Erythritol	307	Unit	217	Unit	35	2
		Malic acid	233	Unit	191	Unit	35	2
		Malic acid	233	Unit	189	Unit	35	2
		Salicylic acid	267	Unit	209	Unit	35	8
		Salicylic acid	209	Unit	91	Unit	35	8
4	10.9	2-hydroxyglutaric acid	350	Unit	204	Unit	50	6
		2-ketoglutaric acid	288	Unit	198	Unit	50	2
		2-ketoglutaric acid	198	Unit	167	Unit	50	2
		2-ketoglutaric acid	198	Unit	154	Unit	50	6
5	11.4	Aconitic acid	375	Unit	285	Unit	20	4
		Aconitic acid	375	Unit	211	Unit	20	4
		Arabinose	307	Unit	217	Unit	20	2
		Arabitol	277	Unit	189	Unit	20	4
		Fucose	277	Unit	189	Unit	20	4
		Rhamnose	277	Unit	189	Unit	20	4
		Ribose	307	Unit	217	Unit	20	2
		Xylitol	307	Unit	217	Unit	20	2
		Xylose	307	Unit	217	Unit	20	2
6	12.6	Citric acid	257	Unit	201	Unit	30	4
		Shikimic acid	255	Unit	239	Unit	30	4
		Isocitric acid	245	Unit	191	Unit	30	4
		Shikimic acid	204	Unit	189	Unit	30	8
		Citric acid	183	Unit	139	Unit	30	4
7	13.0	13C6 Sorbitol	323	Unit	132	Unit	10	8
		2-keto gluconic acid	349	Unit	201	Unit	10	8
		2-keto gluconic acid	349	Unit	186	Unit	10	4
		Fructose	307	Unit	217	Unit	10	2
		Galactitol	319	Unit	129	Unit	10	10
		Galactose	319	Unit	157	Unit	10	4
		Glucose	319	Unit	129	Unit	10	10

(continued)

Table 1
(continued)

Time segment	Time segment start time (min)	Compound name	Precursor ion	MS1 resolution	Product ion	MS2 resolution	Dwell	Collision energy
		Glucuronic acid	364	Unit	160	Unit	10	4
		Mannitol	319	Unit	157	Unit	10	4
		Mannitol	319	Unit	129	Unit	10	10
		Mannose	319	Unit	129	Unit	10	10
		Quinic acid	345	Unit	255	Unit	10	8
		Quinic acid	255	Unit	239	Unit	10	8
		Syringic acid	327	Unit	312	Unit	10	18
8	14.6	Uric acid	456	Unit	441	Unit	30	4
		Uric acid	456	Unit	382	Unit	30	4
		Ferulic acid	308	Unit	293	Unit	30	18
		Ferulic acid	308	Unit	219	Unit	30	4
		Inositol	305	Unit	217	Unit	30	8
		Caffeic acid	219	Unit	191	Unit	30	12
9	15.8	Glucose-6-phosphate	471	Unit	387	Unit	70	4
		Fructose-6-phospate	459	Unit	315	Unit	70	4
10	17.0	Sucrose	361	Unit	169	Unit	50	10
		Trehalose	361	Unit	169	Unit	50	10
		Maltose	361	Unit	169	Unit	50	10
11	21.1	Erlose	361	Unit	169	Unit	50	10
		Melibiose	361	Unit	169	Unit	50	10
		Raffinose	361	Unit	169	Unit	50	10

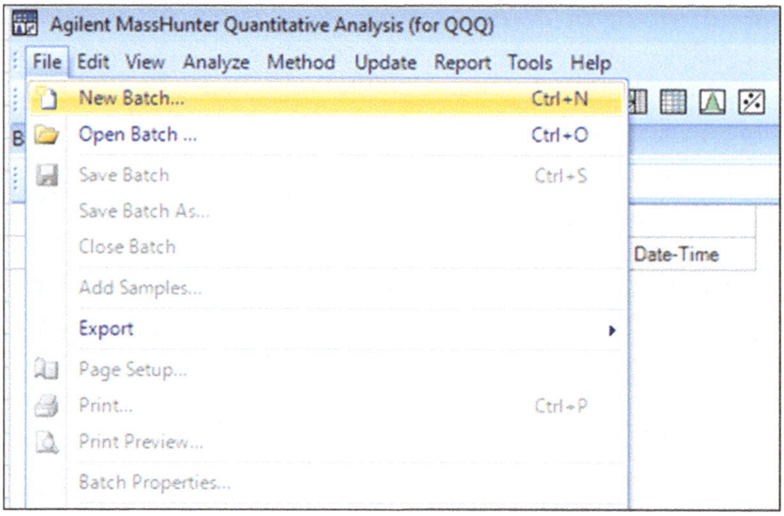

Fig. 4 Creating a new batch

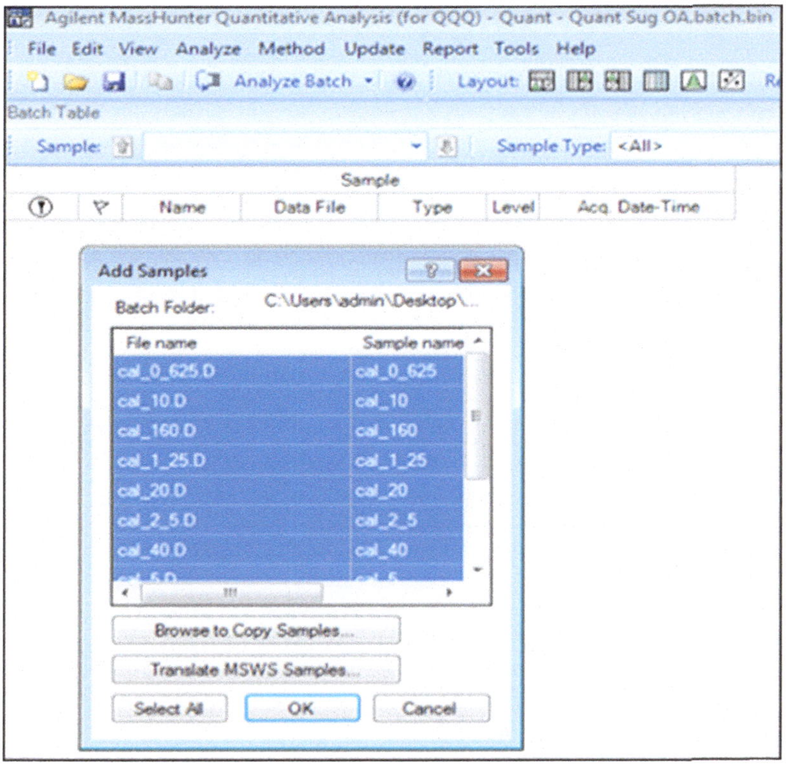

Fig. 5 Adding samples to the batch

		Sample				
ⓘ	▽	Name	Data File	Type	Level	Acq. Date-Time
▶		cal_0_625	cal_0_625.D	Cal	1	6/25/2017 1:15 AM
		cal_1_25	cal_1_25.D	Cal	2	6/25/2017 2:03 AM
		cal_2_5	cal_2_5.D	Cal	3	6/25/2017 2:52 AM
		cal_5	cal_5.D	Cal	4	6/25/2017 3:42 AM
		cal_10	cal_10.D	Cal	5	6/25/2017 4:30 AM
		cal_20	cal_20.D	Cal	6	6/25/2017 5:19 AM
		cal_40	cal_40.D	Cal	7	6/25/2017 6:08 AM
		cal_80	cal_80.D	Cal	8	6/25/2017 6:57 AM
		cal_160	cal_160.D	Cal	9	6/25/2017 7:46 AM
		E_10	E_10.D	Sample		6/25/2017 10:51 AM
		E_20	E_20.D	Sample		6/25/2017 12:29 PM
		E_40	E_40.D	Sample		6/25/2017 2:45 PM

Fig. 6 Setting up sample types and levels

6. Go through "Method set up tasks" in the method view.
7. MRM compound setup. MassHunter software will automatically create a compound list, their RTs and MRMs based on the acquisition method and selected calibration mix data file. However, the user will need to check, verify and change RTs based on the elution order in the total ion chromatogram (TIC)

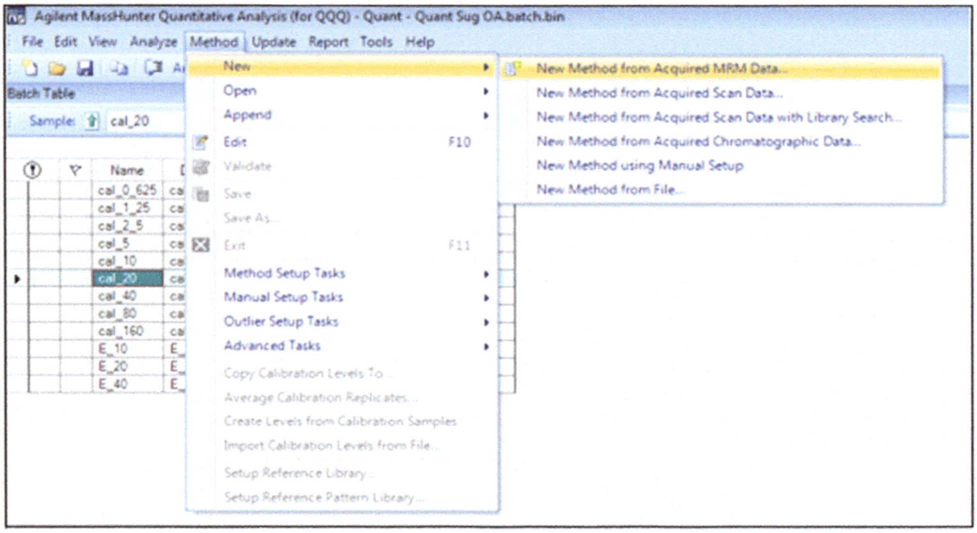

Fig. 7 Creating a new method

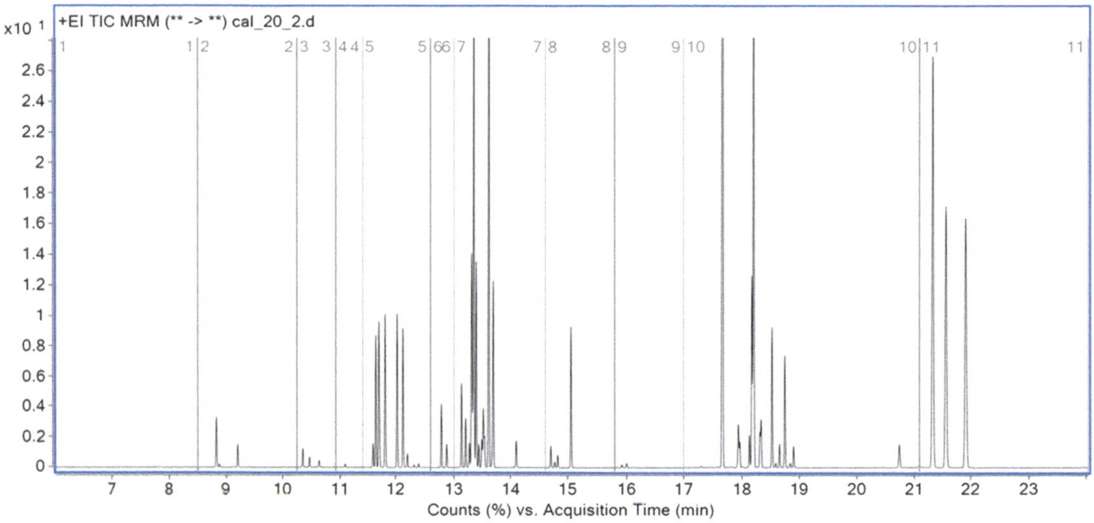

Fig. 8 Total ion chromatogram (TIC) of 20 μM calibration mix

(Fig. 8) (*see* **Note 4**). Order of elution of sugars and organic acids will remain unchanged from run to run (Table 2).

8. Type in "160" as the "Dil. high conc" and insert 1:2 as "Dil. pattern" for all compounds. On the top task bar, under "Method table" in "# of levels" box, specify the number of concentration levels (in this example it is 9). Select the first target compound in the method table and press "Create levels". This will create a "calibration table" with 9 concentration levels for the selected metabolite standard.

Table 2
Order of elution of sugars and organic acids

Name	TS	Transition	Type	Precursor ion	Product ion	RT (min)
Malonic acid	1	233.0 -> 143.0	Target	233.0	143.0	7.85
Maleic acid	2	245.0 -> 217.0	Target	245.0	217.0	8.81
Nicotinic acid	2	180.0 -> 106.0	Target	180.0	106.0	8.83
Succinic acid	2	172.0 -> 113.0	Target	172.0	113.0	8.89
Itaconic acid	2	215.0 -> 133.0	Target	215.0	133.0	9.15
Fumaric acid	2	245.0 -> 217.0	Target	245.0	217.0	9.22
Pipecolic acid	2	156.0 -> 84.0	Target	156.0	84.0	9.41
Malic acid	3	233.0 -> 189.0	Target	233.0	189.0	10.38
Erythritol	3	307.0 -> 217.0	Target	307.0	217.0	10.50
Salicylic acid	3	267.0 -> 209.0	Target	267.0	209.0	10.66
2-hydroxyglutaric acid	4	350.0 -> 204.0	Target	350.0	204.0	11.10
2-ketoglutaric acid	4	198.0 -> 154.0	Target	198.0	154.0	11.13
Xylose	5	307.0 -> 217.0	Target	307.0	217.0	11.68
Arabinose	5	307.0 -> 217.0	Target	307.0	217.0	11.73
Ribose	5	307.0 -> 217.0	Target	307.0	217.0	11.85
Xylitol	5	307.0 -> 217.0	Target	307.0	217.0	12.06
Rhamnose	5	277.0 -> 189.0	Target	277.0	189.0	12.16
Arabitol	5	307.0 -> 217.0	Target	307.0	217.0	12.16
Fucose	5	277.0 -> 189.0	Target	277.0	189.0	12.24
Aconitic acid	5	375.0 -> 211.0	Target	375.0	211.0	12.43
Shikimic acid	6	204.0 -> 189.0	Target	204.0	189.0	12.83
Isocitric acid	6	245.0 -> 83.0	Target	245.0	83.0	12.91
Citric acid	6	183.0 -> 139.0	Target	183.0	139.0	12.92
Quinic acid	7	345.0 -> 255.0	Target	345.0	255.0	13.18
Fructose	7	307.0 -> 217.0	Target	307.0	217.0	13.25
2-keto gluconic acid	7	349.0 -> 201.0	Target	349.0	201.0	13.25
Mannose	7	319.0 -> 129.0	Target	319.0	129.0	13.35
Galactose	7	319.0 -> 157.0	Target	319.0	157.0	13.39
Glucose	7	319.0 -> 129.0	Target	319.0	129.0	13.43
Syringic acid	7	327.0 -> 312.0	Target	327.0	312.0	13.56
Mannitol	7	319.0 -> 129.0	Target	319.0	129.0	13.65

(continued)

Table 2
(continued)

Name	TS	Transition	Type	Precursor ion	Product ion	RT (min)
13C6 Sorbitol	7	323.0 -> 132.0	ISTD	323.0	132.0	13.70
Glucuronic acid	7	364.0 -> 160.0	Target	364.0	160.0	13.72
Galactitol	7	319.0 -> 129.0	Target	319.0	129.0	13.74
Inositol	8	305.0 -> 217.0	Target	305.0	217.0	14.74
Ferulic acid	8	308.0 -> 293.0	Target	308.0	293.0	14.86
Uric acid	8	456.0 -> 382.0	Target	456.0	382.0	15.03
Caffeic acid	8	219.0 -> 191.0	Target	219.0	191.0	15.09
Fructose-6-phospate	9	459.0 -> 315.0	Target	459.0	315.0	15.96
Glucose-6-phosphate	9	471.0 -> 387.0	Target	471.0	387.0	16.05
Sucrose	10	361.0 -> 169.0	Target	361.0	169.0	17.71
Maltose	10	361.0 -> 169.0	Target	361.0	169.0	18.22
Trehalose	10	361.0 -> 169.0	Target	361.0	169.0	18.25
Turanose	10	361.0 -> 169.0	Target	361.0	169.0	18.57
beta-gentiobiose	10	361.0 -> 243.0	Target	361.0	243.0	18.79
Melibiose	11	361.0 -> 169.0	Target	361.0	169.0	20.82
Raffinose	11	361.0 -> 169.0	Target	361.0	169.0	21.41
Erlose	11	361.0 -> 169.0	Target	361.0	169.0	21.65
Melezitose	11	361.0 -> 169.0	Target	361.0	169.0	22.01

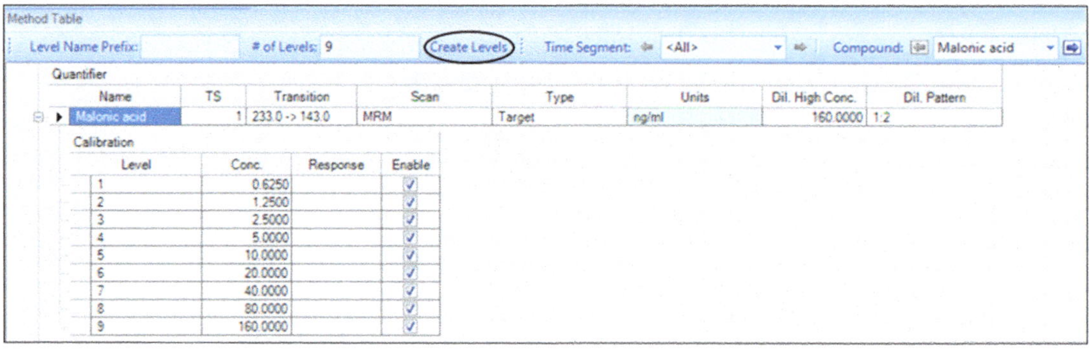

Fig. 9 Creating concentration levels

9. Right-click outside the "calibration table," then: *Copy calibration levels to* → *Select all* → *OK* (Fig. 9)

Quantifier

Name	TS	Transition	Scan	Type	ISTD Compound Name	ISTD Flag	ISTD Conc.	Time Reference Flag
13C6 Sorbitol	7	323.0 -> 132.0	MRM	ISTD	<None>	✓	5.0000	
Malonic acid	1	233.0 -> 143.0	MRM	Target	13C6 Sorbitol		5.0000	
Fumaric acid	2	245.0 -> 217.0	MRM	Target	13C6 Sorbitol		5.0000	
Itaconic acid	2	215.0 -> 133.0	MRM	Target	13C6 Sorbitol		5.0000	
Maleic acid	2	245.0 -> 217.0	MRM	Target	13C6 Sorbitol		5.0000	

Fig. 10 Setting up Internal standard settings

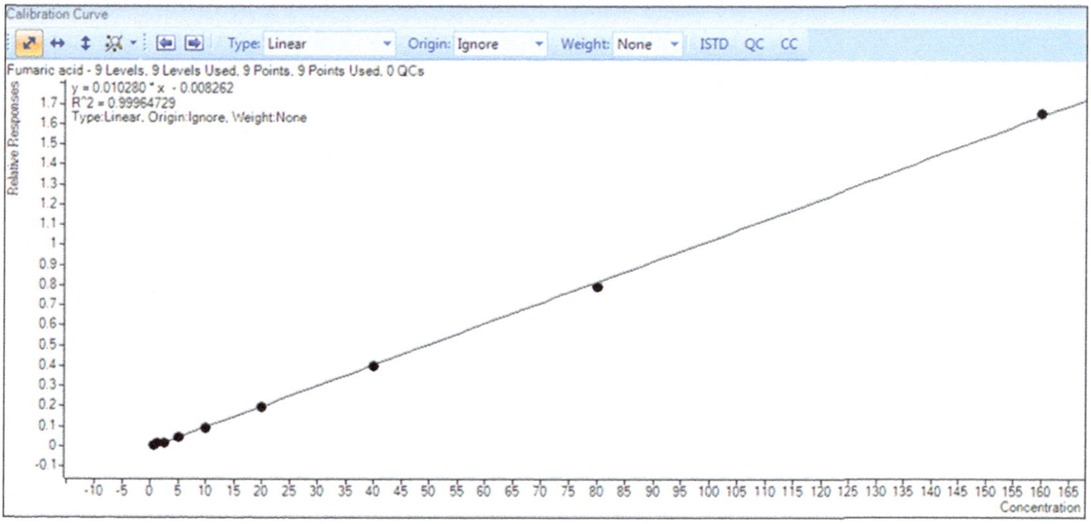

Fig. 11 An example of a calibration curve (fumaric acid)

10. In the "internal standard setup" task, tick the check box under "ISTD flag" for $^{13}C_6$ Sorbitol (*see* **Note 5**) (Fig. 10).
11. Select $^{13}C_6$ Sorbitol as the "ISTD compound name" from drop down box for all compounds except for $^{13}C_6$ Sorbitol.
12. Press "Validate" under method setup tasks to check for errors.
13. Press "Exit" and apply method to batch to process the batch. Interface will be navigated back to batch view.
14. Press "Analyse batch" from the top task bar.
15. Calibration curves for each metabolite will be generated and the concentration levels of each metabolite detected will be determined against the corresponding calibration curve (*see* **Note 6**) (Fig. 11).
16. To review integrations:

    View → Compounds at a glance

    This view allows the overlay of TICs and in return verifies RTs and allows adjusting integrations. If the peaks are poorly integrated, manually correct the integration by locating the peak, then drawing an integration line under the correct peak.

17. After reviewing and editing peak integration go through each calibration curve to find the best fit to achieve linear curve with $R^2 > 0.97$. Finally, press "Analyse batch" again to apply any changes made to the integrations and calibration curves, then Save batch.
18. Calculated concentrations can be finally exported into a spread sheet for final calculations and post statistical analysis:

    File-> Export -> Export table
19. Apply dilution factors, extraction volumes and starting material amount as necessary to express the final concentration in the sample (i.e., picomole/mg of fresh weight).

4 Notes

1. The final concentration of internal standard must be the same between the samples and calibration series after derivatization. Therefore, the volume of 5 μg/mL $^{13}C_6$ Sorbitol stock solution to be added to the calibration series depends on the volume of sample extract containing 5 μg/mL $^{13}C_6$ Sorbitol dried down for analysis.
2. Methoxyamine hydrochloride must be stored in a desiccator and working stock solution of 30 mg/mL in pyridine must be prepared freshly.
3. The final evaporation of samples with methanol is absolutely mandatory. Attention! Do not skip this step [5]. Derivatization reagents and GC-MS instrument is highly susceptible to any residual water and can affect the derivatization and instrument reproducibility and longevity of column.
4. RTs for compounds with the same MRMs needs to be manually checked and verified based on their respective elution order and compared to authentic standards. For example, most of the sugars quantified in this method are structural isomers and therefore form the same MRMs. These sugars must be differentiated based on the RT. Elution order can be established by acquiring individual standards once and recording the RTs in a template. Refer the new RTs back to the order of elution in the template for verification (*see* Table 2).
5. Ideally an internal standard (i.e., stable-isotope-labeled) for each metabolite must be incorporated into the analysis to account for response factors. However, incorporating a mix of stable-isotope-labeled internal standards is costly, therefore is limited to 1 or 2 stable-isotope-labeled internal standards due to feasibility.

6. Abundance of metabolites within a sample is variable. Therefore, certain metabolites may fall below limit of quantification (LOQ) or above linear range. Samples must be concentrated or diluted accordingly, and the analysis must be repeated to determine the accurate levels of these metabolites.

References

1. Hill CB, Roessner U (2013) Metabolic profiling of plants by GC-MS. In: Weckwerth W, Kahl G (eds) The handbook of plant metabolomics: metabolite profiling and networking, 1st edn. Wiley-VCH Verlag GmbH & Co, Weinheim
2. Dias DA, Hill CB, Jayasinghe NS et al (2015) Quantitative profiling of polar primary metabolites of two chickpea cultivars with contrasting responses to salinity. J Chromatogr B 1000:1–13
3. Boughton BA, Callahan DL, Silva C et al (2011) Comprehensive profiling and quantitation of amine group containing metabolites. Anal Chem 83:7523–7530
4. Roessner U, Dias D (2013) A Plant tissue extraction for metabolomics. Methods Mol Biol 1055:21–28
5. De Souza DP (2013) Detection of polar metabolites through the use of gas chromatography-mass spectrometry. Methods Mol Biol 1055:29–37
6. Roessner U, Wagner C, Kopka J et al (2000) Simultaneous analysis of metabolites in potato tuber by gas chromatography-mass spectrometry. Plant J 23:131–142
7. Assche RV, Temmerman L, Dias DA et al (2015) Metabolic profiling of transgenic Caenorhabditis elegans Alzheimer model. Metabolomics 11:477–486
8. Hill CB, Taylor JD, Edwards J et al (2015) Detection of QTL for metabolic and agronomic traits in wheat with adjustments for variation at genetic loci that affect plant phenology. Plant Sci 233:143–154

Chapter 16

Acquisition of Volatile Compounds by Gas Chromatography–Mass Spectrometry (GC-MS)

José G. Vallarino, Alexander Erban, Ines Fehrle, Alisdair R. Fernie, Joachim Kopka, and Sonia Osorio

Abstract

Plants synthesize and emit a large range of volatile organic compounds (VOCs) that play important roles in their interactions with the environment, from attracting pollinators and seed dispersers to protectants such as repellants and pathogen inhibitors. As such, the development of techniques for headspace collection of volatiles in combination with gas chromatography–mass spectrometry (GC-MS) has an important impact on our understanding of the biosynthesis of plant VOCs. Furthermore, knowledge of the plant VOCs can be valuable in relation to plant breeding for improving fruit flavor or enhancing resistance to insects or pathogens. This chapter describes a reliable method for extracting volatile compounds by headspace solid-phase microextraction (HS-SPME), and separate and detect them by GC-MS.

Key words Volatile compounds, VOCs, Plant, Gas chromatography, Headspace solid-phase microextraction, HS-SPME, Mass spectrometry

1 Introduction

Metabolism in plant involves the conversion of high molecular weight precursors to smaller compounds that help to obtain viable seeds and to attract seed-dispersing species. Plants synthesize and emit a large variety of volatile organic compounds (VOCs). Physical properties of these compounds allow them to freely cross cellular membranes and be released into the surrounding environment [1]. These volatiles can be emitted from flowers, leaves, fruits, and roots into the atmosphere or soil, allowing the plant to interact with other organisms. Until now, more than 1700 volatiles have been identified from more than 90 plant families, which contain approximately 1% of all plant specialized metabolites currently known [2]. The significance of VOCs in plant physiology and ecology has been extensively investigated only in the past 15–20 years and has revealed roles for VOCs in the mutualistic interaction of plants with other organism, biotic and abiotic stress

Carla António (ed.), *Plant Metabolomics: Methods and Protocols*, Methods in Molecular Biology, vol. 1778, https://doi.org/10.1007/978-1-4939-7819-9_16, © Springer Science+Business Media, LLC, part of Springer Nature 2018

responses [3]. Probably the most well-known are volatiles emitted from floral tissues attracting pollinators [2]. Biosynthesis of VOCs depends on the availability of carbon, nitrogen, and sulfur, as well as energy provided by primary metabolism. According to the chemical structure, plant volatiles can be classed into hydrocarbons, alcohols, aldehydes, ketones, ethers, and esters. Based on their biosynthetic origin, plant volatiles are generally divided into several classes, including terpenoids, phenylpropanoid/benzenoid, fatty acid derivatives, and amino acid derivatives.

The increasing scientific interest in plant VOCs has led to the development of a variety of systems for the collection and analysis of volatiles [4–6]. All methods for the analysis of plant volatiles attempt to identify the authentic profile of volatile blends emitted by a plant. However, the choice of which system to use in a particular experiment for collection and analysis of plant volatiles dependents on the biological problem and plant material being investigated. Nowadays, VOCs analysis demands sensitive and time-efficient techniques or high-throughput profiling.

VOCs are usually identified and quantitated by extracting them from the airspace (headspace) surrounding aboveground plant parts and passing them through a gas chromatograph with a mass spectrometer detector (GC-MS). When conducting volatile profiling, several approaches are possible for sampling the headspace of plant tissue. Among them, solid-phase microextraction (SPME) has been demonstrated to be capable of isolating a high number of VOCs at detection limits in the ppb (parts per billion by volume). In SPME, volatile compounds are adsorbed on a stationary phase that is coated on a fused silica fiber. The fiber is inserted into the headspace above the sample, which has been allowed to equilibrate at 40–60 °C for 15–30 min, and is exposed for 5–30 min (*see* **Note 1**). The compounds adsorbed on the fiber are then thermally desorbed for 5 min in the GC injector port and then sent through the column and detector (*see* **Note 2**).

Numerous studies have used SPME extraction for the analysis of a broad range of volatile compounds in different matrices, such as whole plant, flower, fruit, air, soil and water samples [7–10]. In some applications, it is desirable to collect volatiles from parts of intact growing plants [9] (Fig. 1c) and others analysis requires homogenized samples in which is frequently added concentrated salt solution (NaCl or $CaCl_2$) to decrease the solubility of volatile compounds and force them into the headspace [10] (Fig. 1b).

As shown in Fig. 2, the total ion current (TIC) chromatograms and thereby the metabolic profiles between the different extraction methods are highly variable.

Some metabolites can be better isolated in in vivo tissues (Fig. 3a, b) and others could be only detected in homogenized tissues (Fig. 3c, d). Thus, it is vital to pay much attention to the extraction protocol design which depends on the target metabolite

Fig. 1 Strategies of plant volatile analysis. (**a**) Blank sample, H_2O or saturated $CaCl_2$. (**b**) Homogenized tissue in H_2O or saturated $CaCl_2$. (**c**) Intact plant tissue with 0.2 mL H_2O. (**d**) Blank sample for intact plant tissue

(s) or the biological questions to be answered. Also, a number of extra TIC peaks are often observed when homogenized samples is used, but not all observed peaks correspond to the detection of a real metabolite coming from the original tissue matrix. Indeed, it could be possible to detect environmental contaminations (Fig. 3e, f) or coming from $CaCl_2$ (Fig. 3g, h). Therefore, for each injection sequence it is a prerequisite to add sample blanks (*see* **Note 3**).

Here, we describe a method based on headspace solid-phase microextraction (HS-SPME) coupled to GC-MS to determinate volatile compounds in a variety of different matrixes (e.g., plant species/tissues). We exemplify it using intact in vivo or homogenized Arabidopsis flowers. An abbreviated scheme of sample preparation and analysis is shown in Fig. 4.

2 Materials

2.1 Sampling and Extraction

1. MilliQ water approx. 0.055 µS/cm.
2. Liquid nitrogen supply.
3. Oscillating ball mill MM200 (e.g., Retsch GmbH) or pestle and mortar.
4. Scalpel blades, spatula, tweezers, scissors.
5. Vortex.
6. Balance.
7. 2.0-mL microcentrifuge tubes (e.g., Eppendorf).
8. 0.2-mL glass vial (e.g., Sci-Vi Crimp Top Vial, Thermo Scientific).
9. 20 mL glass vial with screw neck and magnetic screw cap, septum silicone blue/PTFE white (e.g., Gerstel DHS vials).

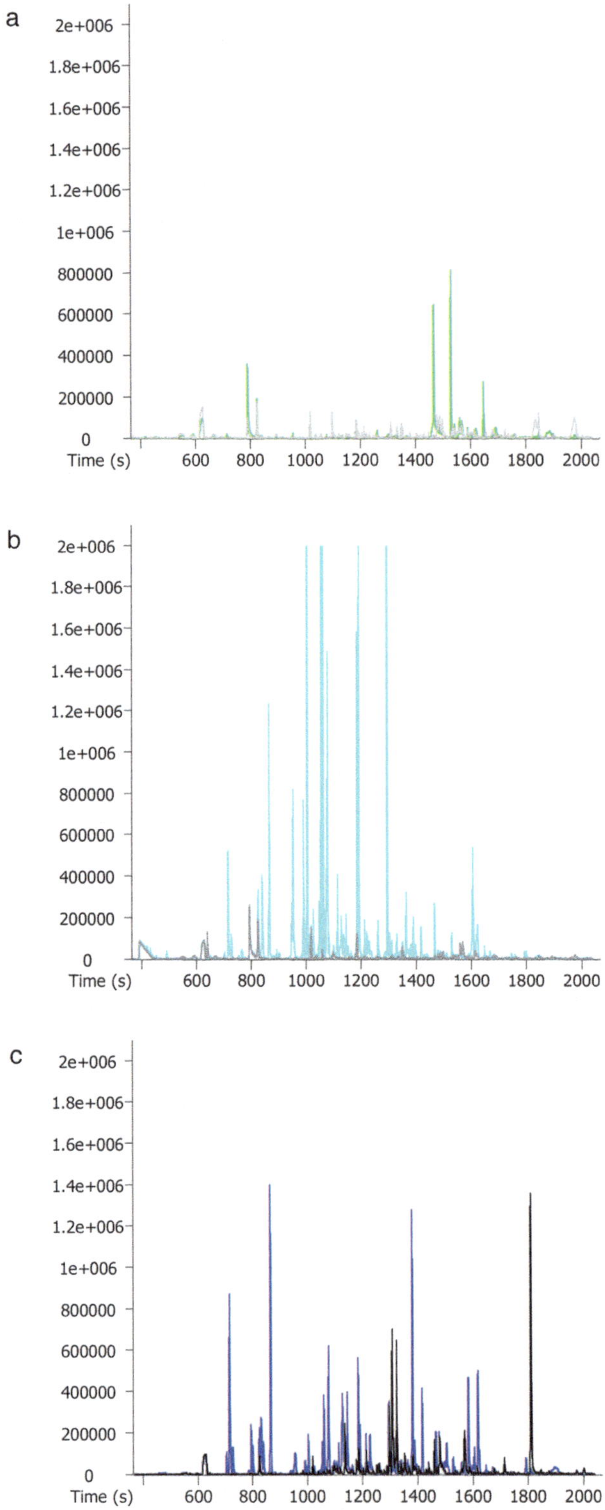

Fig. 2 Total Ion Current (TIC) chromatogram of all samples from Fig. 1. (**a**) TIC chromatogram from samples in Fig. 1c (*green*) and Fig. 1d (*light grey*). (**b**) TIC chromatogram from samples in Fig. 1b (H_2O; *light blue*) and Fig. 1a (H_2O; *light grey*). (**c**) TIC chromatogram from samples in Fig. 1b ($CaCl_2$; *dark blue*) and Fig. 1a ($CaCl_2$; *dark grey*)

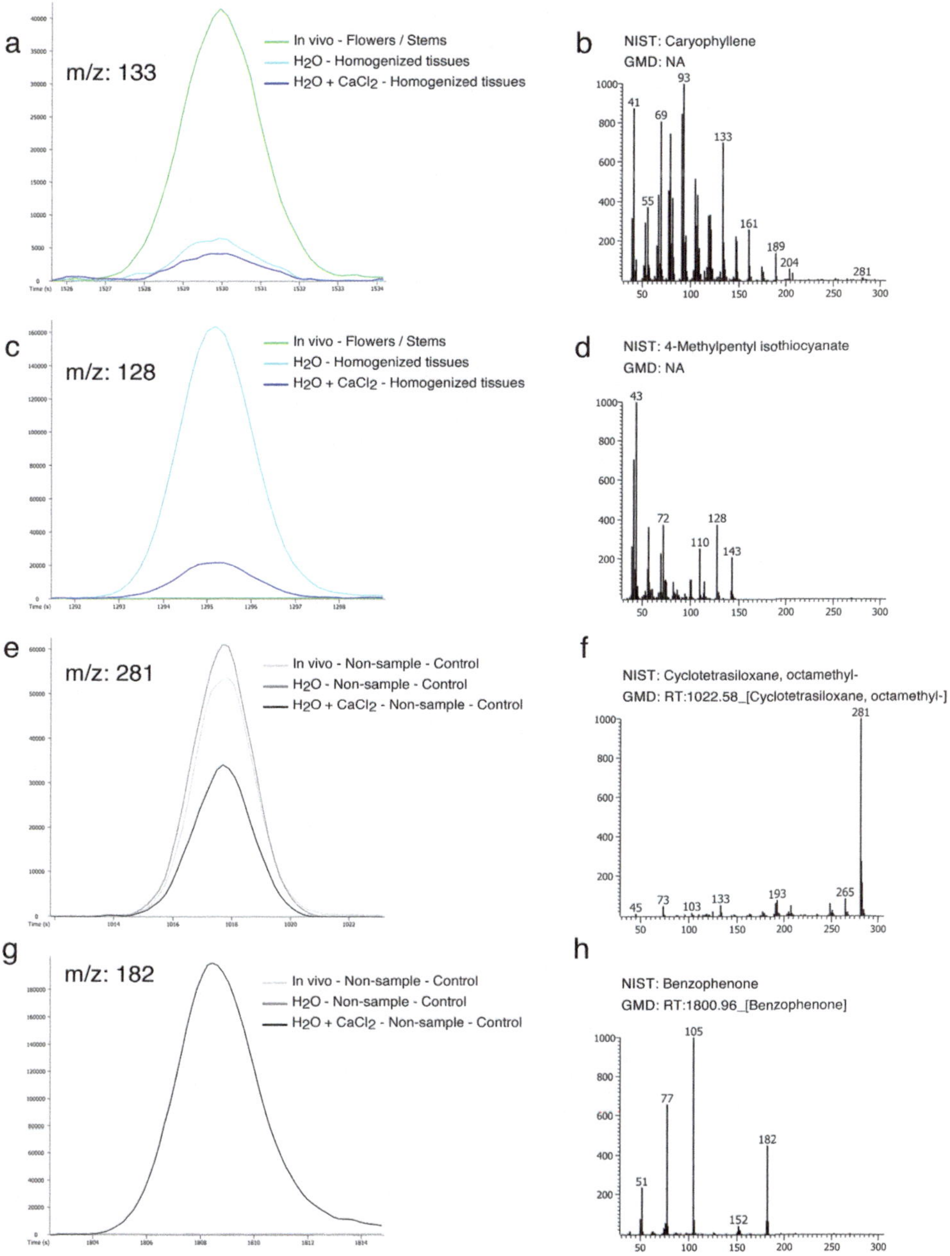

Fig. 3 Tentative annotation of metabolites on the basis of retention time (Table 1) and *m/z* value and MS fragmentation pattern with Golm Metabolome Database (GMD) and/or National Institute of Standards and Technology mass spectral library (NIST). (*Right Panel*; **b, d, f, h**) Mass spectrum used for tentative metabolite annotated. (*Left Panel*; **a, c, e, g**) ion peak from each *m/z* (133, 128, 281, 182, respectively)

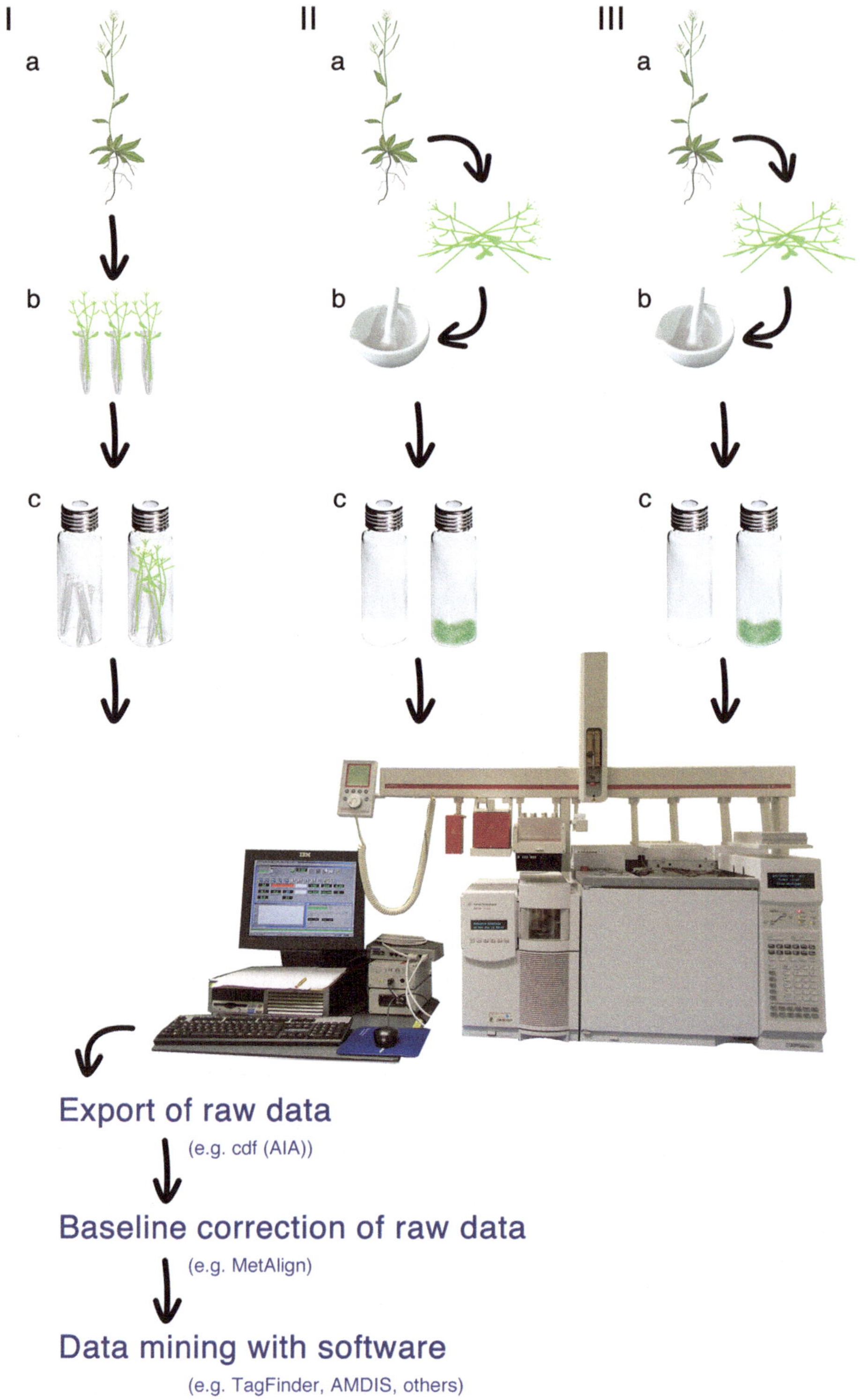

Fig. 4 Scheme of HS-SPME analysis of volatile from intact (**I**) or homogenized (**II**, H_2O and **III**, saturated $CaCl_2$ solution) plant tissue by GC-HS-SPME-MS

10. StableFlex ™ SPME-fiber with 65 μm polydimethylsiloxane/divinylbenzene coating (e.g., Supelco 23Ga, autosampler, Pink) (*see* **Note 4**).
11. Pure standards of all the identified compounds (i.e., Sigma-Aldrich or OldChemin Ltd., Czech Republic).
12. Saturated NaCl or $CaCl_2$.

2.2 GC-HS-SPME-MS

1. Autosampler system, including SPME fiber cleaning and conditioning station and agitator (e.g., Gerstel MPS DHS 2×l Twister).
2. Gas chromatograph with electronic pressure control (Agilent 6890N).
3. Helium carrier gas.
4. Split/Splitless liner (e.g., Agilent liner, splitless, single taper).
5. Capillary column J&W DB-624, 60 m × 0.25 mm × 1.40 μm film thickness.
6. Electron impact ionization mass selective detector (EI/quadrupole MSD, Agilent 5975B).

3 Methods

3.1 Sample Preparation

3.1.1 Volatile Compound Analysis of Intact Plant Tissue (Fig. 4 I)

1. Fill 0.2-mL glass vials with MilliQ water (*see* **Note 5**).
2. Collect plant tissues and quickly put into a 0.2-mL glass vial containing water. Then, put it into a 20 mL vial and close it (Fig. 1c). Consider blank samples (Fig. 1a, d and Fig. 3).
3. Feed the autosampler and keep it at 15 °C.

3.1.2 Volatile Compound Analysis of Homogenized Tissue (Fig. 4 II and III) (See Note 6)

1. Collect the plant material and freeze immediately in liquid nitrogen.
2. Precool the steel cylinders and metal balls to grind the samples in liquid nitrogen. Alternatively, cool the pestle and mortar.
3. Quickly take out two samples and place them into independent steel cylinders together with a metal ball and cover the cylinders. Or use a pestle and mortar to grind the samples.
4. Fix cylinders in the mixer mill and mill at 20 Hz/s for 1 min. Or use a pestle and mortar until a very fine powder is obtained.
5. Quickly take out the cylinders and place back into liquid nitrogen.
6. Transfer the fine powder into a 2–0-mL microcentrifuge precooled tube and keep in liquid nitrogen.
7. Weigh the minimal amount of tissue necessary to obtain quantifiable data by GC-HS-SPME-MS (work quickly to get an exact fresh weight), and keep in liquid nitrogen or store at −80 °C until use.

8. The amount of plant material will vary with plant and tissue type (300 mg of flowers was used to get the data shown in this chapter).
9. Fill each 20 mL vials as follows: (a) Blank and sample vial for homogenized samples in water: 500 μL MilliQ water. (b) Blank and sample vials for homogenized samples in saturated $CaCl_2$: 500 μL of saturated $CaCl_2$.
10. Transfer the frozen material into 20 mL vial, close (Fig. 4 IIc and IIIc) and vortex it. Consider blank samples (Fig. 1a–d and Fig. 3 IIc and IIIc).
11. Feed the autosampler and keep it at 15 °C.

3.2 Data Acquisition

1. Incubate the vials from **step 3** and/or 14 at 50 °C for 10 min with agitation at 250 rpm in autosampler.
2. Insert 24 mm the SPME holder into the vial and keep it for 20 min (extraction time) at 50 °C. Keep the agitation as described in **step 1**.
3. Inject sample in pulsed splitless mode, with the helium carrier gas flow set to 1 mL/min by using the autosampler.
4. Desorb the VOCs by injection of the fiber in the injection port at 250 °C for 1 min.
5. The flow rate is kept constant with electronic pressure control enabled.
6. The injection temperature is set to 250 °C.
7. Injection programs must include fiber cleaning and conditioning steps before and after each injection.
8. Helium flow goes through the fiber at 250 °C for 5 min.
9. Perform chromatography using a 60 m DB-624 capillary column. The temperature program should be isothermal for 2 min at 40 °C, followed by a 10 °C per min ramp to 260 °C, and holding at this temperature for 10 min. Cooling should be as rapid as the instrument specifications allow.
10. Set the transfer line temperature to 250 °C and match ion source conditions.
11. Set the ion source temperature to maximum instrument specifications, 250 °C.
12. The recorded mass range should be m/z 30–300 at 2 scan per s.
13. Proceed the remaining monitored chromatography time with a 360 s solvent delay with filaments turned off.
14. Manual mass defect should be set to 0, filament bias current should be -70 V, and detector voltage should be ~1500–2000 V (*see* **Note 7**).
15. Transfer raw GC-MS profile chromatograms and analyze in comparison to internal standard (Table 1 and Fig. 4) (*see* **Note 8**).

Table 1
List of volatile compounds with retention time based on authentic standard or/and matching to GMD and NIST in our conditions

Name	RT (seconds)	Anotation
Acetate	657.6	Anotated by reference substance
Acetate, hexyl-	1096.74	Anotated by reference substance
Acetate, isopentyl-	943.74	Anotated by reference substance
Acetate, phenethyl-	1380.12	Anotated by reference substance
Acetone	464.04	Anotated by reference substance
Acetophenone	1212.84	Anotated by reference substance
Benzaldehyde	1095.72	Anotated by reference substance
Benzoate	1337.94	Anotated by reference substance
Benzophenone	1800.96	Anotated by reference substance
Benzylalcohol	1184.58	Anotated by reference substance
Benzylnitrile	1309.26	Anotated by reference substance
Butanal, 2-methyl-	680.1	Anotated by reference substance
Butanal, 3-methyl-	670.38	Anotated by reference substance
Butanenitrile, 3-methyl-	807.54	Anotated by reference substance
Butanoate, 3-methyl-, 2-methylpropyl ester-	1084.2	Anotated by reference substance
Butanol	690.9	Anotated by reference substance
Butanol, 1-	699.36	Anotated by reference substance
Butanol, 1-, 2-methyl-	790.56	Anotated by reference substance
Butanol, 3-methyl-	786.06	Anotated by reference substance
Butenal, (E)-2-methyl-2-	801.9	Anotated by reference substance
Butyrate	1079.1	Anotated by reference substance
Butyrate, 2-methyl-, ethyl-	907.68	Anotated by reference substance
Butyrate, ethyl-	845.88	Anotated by reference substance
Butyrolactone, gamma	1105.14	Anotated by reference substance
Damascenone, trans-beta	1499.16	Anotated by reference substance
Decadienal, (E,E)-2.4-	1436.4	Anotated by reference substance
Decanal	1308.42	Anotated by reference substance
Decanoate	1473.06	Anotated by reference substance
Decanoate, ethyl-	1456.02	Anotated by reference substance
Dodecanoate	1653.9	Anotated by reference substance
Dodecanoate, ethyl	1635.48	Anotated by reference substance

(continued)

Table 1 (continued)

Name	RT (seconds)	Anotation
-Ethanol	425.4	Anotated by reference substance
Ethanol, 2-phenyl-	1266.18	Anotated by reference substance
Eugenol	1480.14	Anotated by reference substance
Formate	611.52	Anotated by reference substance
Furan, 2-ethyl-	704.4	Anotated by reference substance
Geranial	1392.18	Anotated by reference substance
Geranylacetone	1532.58	Anotated by reference substance
Guaiacol	1232.76	Anotated by reference substance
Hepten, 5-, 2-one-, 6-methyl-	1088.7	Anotated by reference substance
Heptenal, (E)-2-	1071.84	Anotated by reference substance
Hexadien-1-al, (E,E)-2,4-	1028.16	Anotated by reference substance
Hoxanal	862.98	Anotated by reference substance
Hexanoate	1095.72	Anotated by reference substance
Hexanoate, ethyl-	1079.1	Anotated by reference substance
Hexanoate, isoamyl-	1333.86	Anotated by reference substance
Hexanoate, methyl ester-	999.24	Anotated by reference substance
Hexanoate, 2-ethyl-	1242.24	Anotated by reference substance
Hexanol, 1 -	959.04	Anotated by reference substance
Hexen-1-ol, (Z)-3-	944.7	Anotated by reference substance
Hexenal, (E)-2-	949.32	Anotated by reference substance
Ionone, beta	1602.18	Anotated by reference substance
Isobutylthiazole, 2-	1143.18	Anotated by reference substance
Isobutyrate	847.08	Anotated by reference substance
Isovalerate, ethyl-	912.18	Anotated by reference substance
Lactate, ethyl-	896.7	Anotated by reference substance
Linalool	1203.9	Anotated by reference substance
Mesitaldehyde	1436.52	Anotated by reference substance
Nonanal	1210.26	Anotated by reference substance
Nonanoate	1389.72	Anotated by reference substance
Nonanoate, ethyl-	1371.3	Anotated by reference substance
Octanal	1105.02	Anotated by reference substance

(continued)

Table 1
(continued)

Name	RT (seconds)	Anotation
Octanoate	1293.66	Anotated by reference substance
Octanoate, ethyl-	1280.7	Anotated by reference substance
Octanoate, isoamyl-	1503.06	Anotated by reference substance
Octanoate, pentyl-	1538.58	Anotated by reference substance
Octanol, 1-	1181.28	Anotated by reference substance
Octen. 1-ol, (Z)-3-	1167.24	Anotated by reference substance
Octenal, (E)-2-	1181.94	Anotated by reference substance
Pentanal	727.26	Anotated by reference substance
Pentanoate, pentyl ester-	1270.86	Anotated by reference substance
Pentanol, 1-	826.02	Anotated by reference substance
Penten, 3-ol, 1-	715.26	Anotated by reference substance
Penten, 3-one, 1-	713.16	Anotated by reference substance
Pentenal, (E)-2-	822.54	Anotated by reference substance
Pentylfuran, 2-	1065.96	Anotated by reference substance
Phenylacetaldehyde	1187.34	Anotated by reference substance
Phenylethane, 1-nitro-2-	1466.22	Anotated by reference substance
Pinene, alpha	1008.78	Anotated by reference substance
Propanal, 2-methyl-	536.64	Anotated by reference substance
Propanol, 1-, 2-methyl-	642.84	Anotated by reference substance
Salicylate, ethyl-	1397.7	Anotated by reference substance
Salicylate, methyl-	1331.88	Anotated by reference substance
Succinate, diethyl-	1286.76	Anotated by reference substance
Terpineol	1319.58	Anotated by reference substance
Tolualdehyde, p-	1231.92	Anotated by reference substance
Vanillin	1585.68	Anotated by reference substance
[.alpha.- Terpinolen]	1184.76	Anotated by matching to NIST
[.alpha.-Longipinene]	1465.44	Anotated by matching to NIST
[.alpha.-Phellandrene]	1096.92	Anotated by matching to NIST
[.alpha.-Terpinen]	1110.24	Anotated by matching to NIST
[.alpha.-Thujene]	1095.42	Anotated by matching to NIST
[.beta.-Myrcene]	1062.42	Anotated by matching to NIST

(continued)

Table 1
(continued)

Name	RT (seconds)	Anotation
[.beta.-Phellandrene]	1193.58	Anotated by matching to NIST
[.delta. Nonalactone]	1571.34	Anotated by matching to NIST
[.gamma.-Elemene]	1431.72	Anotated by matching to NIST
[1,2-Bis(trimethylsilyl)benzene]	1622.16	Anotated by matching to NIST
[1,2-Dihydrolinalool]	1247.34	Anotated by matching to NIST
[1-Monolinoleoylglycerol trimethylsilyl ether]	1494.24	Anotated by matching to NIST
[2(3H)-Furanone, dihydro-5-pentyl-9]	1535.82	Anotated by matching to NIST
[2-Octanone]	1100.64	Anotated by matching to NIST
[2-Propanone, 1-(4-methoxyphenyl)-]	1533.12	Anotated by matching to NIST
[3-Hydroxymandelic acid, ethyl ester, di-TMS]	1190.1	Anotated by matching to NIST
[Benzaldehyde, 4-methoxy-]	1425.06	Anotated by matching to NIST
[Benzene, 1,1′-[oxybis(methylene)]bis-]	1798.44	Anotated by matching to NIST
[Benzene, 1,3-dichloro-2-methyl-]	1258.74	Anotated by matching to NIST
[Benzene, 1,4-dichloro-]	1129.86	Anotated by matching to NIST
[Benzene, 1-ethyl-2-(phenylmethyl)-]	1731.78	Anotated by matching to NIST
[Benzene, 1-methoxy-4-(1-propenyl)-]	1373.46	Anotated by matching to NIST
[Benzene, 1-methoxy-4-(1-propenyl)-]	1407	Anotated by matching to NIST
[Benzene, 1-methyl-3-(phenylmethyl)-]	1654.86	Anotated by matching to NIST
[Benzene, 1-methyl-4-(1-methylethenyl)-]	1200.66	Anotated by matching to NIST
[Benzene, 2,4-dichloro-1-methyl-]	1255.02	Anotated by matching to NIST
[Benzene, 2,4-dimethyl-1-(phenylmethyl)-]	1759.8	Anotated by matching to NIST
[Benzene, 2,6-dimethyl-1-(phenylmethyl)-]	1738.62	Anotated by matching to NIST
[Benzene, 2,6-dimethyl-1-(phenylmethyl)-]	1745.28	Anotated by matching to NIST
[Benzyl Benzoate]	1983.96	Anotated by matching to NIST
[Benzyl chloride]	1148.7	Anotated by matching to NIST
[Bibenzyl]	1651.14	Anotated by matching to NIST
[Camphor]	1310.88	Anotated by matching to NIST
[Caryophyllene]	1436.04	Anotated by matching to NIST
[Caryophyllene]	1481.1	Anotated by matching to NIST
[Caryophyllene]	1515.24	Anotated by matching to NIST
[Copaene]	1441.68	Anotated by matching to NIST

(continued)

Table 1 (continued)

Name	RT (seconds)	Anotation
[Copaene]	1471.14	Anotated by matching to NIST
[Cyclohexanemethanol, .alpha.,.alpha.,4-trimethyl-]	1283.64	Anotated by matching to NIST
[Cyclotetrasiloxane, octamethyl-]	1022.58	Anotated by matching to NIST
[Cyclotrisiloxane, hexamethyl-]	828.06	Anotated by matching to NIST
[Diisopropyl ether]	560.94	Anotated by matching to NIST
[Diphenylmethane]	1554.3	Anotated by matching to NIST
[Estragole]	1319.1	Anotated by matching to NIST
[Ethyl Acetate]	603.72	Anotated by matching to NIST
[Eucalyptol]	1140.66	Anotated by matching to NIST
[Formic acid, 1-methylethyl ester]	556.02	Anotated by matching to NIST
[Heptanal]	996.48	Anotated by matching to NIST
[Isoborneol]	1326.6	Anotated by matching to NIST
[Limonene]	1122.6	Anotated by matching to NIST
[Longicyclene]	1495.02	Anotated by matching to NIST
[Longifolene-(V4)]	1461.3	Anotated by matching to NIST
[n-Decanoic acid]	1471.74	Anotated by matching to NIST
[Nonanoic acid, methyl ester]	1311.06	Anotated by matching to NIST
[Ocimene]	1126.32	Anotated by matching to NIST
[Octanoic acid, 2-methyl-]	1347.78	Anotated by matching to NIST
[Octanoic acid, methyl ester]	1215.36	Anotated by matching to NIST
[o-Cymene]	1126.32	Anotated by matching to NIST
[Propanoic acid, 2-methyl-, 1-methylethyl ester]	837.84	Anotated by matching to NIST
[Silicic acid, diethyl bis(trimethylsilyl) ester]	1692.72	Anotated by matching to NIST
[Terpinene, .alpha.]	1149.9	Anotated by matching to NIST
[trans-.beta.-Ocimene]	1109.28	Anotated by matching to NIST
[Undecane]	1161.66	Anotated by matching to NIST

4 Notes

1. Solid-phase microextraction (SPME) fibers can be reused approximately 100 times.
2. SPME is an equilibrium extraction and desorption process, so precise control of sampling conditions is important. Recoveries

may be enhanced by raising the temperature and time, but compounds may be displaced from the absorption sites and compounds that were not originally present in the sample may be formed.

3. To avoid carryover, a sample blank between 10 samples is recommended.
4. There are different types of fiber adsorbents available. By carefully selecting the polarity and thickness of the fiber coating, compounds of different polarity and volatility ranging from high-boiling or semivolatile to volatile compounds can be sampled.
5. In order to ensure the quality of the used water for the extraction, it is recommended to run a blank sample only with H_2O before mixing it with sample.
6. Once the sample has been frozen in liquid nitrogen it must not thaw out even slightly before analysis. Make sure therefore to keep samples at constant freezing temperature (in liquid nitrogen) to avoid degradation of VOCs and precool all components to be used (spatulas, vials, etc.) in liquid nitrogen before they come into contact with the sample.
7. In contrast to LECO's MS, the detector voltage is automatically adjusted in Agilent's MS.
8. For reporting chemical identification, authors should consider references [11–15].

Acknowledgments

We thank the Spanish Ministerio de Ciencia e Innovación, Ministerio de Economía y Competitividad-FEDER, University of Málaga through the Grants AGL2012-40066-CO2-02 and Ramón y Cajal program (RYC-09170). The authors also thank the European Union's Horizon 2020 research and innovation programme (grant agreement number 679303) and the Max Planck Society for financial support.

References

1. Pichersky E, Noel JP, Dudareva N (2006) Biosynthesis of plant volatiles: Nature's diversity and ingenuity. Science 331:808–811
2. Pichersky E, Gershenzon J (2002) The formation and function of plant volatiles: Perfumes for pollinator attraction and defense. Curr Opin Plant Biol 5:237–243
3. Dudareva N, Pichersky E, Gershenzon J (2004) Biochemistry of plant volatiles. Plant Physiol 135:1893–1902
4. Linskens HF, Jackson JF (1997) Modern methods of plant analysis. plant volatile analysis. Springer, New York
5. Millar JG, Sims JJ (1998) Preparation, cleanup, and preliminary fractionation of extracts. In: Millar JG, Haynes KF (eds) Methods in chemical ecology. Kluwer Academic Publishers, Boston, pp 1–37
6. Rambla JL, López-Gresa MP, Bellés JM, Granell A (2015) Metabolomic profiling of

plant tissue. In: Alonso JM, Stepanova AN (eds) Plant functional genomics, Methods in molecular biology. Springer, New York, pp 221–235

7. Flamini G, Cioni PL, Morelli I (2002) Differences in the fragrances of pollen and different floral parts of male and female flowers of Laurus nobilis. J Agric Chem 50:4647–4652
8. Holzinger R, Sandoval-Soto L, Rottenberger S et al (2000) Emissions of volatile organic compounds from Quercus ilex L. measured by proton-transfer-reaction mass spectrometry under different environmental conditions. J Geo- phys Res 105:20573–20579
9. Rohloff J, Bones AM (2005) Volatile profiling of Arabidopsis thaliana-Putative olfactory compounds in plant communications. Phytochemistry 66:1941–1955
10. Zorrilla-Fontanesi Y, Rambla JL, Cabeza A et al (2012) Genetic analysis of strawberry fruit aroma and identification of O-methyltransferase FaOMT as the locus controlling natural variation in mesifurano content. Plant Physiol 159:851–870
11. Sumner LW, Amberg A, Barrett D et al (2007) Proposed minimum reporting standards for chemical analysis Chemical Analysis Working Group (CAWG) Metabolomics Standards Initiative (MSI). Metabolomics 3:211–221
12. Fernie AR, Aharoni A, Willmitzer L et al (2011) Recommendations for reporting metabolite data. Plant Cell 23:2477–2482
13. Dunn WB, Erban A, Weber RJM et al (2013) Mass appeal: metabolite identification in mass spectrometry-focused untargeted metabolomics. Metabolomics 9:S44–S66
14. Rochat B (2017) Proposed confidence scale and ID score in the identification of known-unknown compounds using high resolution MS data. J Am Mass Spectrom 28:709–723
15. Strehmel N, Hummel J, Erban A et al (2008) Retention index thresholds for compound matching in GC-MS metabolite profiling. J Chromatogr B 871:182–190

Chapter 17

Mass Spectrometry Imaging (MSI) for Plant Metabolomics

Berin A. Boughton and Dinaiz Thinagaran

Abstract

Mass spectrometry imaging (MSI) is a developing technique to measure the spatiotemporal distribution of many biomolecules in tissues. Over the preceding decade MSI has been adopted by plant biologists and applied in a broad range of areas including: primary metabolism, natural products, plant defense, plant responses to abiotic and biotic stress, plant lipids, and the developing field of spatial metabolomics. This methods chapter covers preparation of plant tissues for matrix-assisted laser desorption ionization (MALDI)-MSI, including sample embedding and freezing, sectioning, mounting, and matrix deposition using both sublimation and spray deposition prior to MSI analysis.

Key words Mass spectrometry imaging, MSI analysis, Matrix-assisted laser desorption ionization, MALDI, Matrix, Frozen section

1 Introduction

Spatial analysis of plant tissues, including determination of the locations of production, storage and site/s of action of plant natural products, provides fundamental and unique insights into plant biology [1]. Mass spectrometry imaging (MSI) is gaining in popularity and is emerging as one of the leading technologies and has been adopted for the investigation of responses to abiotic and biotic stresses, plant defense mechanisms, beneficial symbiotic relationships, and fundamental ecologically important processes including nitrogen fixation and nutrient cycling [2]. The most popular MSI method is matrix-assisted laser desorption ionization (MALDI), a soft ionization method that is commonly used for direct measurement of molecular ions. MALDI-MSI is dependent upon a matrix layer deposited onto a specimen to facilitate desorption of analytes from solid to gas phase and to promote ionization within the laser ablation plume [3]. Matrix deposition could be performed by sublimation, spraying or spotting [4, 5]. Sublimation is a dry matrix deposition strategy used to introduce a uniform coating of fine matrix crystals over the surface of the specimen,

Carla António (ed.), *Plant Metabolomics: Methods and Protocols*, Methods in Molecular Biology, vol. 1778,
https://doi.org/10.1007/978-1-4939-7819-9_17,

whereas a wet deposition approach is performed using a robot spray to extract analytes from the surface of a specimen which cocrystallize with the sprayed matrix. MALDI-MSI is a popular option due to the availability of commercial instruments and well-developed protocols for mammalian tissues [6–9]. In comparison to mammalian tissues, plant tissues can be more difficult to prepare and require some special considerations for freezing, sectioning and mounting. Prior to analysis, tissues must be collected and stored [10], the steps taken during both tissue collection and storage are critical for successful MSI analysis and often vary depending upon the analyte of interest [1, 11]. Most experiments will have a distinct timing miss-match between collection of sample and analysis, requiring the storage of samples for a period of time prior to analysis and this is usually conducted by freezing the tissues. Typically, to protect delicate tissues, structures and small molecule metabolite distributions, a gentle freezing approach is recommended instead of immersion in liquid nitrogen, methods include freezing over liquid nitrogen in the nitrogen atmosphere or in cold carbon dioxide atmosphere over dry ice, dipped into isopentane–-liquid nitrogen or isopentane–dry ice slurries. Care must be taken to retain the tissue morphology during the freezing process to preserve an accurate representation of the native tissue; soft tissues may deform and take the shape of the container (tube or tray) within which they are frozen. Once frozen, tissues and analytes are generally stable for months to years when stored at −80 °C. In this chapter, we describe a protocol for plant specimen embedding, sectioning and mounting based upon a modified Kawamoto tape transfer method [12–15], followed by standard dry and wet matrix deposition approaches for MALDI-MSI.

2 Materials

All matrix solutions should be made using LCMS (or higher) grade quality solvents. All materials are prepared and stored at room temperature, unless otherwise specified.

2.1 Freezing and Embedding Plant Tissues

1. Stainless steel embedding container (Section-Lab Co. Ltd., Hiroshima, Japan).
2. Embedding solutions:
   (a) Carboxy-methyl cellulose (CMC): 1–5% solution. Add 10–50 g CMC to deionized water. Use rapid stirring and heating to make higher concentration solutions. Addition of CMC powder to a rapid water vortex aids dissolution. Store at room temperature and use within 1 month.
   (b) Super Cryo Embedding Media (SCEM) (Section-Lab Co. Ltd., Hiroshima, Japan).

3. Liquid nitrogen.
4. Isopropanol–dry ice slurry. Add isopropanol slowly to 0.2–0.5 kg of dry ice with stirring until a slurry is formed.

2.2 Sectioning and Mounting Plant Tissue Samples

1. Leica CM1860 cryotome (Leica Biosystems, Mount Waverly, VIC, Australia). Set cryochamber temperature to −20 °C.
2. Feather C35 (Carbon Steel) or N35 (Stainless Steel) disposable razor blades.
3. Optimal Cutting Temperature (O.C.T.) embedding compound (Sakura Tissue-Tek or Scigen Tissue-Plus).
4. Single-sided high-quality carbon steel razor blades.
5. Stainless steel forceps.
6. Kimwipes.
7. 80% ethanol solution.
8. Type 2C(9) cryofilm (Section-Lab Co. Ltd., Hiroshima, Japan).
9. Deer skin cryofilm application tool (Section-Lab Co. Ltd., Hiroshima, Japan).
10. Double-sided conductive carbon tape for SEM: 12 mm or 25 mm width.
11. Glass slides: 25 mm × 75 mm. Menzel Superfrost™ Plus or Superfrost™ Polylysine™ for MALDI sources that do not require conductive glass slides, or Bruker Indium-Tin Oxide (ITO) coated slides for sources that require conductive glass slides such as MALDI-TOF.

2.3 Matrix Deposition via Sublimation

1. Sublimation apparatus. It comprises an Edwards high vacuum pump, bench top hot plate, sand bath, vacuum gauge monitor, glass cold finger, and glass sublimation apparatus.
2. Pure MALDI matrix (powder/crystal): 2,5-dihydroxybenzoic acid (DHB) (positive ion mode) or 9-aminoacridine (9-AA) (negative ion mode).
3. Ice–water slurry.

2.4 Matrix Deposition via TM Sprayer

1. Matrix spray device. This matrix deposition protocol is based on the HTX TM Sprayer (HTX Technologies, LLC, Chapel Hill, NC, USA) robotic spray unit comprised of a Shimadzu LC-20 AD Quaternary Liquid Chromatography (LC) Pump (Shimadzu Scientific Instruments, Ermington, NSW, Australia), HTX TM-Sprayer and an attached laptop with installed operating software.
2. Dissolved matrix solution:
   (a) 2,5-Dihydroxybenzoic acid (DHB) matrix solution: 40 mg/mL. Dissolve 0.4 g DHB in 10 mL 100% acetone (positive ion mode). Use immediately after preparation.

(b) 9-aminoacridine (9-AA) matrix solution: 20 mg/mL. Dissolve 0.2 g 9-AA in 10 mL 100% acetone (negative ion mode) (*see* **Note 1**). Use immediately after preparation.

3. Acid wash solution: 0.5% acetic acid in acetonitrile. Add 5 mL acetic acid to 1 L acetonitrile.
4. Basic wash solution: 0.5% ammonium hydroxide in acetonitrile. Add 5 mL ammonium hydroxide to 1 L acetonitrile.

3 Methods

Carry out all protocols at room temperature, unless otherwise specified.

3.1 Freezing and Embedding Fresh Plant Tissues

The embedding procedure below is based upon fresh tissues. Prefrozen tissues may also be used (*see* **Note 2**).

1. Prefill the bottom of the stainless-steel embedding container with a thin layer of 1-5% CMC.
2. Place the sample in the embedding container and fill with more 1-5% CMC until the sample is completely submerged within the embedding medium (Fig. 1a) (*see* **Note 2**).
3. Place the embedding container containing the sample and 1–5% CMC into the isopropanol–dry ice slurry until the CMC is completely frozen (Fig. 1b) (*see* **Note 3**).
4. Remove the frozen CMC block from the embedding container carefully without fracturing the block and store at −80 °C (Fig. 1c) (*see* **Note 4**).

3.2 Sectioning and Mounting of Tissues

All procedures for tissue sectioning and mounting must be performed in a cryochamber at −20 °C (*see* **Note 5**). Always clean the surface of the entire inner and outer surfaces of the cryotome and all tools in the cryochamber with 80% ethanol solution before and after use. Follow all waste disposal regulations diligently for all wastes produced. Be careful when handling knife blades and always remove knife blade from the holder after every sectioning event for safety reasons. All tools should be precooled to −20 °C in the cryochamber prior to use.

1. Place the CMC block in the cryochamber using a pair forceps to allow equilibration in temperature for 15–30 min.
2. Carefully trim the CMC block to depth with a precooled flat razor blade if the block is too large (*see* **Note 6**).
3. Mount the specimen disc then spread a minimal amount of OCT onto the surface of the specimen disc then quickly adhere the CMC block onto the specimen disc with a pair of forceps (Fig. 2a) (*see* **Note 7**).

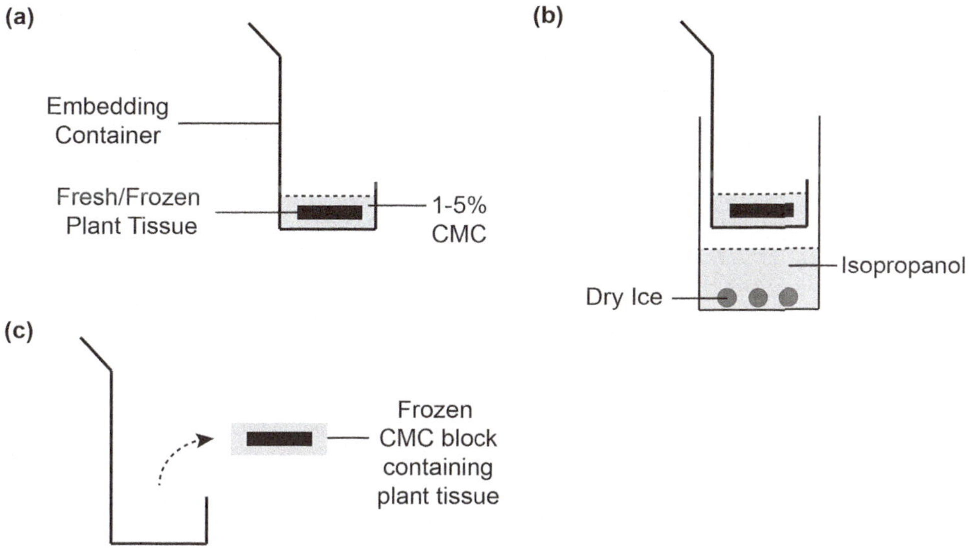

Fig. 1 Freezing and embedding of frozen plant tissue. (**a**) Fresh or frozen plant tissue is placed in an embedding container filled with 1–5% CMC. (**b**) Embedding container is placed in an isopropanol–dry ice slurry until 1–5% CMC is completely frozen. (**c**) Frozen CMC block is then carefully removed from the embedding container for sectioning

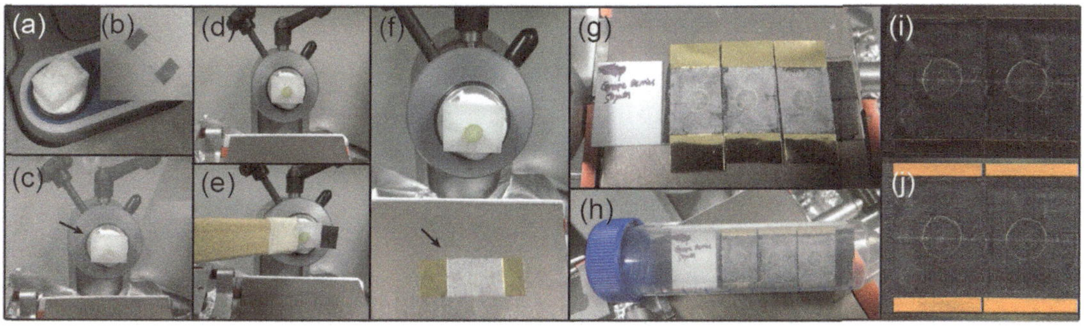

Fig. 2 Preparation of plant tissue sections and mounting onto glass slides. (**a**) A CMC block is mounted onto the specimen disc with a minimal amount of OCT. (**b**) A cryofilm is pretrimmed into the size of the CMC sectioning surface. (**c**) The shaft of the specimen disc is placed into the location hole of the orientable specimen head in the cryochamber (*black arrow*) at a desired sectioning angle. (**d**) The CMC block is sectioned through to a point where a tissue section is desired. (**e**) A pretrimmed cryofilm is gently adhered onto the exposed cut surface of the CMC block and pressed tightly using the deer skin application tool. (**f**) A section is obtained from the CMC block (*black arrow*) with the sectioned side facing upward. (**g**) The collected film section is adhered onto a precooled glass slide with double-sided conductive carbon tape. (**h**) Glass slide with the adhered film sections is then transferred into a chilled empty 50 mL tube, loosely capped and freeze dried. (**i**) Freeze-dried plant tissue specimen prior to matrix deposition. (**j**) Sectioned plant specimen deposited with DHB matrix

4. Place the shaft of the specimen disc into the location hole of the orientable specimen head in the cryochamber, then orient the specimen surface with the lever to a desired sectioning angle. Tighten the screw to prevent any sample movement during sectioning (Fig. 2c) (*see* **Note 8**).
5. Set the desired sectioning thickness on the cryotome control panel (*see* **Note 9**).
6. Remove a new knife blade from the knife magazine and carefully clean the knife blade with 80% ethanol. Wipe the knife blade dry with Kimwipes before placing it into the knife holder (*see* **Note 10**). Lock the knife blade into place by gently closing the blade locking clamping lever.
7. Unlock and gently rotate the handwheel toward the rear of the cryotome to start sectioning (*see* **Note 11**). Section through the CMC and tissue sample block to a point where a tissue section is desired (Fig. 2d).
8. Clean the sides of the sample block with a brush to get rid of any unwanted material.
9. Gently adhere a piece of pretrimmed cryofilm (type 2C(9)) (Fig. 2b) to the exposed cut surface of the block and press it tightly with a deer skin application tool (Fig. 2e).
10. Obtain a section by rotating the handwheel. Keep the section side upward (Fig. 2f; black arrow) (*see* **Note 12**).
11. Place the sectioned-side of the collected film section upward using a pair of forceps and adhere it onto a precooled glass slide with double-sided conductive carbon tape (Figs. 2g and 3) (*see* **Note 13**).
12. The glass slide with the frozen section is transferred into a chilled empty 50 mL tube (Fig. 2h), then freeze dried overnight at −55 °C at 1 mBar (*see* **Note 14**).

3.3 Matrix Deposition

Matrix deposition could be performed by either using a wet or a dry deposition approach.

3.3.1 Dry Matrix Deposition via Sublimation

It is recommended to assemble and operate the sublimation apparatus within a fume hood or vent exhaust gases into the fume hood when using toxic matrices. Always inspect the sublimation

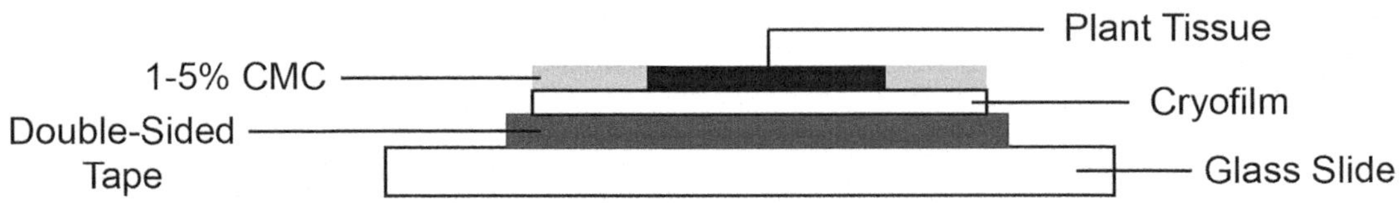

Fig. 3 Side-view of a glass slide after mounting of a sectioned plant tissue sample. A collected film-section is adhered onto a precooled glass slide with double-sided conductive carbon tape, with the section-side facing upward

apparatus for damage before each use. Do not use the sublimation apparatus if the glass sublimation jar or cold finger is cracked or damaged.

1. Turn on the hot plate and allow the sand bath temperature to equilibrate to about 130–140 °C.
2. Preweigh the sample slide prior to deposition (*see* **Note 15**).
3. Mount slide to the bottom side of the cold finger using laboratory tape ensuring that the slide is evenly attached (*see* **Note 16**).
4. Place approximately 0.2–0.3 g of matrix powder into the bottom of the sublimation jar and spread evenly.
5. Assemble sublimation apparatus and add ice–water slurry to the cold finger to prechill the sample slide (Fig. 4) (*see* **Note 17**).
6. Arrange the sublimation apparatus on the heated sand bath and preheat the base for 0.5–1 min to allow even temperature (*see* **Note 18**).
7. Connect the vacuum tubing to the sublimation apparatus and turn on the vacuum pump. Allow matrix deposition for 2–5 min or until a satisfactory amount of fine matrix coating is applied onto the specimen glass slide (*see* **Note 19**).
8. Once complete, gently release the vacuum, disconnect the apparatus from the vacuum pump tubing and remove it from the heat source. It is essential to carefully release the vacuum within the sublimation jar to ensure that air-currents do not stir matrix from the base of the apparatus.
9. Remove specimen slide from the sublimation apparatus and weigh to determine amount of matrix deposited.

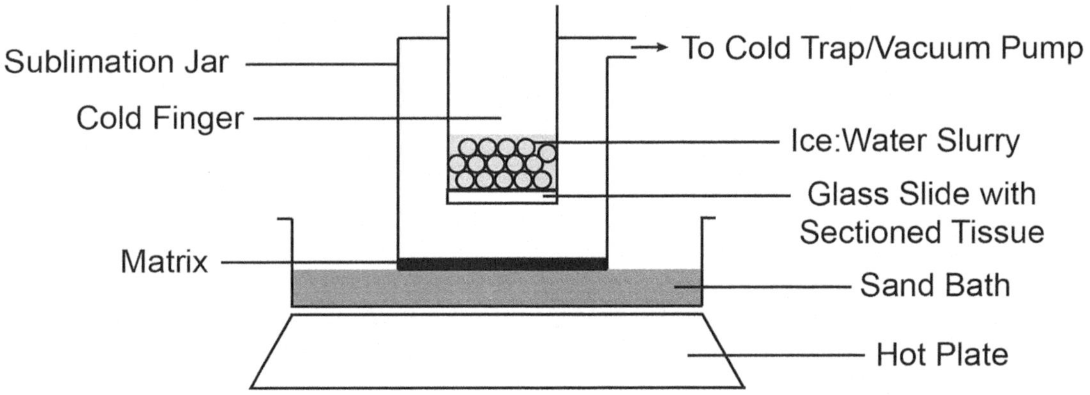

Fig. 4 Assembly of a sublimation apparatus. A sublimation jar is placed on a hot sand bath heated on a hot plate. The sublimation jar contains a glass slide with sectioned plant tissue adhered on the bottom side of the cold finger while matrix is placed into the bottom of the sublimation jar. Ice–water slurry is added into the cold finger of the jar

3.3.2 Wet Deposition via Matrix Spray

The following spray deposition protocol is based on the TM Sprayer unit.

1. Power on the TM Sprayer.
2. Ensure gas supply is connected, turned on, and maintain pressure on the TM Sprayer to between 10 and 11 psi.
3. Turn on the Shimadzu LC-20 AD Pump and ensure adequate solvent flow (0.1–0.2 mL/min) (*see* **Note 20**).
4. On the attached PC, load a saved method or set up a new spray method on the TM Sprayer software (*see* **Note 21**).
5. Set flow path to "Load" before slowly charging 6–7 mL of dissolved matrix solution into the loop via a Luer-lock syringe (*see* **Note 22**).
6. Select matching solvent(s) on the LC pump.
7. Remove the front glass cover and place a slide of interest into the TM Sprayer. Secure the slide in an appropriate position with tape. Replace the front glass cover (*see* **Note 23**).
8. Set flow path to "Spray" on the TM Sprayer and wait approximately 1 min to allow matrix to flow through (*see* **Note 24**).
9. Click on "start" on the TM Sprayer operating software to begin matrix deposition. Once it is completed, switch the flow path back to "Load." Wait for several minutes before opening the front glass cover to allow any aerosol toxic matrix to be drained out of the system. Remove glass slide and store it in a vacuum desiccator until measurement.
10. Flush matrix loop with wash solution. If an acidic matrix (DHB) is used, flush the matrix loop with the following in a sequential manner: 6 mL basic wash solution, 6 mL acetone, 6 mL acidic wash solution, and 6 mL acetone. If a basic matrix is used, start out with 6 mL acidic wash solution, followed by 6 mL acetone, 6 mL basic wash solution, and 6 mL acetone (*see* **Note 25**).
11. Flush nozzle for several minutes by running through solvents at 1–2 mL/min for 5–10 min to ensure nozzle is not blocked. Clean nozzle head with 80% ethanol in a spray bottle. Clean the spray chamber of residual matrix and replace the aluminum foil over the spray area.
12. Turn down temperature to 30 °C if a higher temperature was used for matrix deposition. Turn off LC pump and gas at regulator and power down TM Sprayer when not in use (*see* **Note 26**).

4 Notes

1. It is best to use between 5 and 40 mg/mL of matrix dissolved in an appropriate solvent at room temperature such as water, ethanol, methanol, isopropanol, acetone and acetonitrile optimized to extract analytes of interest. A good alternative to acetone are water/methanol mixtures. Take care when heating solvents to dissolve matrix. If a saturated solution is loaded into the instrument, there are high possibilities for crystallization within the lines. This will lead to blockages in the lines of the instrument or blockage of the spray nozzle. There are multiple matrices available that may be used for different purposes. The matrices listed here, DHB and 9AA, are two common matrices reported in the literature for positive and negative ionisation mode analysis respectively.
2. Fresh or frozen plant tissue samples can be subjected to embedding. If using prefrozen tissues, it is recommended to gently freeze the tissues in cold atmosphere above dry-ice or liquid nitrogen. Always ensure that the embedding container is large enough to fit the entire plant tissue sample.
3. Alternatively, the sample may be held above liquid nitrogen in cold atmosphere until the sample has completely frozen. Dipping the container in liquid nitrogen rapidly freezes the sample risking fractures to the CMC block.
4. Fractures in the CMC block may damage the embedded plant sample and it would be harder to produce sections with a consistent thickness.
5. Always ensure cryotome is at desired temperature before use and that the handwheel is locked whenever the cryotome is not in use or changing between samples.
6. Care must be taken to not remove any of the sample tissue during trimming.
7. Use the minimal amount of OCT required to adhere the block ensuring no OCT will be cut during sectioning [5].
8. Ensure that the shaft of the specimen disc is completely inserted. Make sure to lock the handle of the handwheel to prevent it from movement of the specimen head during mounting of specimen disc. If the blade is in place, use the knife guard to cover the knife edge to prevent any cutting injury.
9. Due to the nature of frozen plant tissues which tend to fracture and leave gaps during sectioning, sample thickness of 25–50 μm is recommended.

10. Failure to clean knife blades prior to use would introduce unwanted compounds into downstream mass spectrometry analyses. Always use different blades for different samples to avoid cross-contamination.
11. If the sample head is too far away from the knife edge, slowly advance the specimen head closer to the blade using the specimen head motor advancement control panel on the cryotome. Trim the sample block to the desired depth using a thicker section size of 50–100 μm. When approaching the desired area for sectioning set the section thickness to 35 μm and take at least one section before collecting sections.
12. Fold up the edges of the cryofilm using a pair of forceps to ensure they are not cut during the sectioning process.
13. Prepare glass slides by cutting double-sided tape to size and adhere to the surface. Ensure there are no air bubbles or ripples in the tape surface. When mounting the film-section onto the double-sided tape, ensure no air bubbles are introduced between the cryofilm and double-sided tape. The film-section may be gently compressed using a Teflon-coated glass slide, but be careful to not crush the sample. Keep glass slide within the cryochamber and away from the glass cover of the cryotome as specimen may defrost.
14. Freeze-drying ensures no delocalization of metabolites when the tissue is thawed.
15. Preweighing the sample slide prior to deposition allows calculation of total amount of matrix deposited on the slide after sublimation. Measurement of the area covered allows calculation of the density of matrix (mg) per unit area (mm^2). Alternatively the density of the matrix per unit area (mg/mm^2) could be determined using the formula (*see* www.htximaging.com technical note #31):

$$\frac{\text{no.of passes} \times \text{matrix concentration (mg/mL)} \times \text{solution flow rate (mL/min)}}{(\text{nozzle velocity (mm/min)} \times \text{track spacing (mm)})}$$

16. Careful application of tape to the ends of the slides covering the sample label masks these areas from deposition. Once removed from the sample slide the masked off areas allow safe handling of the slide and leave the sample label area clear.
17. Prechilling the cold finger and slide allows even and rapid deposition of the matrix.
18. Preheating the bottom of the sublimation apparatus ensures an even temperature when vacuum is applied, decreasing variability in time for deposition.

19. A suitable amount of matrix must be determined empirically for each tissue type and compound class of interest.
20. Before usage of TM Sprayer, be sure to purge all system lines by opening the purge valve and setting flow rate at 5 mL/min to get rid of visible bubbles within the lines (if any). Next, flush the system at both "Load" and "Spray" positions at 1–2 mL/min to ensure that the system is not blocked with matrix from the previous runs. If the system is blocked or the pressure is too high (>300 psi), flush system using isopropanol or 100% water depending on prior matrix used. Alternatively, acid and base wash solutions may be used to remove stubborn matrix contamination. For an acidic matrix, use basic wash solution to dissolve any matrix. For basic matrices, use the acid wash solution.
21. The number of passes, speed of pass, track spacing, solvent selection (mix) and solvent flow rate must be optimized to create the best coating (size of crystals = small, coating thickness "Goldilocks Zone") for each sample type. As a general guide, flow rate: 0.1–0.2 mL/min; nozzle temperature: 30 °C; number of passes: 4–8; speed of pass: 1200–1300 mm/min; track spacing: 2.0 mm; track offset: 1.0 mm. Higher concentrations of matrix will require less numbers of passes to deposit a suitable amount of matrix. It is recommended to select a region greater than the size of the slide (or tissue) to spray to ensure an even deposition across the surface.
22. This ensures the 5 mL sample loop is filled with matrix with a 1–2 mL overflow. Ensure that the syringe remains inline to prevent matrix leakage from the loop.
23. Do not operate with the front glass open during the main spraying procedure as this reduces the risk of inhalation of aerosolized chemicals.
24. If a nontoxic matrix is used, a clean glass slide could be held under the spray nozzle to visualize if any matrix is being sprayed.
25. Basic wash solution is used to neutralize any leftover acidic matrix within the loop and vice versa.
26. Only turn off LC pump when nozzle temperature is below 50 °C.

Acknowledgment

The University of Melbourne node of Metabolomics Australia, a National Collaborative Research Infrastructure Strategy (NCRIS) initiative under Bioplatforms Australia (BPA) Pty Ltd.

References

1. Boughton BA, Thinagaran D, Sarabia D et al (2016) Mass spectrometry imaging for plant biology: a review. Phytochem Rev 15:445–488
2. Bjarnholt N, Li B, D'Alvise J, Janfelt C (2014) Mass spectrometry imaging of plant metabolites - principles and possibilities. Nat Prod Rep 31:818–837
3. Dreisewerd K (2003) The desorption process in MALDI. Chem Rev 103:395–426
4. Hankin JA, Barkley RM, Murphy RC (2007) Sublimation as a method of matrix application for mass spectrometric imaging. J Am Soc Mass Spectrom 18:1646–1652
5. Schwartz SA, Reyzer ML, Caprioli RM (2003) Direct tissue analysis using matrix-assisted laser desorption/ionization mass spectrometry: practical aspects of sample preparation. J Mass Spectrom 38:699–708
6. Caprioli RM, Farmer TB, Gile J (1997) Molecular imaging of biological samples: localization of peptides and proteins using MALDI-TOF MS. Anal Chem 69:4751–4760
7. Norris JL, Caprioli RM (2013) Analysis of tissue specimens by matrix-assisted laser desorption/ionization imaging mass spectrometry in biological and clinical research. Chem Rev 113:2309–2342
8. Spengler B (2015) Mass spectrometry imaging of biomolecular information. Anal Chem 87:64–82
9. Boughton BA, Hamilton B (2017) Spatial metabolite profiling by matrix-assisted laser desorption ionization mass spectrometry imaging. Adv Exp Med Biol 965:291–321
10. Dong Y, Li B, Malitsky S et al (2016) Sample preparation for mass spectrometry imaging of plant tissues: a review. Front Plant Sci 7:60
11. Andersen TB, Martinez-Swatson KA, Rasmussen SA et al (2017) Localization and in-vivo characterization of Thapsia garganica CYP76AE2 indicates a role in thapsigargin biosynthesis. Plant Physiol 174:56–72
12. Kawamoto T (2003) Use of a new adhesive film for the preparation of multi-purpose fresh-frozen sections from hard tissues, whole-animals, insects and plants. Arch Histol Cytol 66:123–143
13. Gorzolka K, Kolling J, Nattkemper TW, Niehaus K (2016) Spatio-temporal metabolite profiling of the barley germination process by MALDI MS Imaging. PLoS One 11: e0150208
14. Jarvis DE, Ho YS, Lightfoot DJ et al (2017) The genome of Chenopodium quinoa. Nature 542:307–312
15. Jarvis DE, Ho YS, Lightfoot DJ et al (2017) Corrigendum: The genome of Chenopodium quinoa. Nature 545:510

Chapter 18

Laser Ablation Electrospray Ionization-Mass Spectrometry Imaging (LAESI-MS) for Spatially Resolved Plant Metabolomics

Desalegn W. Etalo, Carmen Díez-Simón, Ric C. H. deVos, and Robert D. Hall

Abstract

There is increasing demand to bring the level of metabolomics analyses down to the tissue or cell level. Significant progress has been made involving the use of in situ metabolomics imaging techniques where no tissue collection or extraction is needed prior to analysis. In this chapter we describe a relatively new method which is simple and easy to use. No ectopic matrix or vacuum is required, and analyses are performed with living plant materials directly from (or even still attached to) the plant. Although relatively straightforward, there are still a few caveats as regards this method which are described at the end of the chapter.

Key words LAESI, Laser ablation ESI-MS, In situ metabolomics, Spatial resolution, Plant metabolomics, Imaging

1 Introduction

Metabolomics has already become an important tool to study the chemically diverse metabolites that are accumulated by plants. A main underlying goal of many metabolomics studies is to be able to define specific metabolic pathways, investigate their interconnectivity, identify their control points and follow their dynamism. However, plant systems are a technological challenge for metabolomics, both in terms of their huge chemical diversity and the very strong temporal and spatial heterogeneity that is characteristic of their biosynthesis and accumulation.

Standard metabolomics approaches are generally based upon the extraction, separation and (if possible) identification of the detectable metabolites in a particular tissue. A widely used detection method in plant metabolomics is mass spectrometry (MS) due to its sensitivity, high resolution, speed, and broad application

Carla António (ed.), *Plant Metabolomics: Methods and Protocols*, Methods in Molecular Biology, vol. 1778,
https://doi.org/10.1007/978-1-4939-7819-9_18, © Springer Science+Business Media, LLC, part of Springer Nature 2018

[1]. GC (gas chromatography), LC (liquid chromatography), and CE (capillary electrophoresis) are the separation techniques of choice when coupled to the MS. Other approaches focus on NMR [2], and these methods are receiving growing attention as increasing instrument sensitivity is enabling increasing numbers of metabolites to be detected and directly identified. Nevertheless, all of these approaches suffer from a significant, common drawback. Prior to analysis, plant material must be collected, usually pooled and then extracted to provide a robust, biologically relevant sample with minimal technical "noise." Plant tissues are, however, highly heterogeneous, comprising multiple cell types, each of which has its own specific biochemical profile. This heterogeneity, which is of high importance for the structure and function of the plant, is therefore lost and we have to make do with the "law of averages." While in many cases, such an average "global overview" is perfectly adequate to meet our needs, when we wish to delve deeper into the precise metabolic heterogeneity and chemical diversity of plants and try to understand how these are organized at the cellular and molecular level, this loss of spatial heterogeneity becomes a serious limitation.

Analytical approaches enabling analysis at the cellular level are therefore required. Over the last ten or more years, different inroads have been taken to allow us to "look into" plant tissues and perform in situ chemical profiling. These so-called metabolite "imaging" approaches (MSI, mass spectrometry-based imaging) are gaining in popularity for tackling specific metabolomics questions and a number of contrasting technologies have been developed. Significant progress has been made and MS imaging applications have been employed in the animal [3], human [4], microbial [5], and plant sciences [6]. Perhaps the longest-established and most broadly adapted methods are based on MALDI-MS (matrix-assisted laser desorption ionization) technologies [7]. Direct analysis in real time (DART) and desorption electrospray ionization (DESI) approaches [8, 9] provide alternative methods which have also received attention. Each method has its particular advantages and disadvantages (requirement for an applied ionization matrix, use of a vacuum during analysis, loss of spatial resolution, etc.). All these approaches look at the natural or exposed surface of a plant and following metabolite vaporization and ionization of the constituent molecules, use MS for detection. Such direct imaging approaches potentially avoid extraction-related artifacts but in contrast, do carry certain potential limitations, e.g., a greater risk of matrix/ion suppression effects and a lack of resolution through the absence of separation techniques [10].

Both the most widely used MSI techniques, secondary ion mass spectrometry (SIMS) [11, 12] and MALDI [13–15], usually require measurements to be made under vacuum conditions to obtain the best results. SIMS is characterized by a very high spatial resolution (sub-μm), which can detect inorganic compounds at relatively low masses, whereas MALDI is able to detect small to large molecules at a spatial resolution down to ca. 10 μm [16]. An atmospheric pressure (AP)-MALDI technique has been developed which allows for sampling in an ambient environment. However, this approach suffers from significant ion losses between the source and the mass analyzer [17]. Nevertheless, a particular advantage of MALDI is that, as well as small molecules, it can also very effectively be used to image macromolecules such as proteins.

From 2007, ambient ionization techniques for MSI were starting to be used which involve no sample pretreatment and no need for a vacuum. Desorption electrospray ionization (DESI) is especially useful for ionizing polar compounds, but more hydrophobic compounds can also be detected using appropriate solvents [18, 19]. Later, laser ablation electrospray ionization (LAESI) was also developed which had the advantage that it can separate the desorption and ionization process into two different steps [20] while still operating at atmospheric pressure [21]. The typical spatial resolution in a LAESI experiment is 200–300 μm although performing in situ analysis of single plant cells at a lower spatial resolution has been demonstrated using onion epidermis cells [22]. Furthermore, LAESI is one of the first ionization techniques that enables 3D profiling since the laser can gradually remove a tissue, layer by layer [23–25].

LAESI mass spectrometry imaging (MSI) uses a mid-IR laser beam to excite the hydroxyl vibrations of water molecules in the sample to produce a plume of neutral ion species which are then ionized by charged solvent droplets from a nanoelectrospray. These charged ions then directly enter the MS mass analyzer for mass analysis (Fig. 1). To perform LAESI-MS the fresh/living sample is simply mounted flat onto a sample holder and positioned at the focal point of the mid-infrared light ablation beam. The molecular image of a sample is reconstructed by representing the intensity of the total or specific ion signal on a false color scale and correlating this with the coordinates of each pixel of the analyzed area (Figs. 1 and 2, previously unpublished example). The use (and limitations) of LAESI-MS imaging is nicely demonstrated in our recent paper where LAESI/visible mapping of visible pigments in *Phalaenopsis* orchid petals was performed to provide a clear proof of concept [26]. This is the paper upon which this chapter is based.

In this chapter we provide a step-by-step guide to performing a LAESI-MS-based analysis on *Phalaenopsis* petals which are easily obtained at local garden centers and which should provide a straightforward introduction to the technology and its potential.

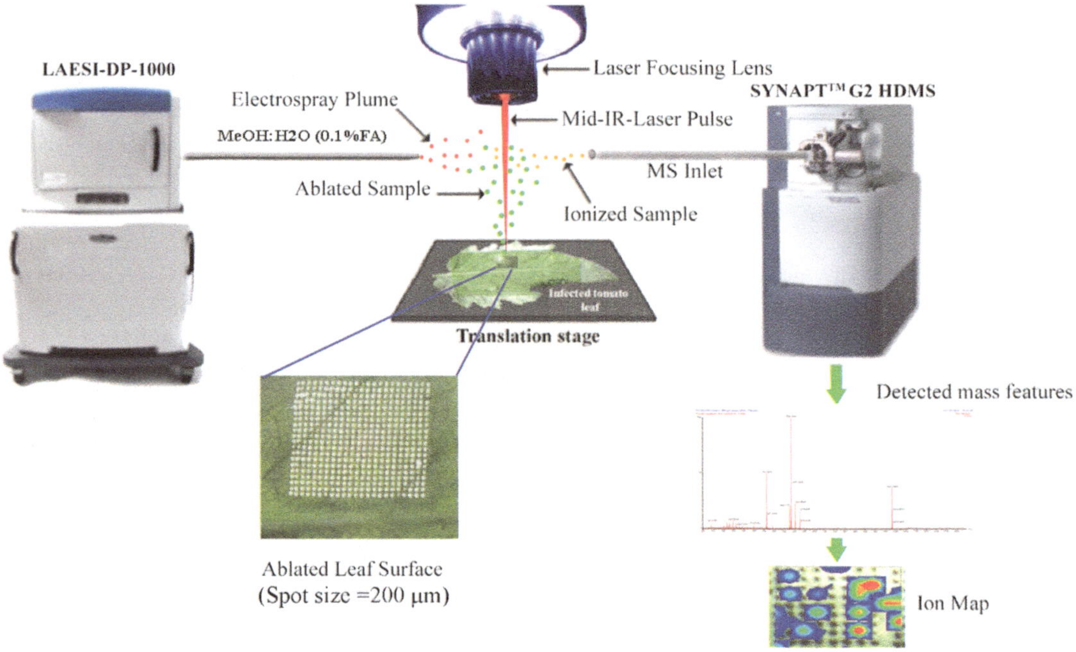

Fig. 1 A schematic representation of LAESI-MS. Under ambient atmospheric conditions, water-containing tissues such as plant leaves are subjected to mid-IR (2940 nm) laser pulses. The absorption of such an intense laser pulse by the sample produces small bubbles within the sample without ion formation. When the bubble explodes, a jet of vaporized material (ablated sample) is ejected above the sample surface without ionization. The ablated sample then intersects a stream of ions from an electrospray source (electrospray plume), which ionizes the molecules of the jet (ionized sample). The ionized molecules subsequently are directly drawn into the mass spectrometer for analysis. Using Protea Plot software for data analysis, the detected mass features can subsequently be converted into two-dimensional or three-dimensional ion maps and can be overlaid on the original sample image, reproduced from [26]

2 Materials

2.1 Plant Tissue

Phalaenopsis orchid petals from well-watered, healthy plants.

2.2 Chemicals

1. Ultra LC-MS-grade methanol (MeOH).
2. Ultrapure water (Actu-All Chemicals).
3. Formic acid (FA) 98–100% (Biosolve Chemicals).
4. Sodium formate suitable for MS calibration over the mass range of 50–1500 Da.
5. Leucine enkaphaline (LE, ≥ 95% purity) standard for lock-mass correction.

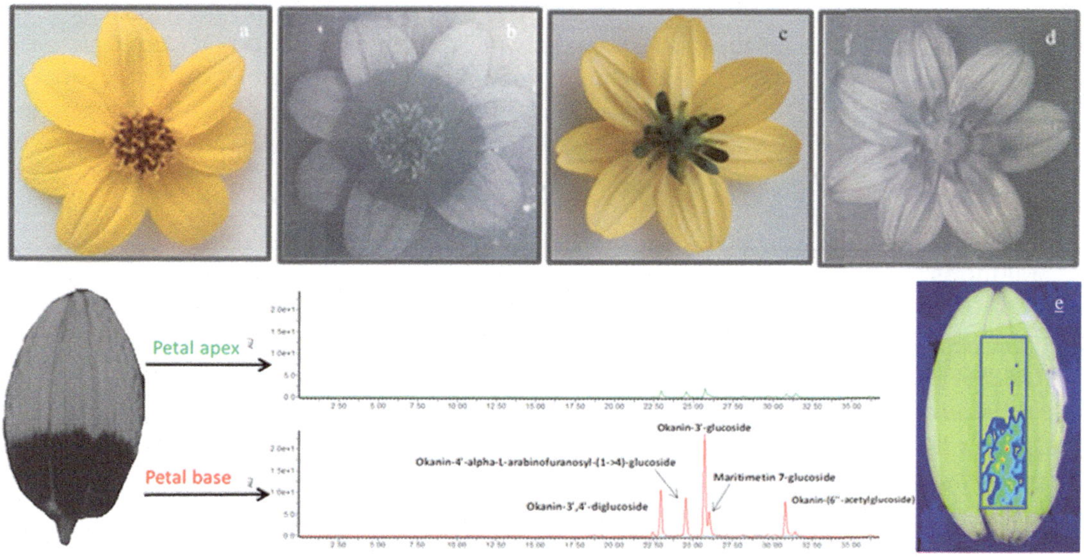

Fig. 2 Analysis of flowers of the garden plant *Bidens ferulifolia*. *Bidens ferulifolia* flower as seen from above in daylight (**a**) and after UV photography (**b**) and from underneath in daylight (**c**) and after UV photography. The presence of okanin derivatives are only visible under UV light and are strictly located in the lower regions of the petal. LC-MS analysis of petal extracts and LAESI imaging (**e**) both confirm their chemical location

2.3 Reagents and Solvents

1. Electrospray solution: MeOH–H_2O (50:50, v/v) with 0.1% (v/v) FA.
2. MS calibration solution: prepare a sodium formate solution 10% FA–0.1 M NaOH–acetonitrile (1:1:8, v/v/v).
3. Lock-mass correction solution: Leucine enkaphaline (LE, *m/z* 556.2771) in acetonitrile–ultrapure water (50:50, v/v).

2.4 Equipment and Software

1. 4.0-mL glass tubes to store electrospray solution.
2. Razor blade.
3. Double-sided sticky tape.
4. Standard glass slides for mounting samples (Protea).
5. Syringe ID (4.6 mm diameter, 100-μm internal diameter stainless steel nanospray emitter).
6. Protea Biosciences LAESI model DP-1000 LAESI system with LAESI desktop Software™ (LDS) to control all on-board LAESI DP-1000 source hardware components, contact closure for triggering of mass spectra data acquisition, and synchronization of LAESI laser with raw data capture in the MS for imaging location and spatial analysis.
7. Waters mass spectrometer model Synapt G2-S with Masslynx software (version 4.1).
8. Protea Plot v.2.0.1.3 (Protea Biosciences) software for generation of ion maps, extraction of mass spectral data from vendor

files, viewing of TIC, base peak plots, and individual or averaged mass spectra, generation of 2D and 3D contour plots (ion/heat maps) for mass spectral imaging studies, export of mass spectral data, 2D/3D heat maps, sample images and automated baseline subtraction, and differential sample-zone analysis features.

3 Methods

3.1 Connecting the LAESI DP-1000 Direct Ionization System to a Waters Mass Spectrometer (Synapt G2-S)

1. Calibrate the mass spectrometer using the MS calibration solution following the procedure provided by the vendor.
2. Check for good performance in terms of mass accuracy and resolution before connecting it to the LAESI-DP-1000 system. A cluster of 21 reference peaks with mass range of 50–1500 Da should be detected.
3. Operate the time-of flight (TOF) mass analyzer of the Synapt G2S in the V-reflectron mode at a mass resolution of 20,000 (full width at one-half maximum).
4. Set the MS parameters as follows: source temperature 150 °C, desolvation temperature 150 °C, sampling cone voltage 30 V.
5. Acquire the positive ions in the range of *m/z* 90–1500.
6. Use LE (*m/z* 556.2771), which was added to the spray solvent, to lock-mass correct the TOF-MS post-data acquisition data.
7. To prepare the LAESI DP-1000 System for use, first insert the LAESI extension tube into the mass spectrometer and fit it to the MS Adapter Plate onto the MS being used.
8. Roll the LAESI system in front of the mass spectrometer by adjusting the vertical height of the LAESI DP-1000 system as needed using the 8 mm hex shaft on the lower right corner of the unit. After the vertical height has been adjusted, roll the LAESI DP-1000 system onto the MS adapter Plate ensuring not to bend the extension tube as it is guided through the rear through-hole in the back of the unit.
9. Connect the female end of power cord to the LAESI DP-1000 system at the lower right corner of the unit and then plug the male end of the power cord into a standard outlet. Connect the B end of the USB cable to the LAESI DP-1000 system at the lower right corner of the unit, and then plug the A end of the USB cable into the computer that will control the LAESI DP-1000 system. The MS computer may be used to control the LAESI DP-1000 system.
10. Connect the MS Trigger Cable to the LAESI DP-1000 system at the lower right corner of the unit. Connect the prewired terminal block to the "External Connections 1" port at the rear

of the MS. Also connect the other prewired terminal block to the Port on the eSAT/IN kit provided.

11. Connect the MS Sync Cable to the LAESI DP-1000 system at the lower right corner of the unit. Connect the ethernet cable to Channel 1 of the eSAT/IN kit.
12. Connect the power cord to the eSAT/IN kit and plug the power cord into a standard outlet.
13. Also connect the eSAT/IN kit to the network switch connecting the MS and the MS computer via the ethernet connection. At this time cycle the power on the computer and the electronics on the MS, ensuring that the MS does not vent.
14. Once the computer and MS are reset, and communication between the MS and computer has been established, open MassLynx and open the Inlet window.
15. In the Inlet window, click on the "Tools" menu and navigate to "Instrument Configuration." Once the configuration page has opened, click the "Configure" button then "Next."
16. The first option will be to configure the LC pump: click "None" and then "Next."
17. The next option will be to configure the autosampler: click "External" and then "Next."
18. On the next page, scroll toward the bottom and click on the eSAT/IN kit and add it to the "Configured Detectors," and then click "Next."
19. On the next page, click the "Scan for Instruments" button; this will scan the system for all available Waters hardware. When it completes click the "Ok" button.
20. On the right, next to "Detector Waters ESATIN," click the drop-down button next to the eSAT/IN kit, and select the serial number that appears for the eSAT/IN kit.
21. Click "Next" followed by "Finish" on the subsequent page. A message should appear saying that the autosampler has been set to external, click "Ok."
22. At this point, the e-SAT/IN kit should appear under the "Detector 1" location in the "Selected Configuration Addresses" box. If it does not, click the "Configure" button again, and follow the steps above again until the eSAT/IN kit appears under the "Detector 1."
23. Next, click the "Events and Triggering" button followed by "Next." Enable the "Event In" Channel 1 option, and click "Next."
24. Under "Detector," click the "Trigger by Contact Closure" option, followed by "Next" and "Finish."

25. On the left of the Inlet window, a "Waters eSAT/IN" icon should appear. Double-click on it to open the options page. "Channel 1" should be enabled and in the "Description" area, type "LAESI Analog Signal."
26. Change the "Sampling Rate" to 10, the "Units" to V, and the "Scale Factor" to 1. It is important to note that in the upper right-hand corner, there is a "Run Time" box. The user MUST specify a run time in this box prior to starting a LAESI-MS analysis. Consequently, a different run time will need to be entered and the "Inlet file" will need to be saved for every LAESI-MS analysis. If a time is not specified, the eSAT/IN kit will not record any data. Also, if the time specified is too long, the eSAT/IN kit will continue to record, even if the MS has stopped. Under this condition, the eSAT/IN kit will not be ready for a consecutive run.
27. Once the Inlet file has been modified, save the "Inlet file."

3.2 Mounting the Plant Tissue and Preparing for Measurement

1. Clean the syringe of the stainless steel nanospray emitter thoroughly with MeOH and fill it with degassed electrospray solution. While filling the syringe, ensure that no air bubbles are trapped in the syringe. Next, connect the syringe to the tubing connected to the "ESI Emitter Assembly," and prime the ESI emitter to ensure that all air has been removed from the ESI emitter tubing (*see* **Notes 1** and **2**).
2. Detach the *Phalaenopsis* spp. petal using a razor blade and mount the tissue as flat as possible using double-sided sticky tape with the adaxial side exposed (upwards) onto a standard sample slide (e.g., chemically precleaned glass slide) (*see* **Note 3**).
3. To perform an imaging analysis of the petal, first turn on the LAESI DP-1000, open the LDS software, and enter the "Interactive Mode." Open the "High Voltage Control" and "Syringe Pump Control" subsystem panels to initiate the electrospray to allow time for stabilization. In the "Syringe Pump Control" panel enter the appropriate syringe ID (4.6 mm diameter) and press "Apply." Next, enter the desired electrospray flow rate and then press "Apply." Typically, a 2 μL/min flow rate works well for syringe of 4.6 mm diameter (*see* **Note 4**). Click the "On/Off" toggle and the syringe pump should start pumping. Additionally, a green "Syringe Pumping" note will appear in the Status Panel in the lower right hand corner of the LDS screen. In the "High Voltage Control" panel choose the desired polarity (positive or negative; in our case, most of the analysis was performed in positive ionization mode) and then enter the electrospray voltage (3800 V was used in our analysis). After entering the voltage, click "Apply" and click the

"On/Off" toggle to apply the high voltage to the emitter. After enabling the voltage, an orange "Voltage On" note will appear in the Status Panel in the lower right hand corner of the LDS screen. This will also engage the door lock and illuminate the orange "High Voltage" LED on the front panel of the LAESI DP-1000 system to indicate the high voltage is on. After a short amount of time, the electrospray background signal should be apparent in the mass spectrum assuming the mass spectrometer has been enabled and is scanning.

4. Once the electrospray has stabilized, a sample can be introduced for analysis. Click the "Load" button in the "Stage Control" panel to move the Sample Nest forward. Next, click the "On/Off" toggle in the "High Voltage Control" panel to disable the high voltage this will unlock the door, allowing the door to be opened. Keep the Peltier-cooled motorized *x-y-z* sample stage at 25 °C. Once the Sample Nest is accessible, remove the Sample Tray, place the sample slide onto the aluminum slide holder, and replace the Sample Tray in the Sample Nest. Ensure that the Sample Tray is seated correctly through the illumination of the green "Sample Loaded" LED. After the tray is loaded, click the "Return" button in the "Stage Control" panel to return the Sample Nest to its previous location. Pressing the "Return" button WILL NOT enable the high voltage or the syringe pump; this must be done manually in their respective subsystem panels.
5. To start an imaging analysis, exit the interactive mode by clicking "Back" and create a "New Project." Type a name for the project and navigate to where the project will be saved on the hard drive. Once this is done, click "Create"; this will make the new project and open the "Project Summary" page. Enter any desired meta-data, such as user name, sample description, and comments.
6. For an imaging experiment the analysis pattern and location (s) must be defined. Before this can occur, a picture of the sample area must be acquired to view the sample. Click the "Modify Analysis" button to enter the analysis definition screen. Chose a "2D Imaging Analysis" from the dropdown menu in the upper left corner of the screen. Open the Camera window and press the "Move" button to move the Sample Nest to the image acquisition location. While the Sample Nest is moving press the "W" button to activate the wide-angle camera and then turn on the camera by pressing the power button. Enable the "Autogain" and "Auto White Balance" in the "Camera Settings" dialog box, if not already enabled. After the stage has reached the image acquisition location and stopped, the "Image Acquisition" button will become active. To acquire an image of the sample area, press

the "Image Acquisition" button. If the image is not suitable, open the "Camera Settings" dialogue box and make adjustments to the live image. Once the adjustments have been made, reacquire the image. To save the image, right click on the image, then navigate to "View" >> "Save Scanned Image." Then type in a name for the image and navigate to where the image will be saved. Once the image has been acquired and saved, press the "Return" button to move the Sample Nest back to its previous location.

7. Next, define the analysis pattern by clicking on the "Zoom Mode" button and use the scroll wheel on the mouse to zoom in on the image to see the sample of interest. Next, click on the "Grid" button to define a grid pattern on the image of the sample (*see* **Note 5**). Move the mouse arrow to the desired location of the first analysis point and left-click. Move the mouse arrow to the end of the analysis such that the grid is completely covering the sample in the image. Before left-clicking again, use the scroll wheel on the mouse to change the grid mesh, or the pixel center-to-center distances (mesh interval of 0.5 mm was used for analysis of orchid petals). Once the desired grid mesh has been achieved, left-click to finalize the grid selection. Using the "Select Mode" function, the grid can be moved as needed to better cover the sample in the image. In the upper left corner in the grid coordinate section, choose either a "Typewriter" or "Serpentine" analysis pattern; typically, serpentine is faster. Finally, enter the correct Z and L positions to ensure that the Sample Nest is the appropriate distance from the ESI emitter and that the focusing lens will be in focus at the specified Z position. In our trial with orchid petals often the focusing lens L value and the sample stage Z value were typically in the order of 4.3 and 20.8 mm, respectively. After entering the Z and L coordinates, click the "Back" button to navigate back to the "Project Summary Page."

8. Next, determine the laser parameters by pressing the "Configure Laser" button and enter the laser settings screen. The machine is equipped with a 2940-nm mid-infrared laser. First, choose the laser output energy by entering the percent output, typically 30% was used in our analysis. Next, choose the number of laser pulses at each location, again 10 pulses per location is suitable for *Phalaenopsis* spp. petal analysis (*see* **Note 6**). Then, enter the frequency of the laser pulses; typically, a frequency of 10 Hz is used. Finally, determine the dwell time, or the time the stage will remain at each location after the last laser pulse has been fired, before moving to the next location. We found 2 s of dwell time to be optimal. To help visualize where the laser pulses occur relative to the MS scans, enable the "MS Scan Visualization" function and enter the approximate scan

time of the mass spectrometer (1 s). Please be aware that this function is only for visualization and DOES NOT control the mass spectrometer scan time. After all the settings are entered and satisfactory, press the "Back" button to navigate back to the "Project Summary Page."

9. Now set up the electrospray parameters by clicking the "Configure Electrospray and Cooling" button and enter the electrospray settings screen. Enter the correct syringe ID (4.6 mm diameter) by either typing in the ID or using the slide bar. Next, enter the desired flow rate (2 μL/min). Next, choose the desired polarity of the electrospray (We used positive ionization mode) and then enter the desired electrospray voltage (3800 V) (*see* **Note 7**). If these settings were enabled in interactive mode they will remain on while the LAESI method is being constructed. Once all settings have been entered, click the "Back" button to return to the "Project Summary Page".
10. Next, set up the mass spectrometer options by clicking the "Configure MS" button to enter the MS options screen. First, choose the type of mass spectrometer being used. Next, enable the start trigger by clicking the toggle. Finally, choose whether the MS data will be high resolution or low resolution. After, these settings are established click the "Back" button to navigate back to the "Project Summary Page."
11. Save the changes to the LAESI method by clicking the "Save" button.
12. Now enter the Run Screen by clicking the "Run Screen" button. Type a name for the LAESI-MS Run, and any other desired comments. Next, click the "Create Folder" button; this will create a new folder with the analysis name inside the LAESI project folder. Next turn on the Inline Camera by pressing the power button.
13. Before the LAESI analysis can be started the MS settings must be established.

    For a Waters mass spectrometer, the data must be acquired by creating a sample list in Masslynx. First, type in a file name for the LAESI data. Next, choose a "MS Method" and an "Inlet" file for the data acquisition. For a Waters mass spectrometer, the MS method and Inlet files must have a specified end time, an estimated analysis time can be found on the "Project Summary Page" of the LDS and this values can be entered for the end time. Finally, specify any additional files or other information in the sample list and save the sample list. Some of the important TOF MS function parameters include acquisition mass range (90–1500 Da), scan time (1 s), interscan time (0.015 s), Data Format (continuum), and Analyzer (resolution) (*see* **Note 8**).

14. Before the LAESI analysis is started, first start the MS, which will ready the MS for the trigger signal from the LAESI DP-1000 unit.
15. Now the LAESI-MS analysis can be started by pressing the "Start" button on the Run Screen in the LDS. This will start the analysis and the LAESI method will be uploaded from the PC to the LAESI DP-1000 unit. After the upload is complete, the contact closure will be triggered and a laser warm up and background acquisition time will start for 45 s. After the warm up time, the sample nest will move to the first location and the LAESI analysis will start. Attention! If necessary, the LAESI analysis can be paused by clicking the "Pause" button. In addition, the ESI can be disengaged by pressing the "Unlock" button; this will also unlock the door. After any corrections are made, the ESI can be reengaged by pressing the "Lock" button, which will reapply the high voltage and lock the door. After pressing the "Lock" button the high voltage will be activated and the ESI will stabilize for about 10 s. After this, the "Start/Resume" button will become active and the LAESI analysis can be completed.
16. After the analysis is complete, the ESI is automatically deactivated. Also, stop the MS data acquisition if an automatic stop time was not used.
17. Next, acquire a post-acquisition image of the sample by clicking the "Back" button to return to "Project Summary Page." Reenter the Modify Analysis screen, open the camera window, and press the "Move" button to move the Sample Nest to the image acquisition location. While the Sample Nest is moving press the "W" button to activate the wide-angle camera and then turn on the camera by pressing the power button. After the stage has reached the image acquisition location and stopped, the "Image Acquisition" button will become active. To acquire an image of the sample area, press the "Image Acquisition" button. If the image is not suitable, open the "Camera Settings" dialog box and make adjustments to the live image. Once the adjustments have been made, reacquire the image. To save the image, right click on the image, then navigate to "View" >> "Save Scanned Image." Then type in a name for the image and navigate to where the image will be saved. Once the image has been acquired and saved, press the "Return" button to move the Sample Nest back to its previous location. After, the stage has returned to the previous location, navigate back to the "Project Summary Page" by pressing the "Back" button.
18. Return to the "Run Screen" and press the "Unload" button to bring the Sample Nest to the front of the LAESI unit to remove the Sample Tray and remove the sample from the slide holder. At this point the LAESI DP-1000 can be shutdown.

19. Check the deviation in mass accuracy of a known metabolite in orchid petal. This can be performed by comparing the calculated exact mass with the measured exact mass of a reference metabolite (*see* **Note 9**). If there is significant deviation in mass accuracy, perform post accusation correction using MassLynx, following the procedure provided by the vendor.
20. To perform the analysis, use Protea Plot v.2.0.1.3 after importing the MassLynx raw data files. Plot the ion intensity signal for a selected *m/z* value (10-ppm window) against the absolute coordinates of analysis to obtain lateral and 3D molecular images. Ions of potential interest for the generation of accurate mass ion maps can be identified using MassLynx via background subtraction of adjacent BPI chronogram regions from those BPI regions that coincided with the analog signal from the mid-infrared laser pulses (*see* **Notes 10** and **11**).

4 Notes

1. Run the electrospray continuously for few hours to check its stability. Two of the most common causes of electrospray instability are clogging of the needle and loose attachment of the needle to the pump. When clogging is encountered, clean the syringe with MeOH. If the problem persists, sonication of the needle in an organic solvent such as MeOH has proven to be effective. Small particles responsible for clogging of needles is often encountered if dirty glassware was used to store the electrospray solvent. Use clean glassware, preferably new 4.0-mL glass tubes to store the electrospray solution.
2. Various electrospray solvents can be used depending on the type of metabolite classes of interest associated with the tissue.
3. Make sure that the samples are mounted on the slide as flat as possible. Failure to do so will lead to pseudo variability in metabolite abundance.
4. The flow rate and the spraying voltage can be varied to obtain a stable spray.
5. Make sure that the selected area for the analysis has fair distribution of purple and white patches. Few purple spots in combination with wider white patches will give reliable information regarding the metabolome composition of these contrasting tissues.
6. If the metabolome composition, abundance and distribution of the upper most layer of the petal is the primary target, a reduced number of laser pulses per location in combination with lower level of laser energy can be used. Each laser pulse creates a 200 μm wide feature with a depth of 20–30 μm. The

number of pulses needed is also dependent on the thickness of the tissue so thinner petals or other thin tissues may require fewer pulses for complete analysis.

7. The choice of the mass spectrometer polarity depends on the classes of metabolite of interest for the analysis. Some plant metabolites can be ionized easily in positive ionization mode while others are better ionized and detected in negative ionization mode. A majority of the metabolites can be detected in both ionization modes. For analysis of anthocyanins that are primarily associated with the purple patches in the petal positive ionization mode is more robust. Apart from protonated ions, primarily singly charged sodium adducts and, occasionally, singly charged potassium adducts, are commonly encountered when the analysis is performed in positive ionization mode.
8. For the settings given, a total scan time of 20 min is needed for a measurement area of 16 mm^2.
9. Correction on the mass features can be done online or offline. Offline correction corresponds to the post-accusation correction of the detected mass features using the reference (lock) mass leucine enkaphaline (*m/z* 556.2771) added to the electrospray solvent. Post accusation mass correction was primarily employed in our analysis.
10. LAESI analysis of orchid petals produces hundreds of mass features. One of the difficulties during data analysis and visualization is the selection of specific mass features that are unique to either the white or the purple parts of the petal. To resolve this issue, prior analysis of the microdissected purple and white parts can provide essential information on the quantitative abundance of a given metabolite in these contrasting tissues. Based on this information, in the LAESI-MS analysis, the relative abundance and spatial distribution of differentially accumulated metabolites in the contrasting tissues can be monitored and mapped.
11. Ion suppression is a major problem in LAESI-MS and other direct infusion techniques. For quantitative determination of metabolites, LAESI-MS analysis needs to be substantiated by LC-MS analysis of laser/microdissected samples. Purple and white spots on the petal can be laser/microdissected and subjected to LC-MS analysis for this purpose.

References

1. Hall RD (2006) Plant metabolomics: from holistic hope, to hype, to hot topic. New Phytol 169:453–468
2. Kim HK, Choi YH, Verpoorte R (2010) NMR-based metabolomic analysis of plants. Nature Protoc 5:536–549
3. Seeley EH, Caprioli RM (2008) Molecular imaging of proteins in tissues by mass spectrometry. Proc Natl Acad Sci U S A 105:18126–18131
4. Li H, Ha E, Donaldson RP et al (2015) Rapid assessment of human amylin aggregation and

its inhibition by copper(II) ions by laser ablation electrospray ionization mass spectrometry with ion mobility separation. Anal Chem 87:9829–9837

5. Dean SN, Walsh C, Goodman H, van Hoek ML (2015) Analysis of mixed biofilm (Staphylococcus aureus and Pseudomonas aeruginosa) by laser ablation electrospray ionization mass spectrometry. Biofouling: J Bioadhesion Biofilm Res 31:151–161
6. Bjarnholt N, Li B, D'Alvise J, Janfelt C (2014) Mass spectrometry imaging of plant metabolites-principles and possibilities. Nat Prod Rep 31:818–837
7. El-Baba TJ, Lutomski CA, Wang BX et al (2014) Toward high spatial resolution sampling and characterization of biological tissue surfaces using mass spectrometry. Anal Bioanal Chem 406:4053–4061
8. Cornett DS, Reyzer ML, Chaurand P, Caprioli RM (2007) MALDI imaging mass spectrometry: molecular snapshots of biochemical systems. Nat Methods 4:828–833
9. Ifa DR, Wu C, Ouyang Z, Cooks RG (2010) Desorption electrospray ionization and other ambient ionization methods: current progress and preview. Analyst 135:669–681
10. Berisha A, Dold S, Guenther S et al (2014) A comprehensive high-resolution mass spectrometry approach for characterization of metabolites by combination of ambient ionization, chromatography and imaging methods. Rapid Comm Mass Spectrom 28:1779–1791
11. Li Z, Bohn PW, Sweedler JV (2010) Comparison of sample pre-treatments for laser desorption ionization and secondary ion mass spectrometry imaging of Miscanthus × giganteus. BioresTechnol 101:5578–5585
12. Saito K, Watanabe Y, Shirakawa M et al (2012) Direct mapping of morphological distribution of syringyl and guaiacyl lignin in the xylem of maple by time-of-flight secondary ion mass spectrometry. Plant J 69:542–552
13. Sarsby J, Towers MW, Stain C et al (2012) Mass spectrometry imaging of glucosinolates in Arabidopsis flowers and siliques. Phytochemistry 77:110–118
14. Gemperline E, Li L (2014) MALDI-mass spectrometric imaging for the investigation of metabolites in Medicago truncatula root nodules. J Vis Exp 85:e51434
15. Yoshimura Y, Zaima N, Moriyama T, Kawamura Y (2012) Different localization patterns of anthocyanin species in the pericarp of black rice revealed by imaging mass spectrometry. PLoS One 7:e31285
16. Hölscher D, Shroff R, Knop K et al (2009) Matrix-free UV-laser desorption/ionization (LDI) mass spectrometric imaging at the single-cell level: distribution of secondary metabolites of Arabidopsis thaliana and Hypericum species. Plant J 60:907–918
17. Li Y, Shirestha B, Vertes A (2007) Atmospheric pressure molecular imaging by infrared MALDI mass spectrometry. Anal Chem 79:523–532
18. Li B, Knudsen C, Hansen NK et al (2013) Visualizing metabolite distribution and enzymatic conversion in plant tissues by desorption electrospray ionization mass spectrometry imaging. Plant J 74:1059–1071
19. Müller T, Oradu S, Ifa DR et al (2011) Direct plant tissue analysis and imprint imaging by desorption electrospray ionization mass spectrometry. Anal Chem 83:5754–5761
20. Nemes P, Vertes A (2007) Laser ablation electrospray ionization for atmospheric pressure, in vivo, and imaging mass spectrometry. Anal Chem 79:8098–8106
21. Bartels B, Svatos A (2015) Spatially resolved *in vitro* plant metabolomics by laser ablation-based mass spectrometry imaging (MSI) techniques: LDI-MS and LAESI. Front Plant Sci 6:471
22. Shrestha B, Patt JM, Vertes A (2011) In situ cell-by-cell imaging and analysis of small cell populations by mass spectrometry. Anal Chem 83:2947–2955
23. Nemes P, Barton AA, Li Y, Vertes A (2008) Ambient molecular imaging and depth profiling of live tissue by infrared laser ablation electrospray ionization mass spectrometry. Anal Chem 80:4575–4582
24. Nemes P, Barton AA, Vertes A (2009) Three-dimensional imaging of metabolites in tissues under ambient conditions by laser ablation electrospray ionization mass spectrometry. Anal Chem 81:6668–6675
25. Nielen MWF, van Beek TA (2014) Macroscopic and microscopic spatially-resolved analysis of food contaminants and constituents using laser-ablation electrospray ionization mass spectrometry imaging. Anal Bioanal Chem 406:6805–6815
26. Etalo D, de Vos RC, Joosten J, Hall RD (2015) Spatially-resolved plant metabolomics: some potentials and limitations of Laser-Ablation Electrospray Ionization (LAESI) mass spectrometry metabolite imaging. Plant Physiol 169:1424–1435

Chapter 19

Live Single-Cell Mass Spectrometry (LSC-MS) for Plant Metabolomics

Keiko Masuda, Yasmine Abouleila, Ahmed Ali, Toshio Yanagida, and Tsutomu Masujima

Abstract

Live single-cell mass spectrometry (LSC-MS) allows for the detection of hundreds to thousands of metabolite peaks acquired from a single plant cell within a few minutes. Plant cells are first observed under a stereomicroscope, a cell of interest is chosen, and then sampled using a metal-coated glass microcapillary for subsequent analysis. A few microliters of ionization solvent is then added to the rear end of the capillary followed by the introduction of the capillary's content directly into the mass spectrometer. High voltage is applied between the capillary and the mass spectrometer inlet to induce nanospray ionization. Metabolite structural confirmation is performed using tandem mass spectrometry analysis (MS/MS) and fragments are matched with MS/MS databases to predict metabolic pathways. This method enables swift and direct molecular detection and identification of specific metabolites from a single plant cell along with their localization within the cell, which will allow for comprehensive understanding of plant metabolomics on a single cell level.

Key words Plant metabolomics, Single-cell analysis, Mass spectrometry, Nanospray ionization, Live single-cell mass spectrometry, LSC-MS, Tandem mass spectrometry, Differential mobility spectrometry

1 Introduction

1.1 Plant Metabolomics

Metabolomics is the study of low molecular weight (generally less than 2000 Da) metabolites inside the cell using specific analytical approaches, with the broader aim of giving an instantaneous snapshot of complex molecular interactions within the cell. The cell metabolome provides a unique chemical fingerprint which is specific for each cellular process. The number of metabolites that have been elucidated in plants so far is about 200,000 [1, 2]. In contrast to other omics studies, metabolomics provides additional large-scale data that enhances our understanding of plant growth, signaling, development, productivity and biochemical processes by means of adding an extra dimension to the data provided from

Carla António (ed.), *Plant Metabolomics: Methods and Protocols*, Methods in Molecular Biology, vol. 1778,
https://doi.org/10.1007/978-1-4939-7819-9_19,

protein, mRNA, and DNA analysis. Whereas proteins and DNA are subject to cellular processes such as posttranslational modifications and epigenetic regulation, metabolites act as specific signatures of biochemical and physiological activities which allow easier correlation with phenotypes [2–4]. The study of plant metabolomics allows a deeper understanding of plant systems through the elucidation of molecular pathways in specific plant cells and may lead to a better understanding of the biochemical processes that control and regulate such processes.

1.2 Significance of Single Plant Cell Analysis

Previous plant metabolomics studies utilized extracts of plant tissue which represents and average of cells as the main source of information regarding plant cell biology. This approach, however, ignores the individuality of each single cell in plant tissues. Moreover, single-cell analysis studies have revealed that averaged population data might compromise fundamental information about cell-to-cell variation and behavior [5]. Cellular heterogeneity arises from numerous factors such as genetic or epigenetic differences, morphological or environmental changes, and mutations [6]. This means that each individual cell has its own specific features and unique metabolic profile that differ from the neighboring ones, highlighting the necessity to carry out single cell analysis [7].

There are a certain set of challenges that are specific to single cell plant analysis. Some of these challenges are caused by plant biology itself, such as the presence of thick cellulose walls, the irregularly rough surfaces of plant tissues and the higher possibility of biomolecular dilution inside the cells [8]. Other challenges are caused by the metabolites themselves, in which some of them are prone to oxidation and are thus highly labile. Moreover, cell size plays an important factor in the challenges faced since most subcellular sample volumes are in the picoliter scale [5]. In addition, technical difficulties in sampling, preventing contamination and enrichment are more pronounced versus usual population averaging studies. Because of these challenges, the choice of extraction, sample preparation and analytical techniques may greatly influence the outcome of the analytical results. Hence there is a need to develop a sensitive, comprehensive, and reproducible analytical protocol for the analysis of the plant systems on a single-cell level.

1.3 Application of Live Single-Cell Mass Spectrometry (LSC-MS) in Plant Metabolomics

Previously, the most commonly used analytical method for plant metabolite profiling was nuclear magnetic resonance (NMR), which provides structural elucidation of the metabolites and metabolic fingerprints [9]. However, due to its relatively low sensitivity in comparison with mass spectrometry (MS), NMR allows for only the identification and quantification of limited number of metabolites [1, 4]. Recently, MS became the analytical technique of choice and one of the most powerful tools in plant metabolomics studies [10–13]. MS is an analytical technique that is used to measure mass to charge ratio of a compound's molecular ion and may also provide elemental composition and isotopic signature. A mass spectrometer has three main components: first compounds are ionized in an ionization source; then ions are separated according to their mass to charge ratio via a mass analyzer; and finally ions are detected in proportion to their abundance [1, 4]. For structural elucidation, tandem mass spectrometry (MS/MS) is often used, where a parent or precursor ion is fragmented via collision induced dissociation (CID) to generate fragment or product ions which are then used to aid in the determination of molecular structure. We established the "Live Single-Cell Mass Spectrometry (LSC-MS)" technique [5, 8, 14, 15] that utilizes nanospray ionization (NSI) which is a variant of electrospray ionization (ESI) that has higher sensitivity and reduced matrix effect, and is consequently more suitable for single-cell analysis.

The LSC-MS protocol is simple in principle. First, cells are observed under video microscopy, and then, a single cell or part of the cell is sampled using a hollow, tapered and conductive glass microcapillary controlled by a micromanipulator. This is followed by the addition of an ionization solvent to the rear end of the capillary. Finally, direct introduction of the capillary's content into the mass spectrometer inlet is done by supplying a voltage differential between the mass spectrometer and the capillary. In addition, separation of isomers can be done using an ion mobility module connected to the mass spectrometer. Mass spectral data are processed using statistical software to search for specific metabolic peaks, and the identification of metabolites is accomplished using metabolic maps and metabolomics databases. This integration of video-microscopic observation with MS provides direct, swift, versatile and exhaustive analysis of various plant tissues on a cellular or subcellular level.

LSC-MS is one of the methods that pave the way for sensitive, high-resolution single-cell plant analysis with minimal pretreatment and cell damage, and will advance our understanding of the cell mechanisms on a single-cell level.

2 Materials

2.1 Reagents

1. Ultrapure water.
2. Acetonitrile or methanol absolute, LC-MS grade.
3. Formic acid or ammonium formate, LC-MS grade.
4. 1, 3, 6 poly-tyrosine mixture (Pierce® Triple Quadrupole Calibration Solution, Thermo Scientific) for calibration of the mass spectrometer.

5. Nitrogen gas for the collision cell and curtain gas (AB Sciex) of the mass spectrometer.
6. Helium gas, 99.999% pure for ion trapping and collision-induced dissociation (CID) experiments.

2.2 Equipment

1. Stereomicroscope.
2. Micromanipulator (MHW-103 on MMN-1, NARISHIGE).
3. Nanospray tips (Cellomics tip, 1 μm of tip bore diameter, HUMANIX).
4. Nanospray tip holder (IM-H1, NARISHIGE) for sampling.
5. Nanospray tip holder (ES286, Thermo Scientific) for microcentrifugation.
6. Gel loading tips.
7. Hybrid ion trap-orbitrap mass spectrometer (LTQ XL, Velos Pro, Thermo Fisher Scientific) or hybrid quadrupole-orbitrap mass spectrometer (Q-Exactive, Thermo Fisher Scientific) and nanospray ion source (Nanospray Flex, Thermo Fisher Scientific).
8. Triple quadrupole mass spectrometer (QTRAP 5500, AB Sciex) equipped with the ion mobility attachment (SelexION, AB Sciex) and nanospray ion source (Nanospray III, AB Sciex) (*see* **Note 1**).

2.3 Software and Database

1. XCalibur Qual Browser (Thermo Scientific) and Tune (Thermo Scientific), or Analyst (AB Sciex) for MS data acquisition and management.
2. MarkerView (AB Sciex) for alignment of the mass spectra and statistical analysis.
3. MetFrag for MS/MS spectra analysis.
4. Kyoto Encyclopedia of Genes and Genomes (KEGG) for metabolite mapping.

3 Methods

The LSC-MS is divided into three main sections, namely (a) sampling, (b) MS analysis, and (c) data analysis. The general procedure is shown schematically in Fig. 1.

3.1 Sampling of Single Plant Cell

1. Set the three-dimensional micromanipulator and the nanospray tip holder which is connected to the syringe for sample sucking by negative pressure next to the stereomicroscope (Fig. 2).
2. Attach the nanospray tip to the tip holder and set it to the micromanipulator. Locate the top of the tip in the middle of high magnification visual field which will be suitable for

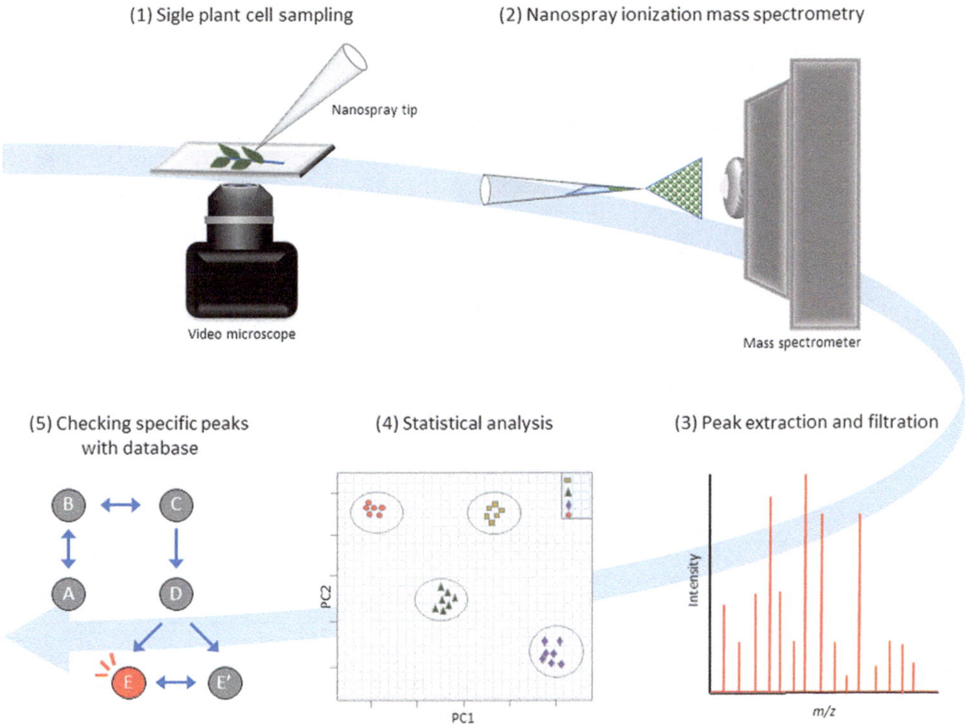

Fig. 1 Schematic general procedure of live single-cell mass spectrometry (LSC-MS)

observing and sampling. Also, confirm that the tip is not broken. Then, just lift it up in a Z-axis direction to make space to put the plant sample on the stage. Never move it in X-axis or Y-axis direction in order not to lose sight of the tip.

3. Cut the plant sample into slices of ~1 mm with a razor, and put it on a glass slide. For a leaf sample, it is recommended to peel the epidermal layers away using adhesive tape or tweezers, or cut the leaf at a slant to expose the cells.
4. Set the sample on the stage of stereomicroscope, adjust the magnification and focus, select the target cell and move it into the center of view. Then, lift the nanospray tip down carefully not to break it, and the tip will come into focus gradually.
5. Touch the target single cell with the top of the tip, and push it through the cell wall. When the tip penetrates the cell wall, a small amount of cell content will automatically go or enter inside the tip by pressure difference and capillarity phenomenon. Suck the cell content more (hundreds of femtoliters) by squeezing the syringe (Fig. 3). Record this procedure by photo or by video to check the timing and the sucked location of the cell precisely, if necessary.

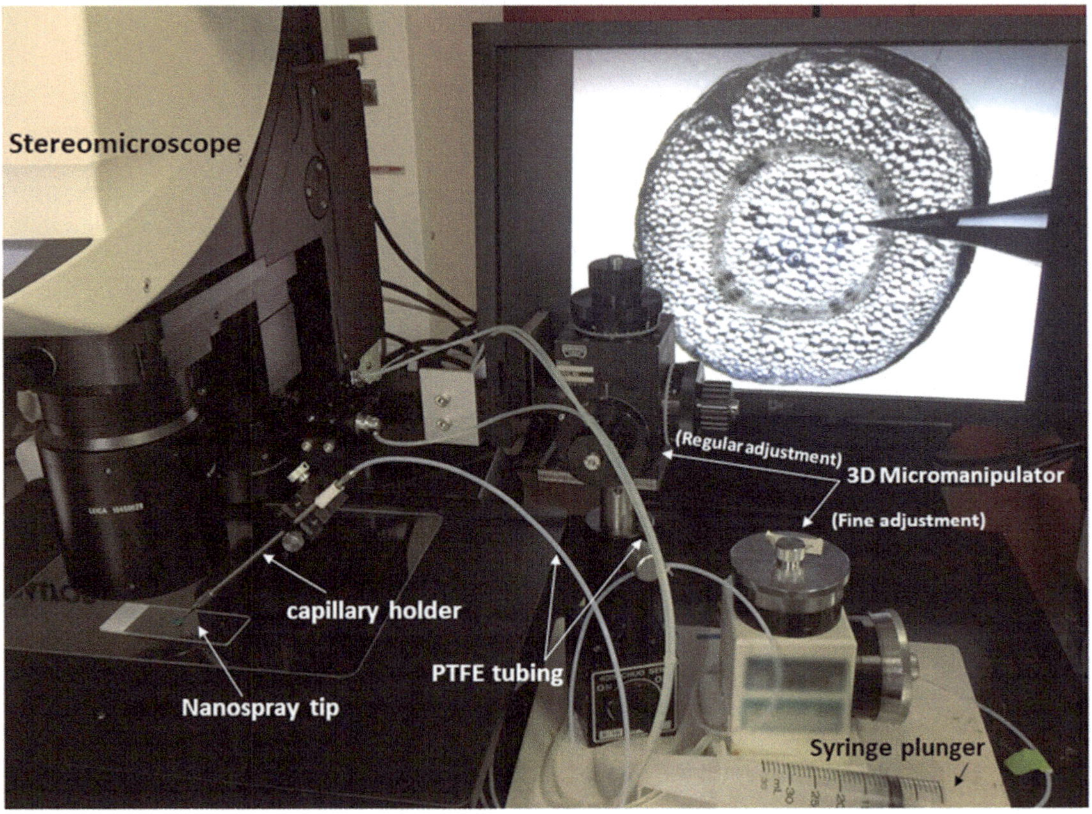

Fig. 2 Equipment setup for single plant cell sampling. Three-dimensional micromanipulators are set on magnetic stand. The nanospray tip holder is connected to the syringe for sample sucking by negative pressure

6. Remove the tip from the cell and then remove it from the tip holder. Pay full attention not to break the tip. Sample can be stored at −80 °C until MS analysis.

3.2 Preparation for LSC-MS Analysis

This step should be performed after each instrument preconditioning as described in Subheadings 3.3 and 3.4, and immediately before each MS analysis.

1. For instrument calibration, we usually use 1, 3, 6 poly-tyrosine. Add 2 μL of the calibration solution into the rear end of the nanospray tip using a gel loading tip. Check for the presence of the following peaks, in positive mode (182, 508, and 997) and in negative mode (180, 506, and 995) then proceed with the calibration.
2. For analysis, add 2 μL of the ionization solution into the rear end of the nanospray tip containing plant sample. The ionization solvent used is usually 50–90% methanol or acetonitrile with 0.1% formic acid for analysis in the positive mode, and 1 mM ammonium formate for positive and negative mode. A molecule with a known m/z value and concentration can be

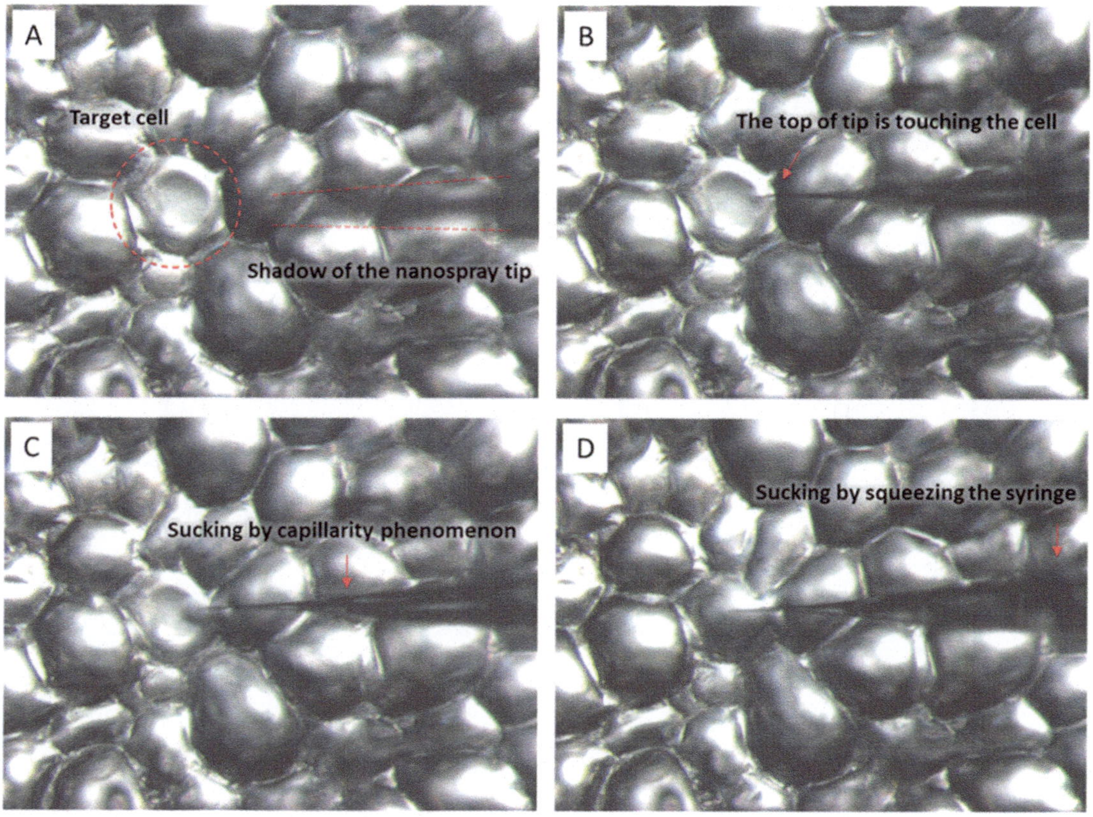

Fig. 3 Single plant cell sampling using the nanospray tip. (**a**) Lift the nanospray tip down carefully. (**b**) Penetrate the target single cell wall using the tip. (**c**) After penetration, a small amount of cell content will automatically go inside the tip by pressure difference and capillarity phenomenon. (**d**) Suck the cell content more by squeezing the syringe

added as an internal standard to improve the accuracy of the analysis (*see* **Note 2**) or to normalize the intensity of the multiple samples (*see* Subheading 3.5). In case of differential mobility spectrometry (DMS) analysis (*see* Subheading 3.4), dip the sharp end of the tip in 50% 2-propanol solution (*see* **Note 3**) then introduce the organic solvent from the other end of the tip.

3. To remove air bubbles and to push the solvent to the sharp end of the capillary, forcibly tap the capillary in the direction of the sharp end, or place the capillary into a microcentrifuge for a few seconds.
4. Set the nanospray tip in the nanospray ion source and adjust its position till it is approximately 3 mm away from the inlet of the mass spectrometer (Fig. 4).
5. Start the analysis and check the total ion count (TIC) to confirm whether a stable spray is properly generated or not. If

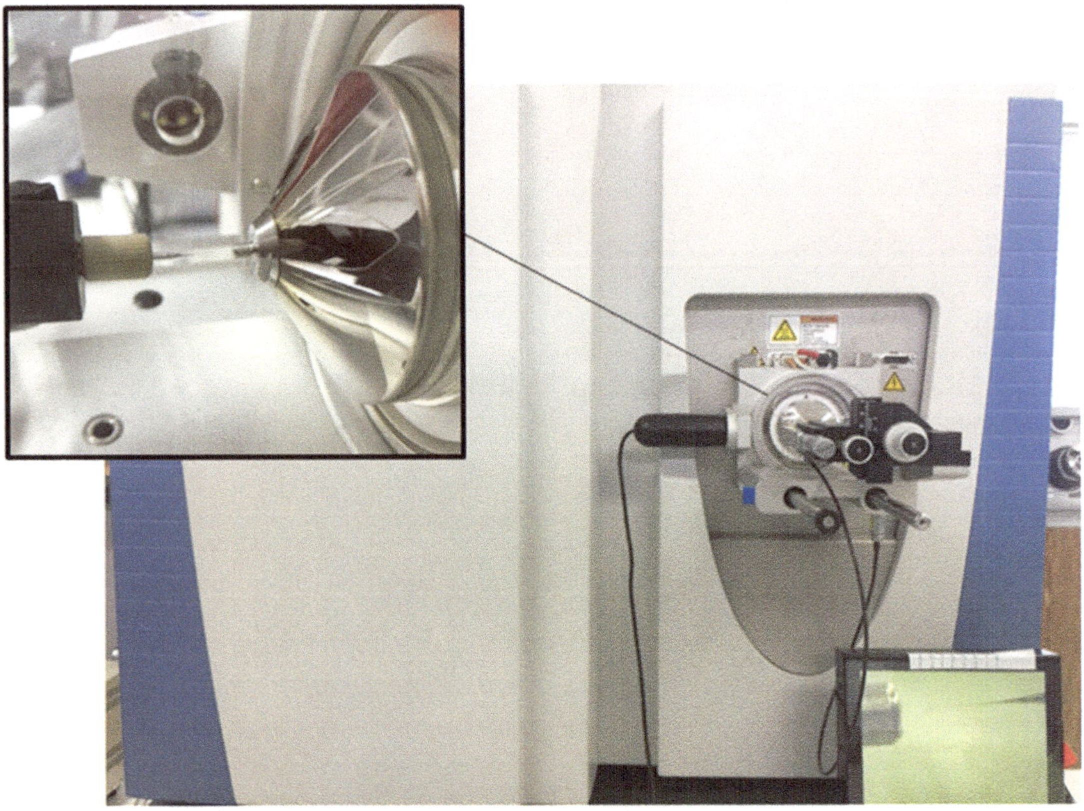

Fig. 4 Nanospray ion source for orbitrap instrument. Approximately 3 mm away from the inlet of the mass spectrometer gives stable ion introducing

the spray is unstable, adjust the position of tip to get the most stable and highest TIC. Check the TIC variation from one scan to another (should be less than 15% variation). If TIC variation is high, check the capillary for the presence of air bubbles and remove it by tapping or centrifugation.

3.3 Orbitrap Preconditioning and MS Analysis

Orbitrap instruments are suitable not only for targeted analysis but also for untargeted analysis due to their relative high mass accuracy, resolution, and sensitivity when compared to triple quadrupole or time of flight mass spectrometers. High mass accuracy and resolution are especially important for the LSC-MS since there is no separation step before MS analysis. Adjust the mass spectrometer settings as follows:

1. Inlet capillary temperature: 200 °C; Resolution: 100,000 FWHM for LTQ XL, Velos Pro and 140,000 FWHM for Q-Exactive (*see* **Note 4**); Ion source voltage: 1.0–1.5 kV for positive mode, and −0.8–−1.0 kV for negative mode (allow for 90 min preconditioning when switching polarities in LTQ XL and Velos Pro).

2. Place the nanospray tip containing the sample and organic solvent on the nanospray ion source and start the analysis with the desired mass range.
3. For MS/MS analysis, collision induced dissociation (CID) is recommended first on LTQ XL and Velos Pro instruments since it allows for subsequent fragmentation if needed. However, high-energy collision dissociation (HCD) can generate a higher number of product ions albeit at the expense of multiple fragmentation steps (only as a final step of MS^n).

3.4 Triple Quadrupole Preconditioning and MS Analysis

Triple quadrupole instruments are used for targeted analysis by running multiple reaction monitoring (MRM) experiments. In MRM, the target metabolite (precursor ion) is isolated in the first quadrupole (Q1) of the instrument. The parent ion is guided afterward to the second quadrupole (q2) where it is fragmented and from the resulting fragments a product ion is selected to pass through the third quadrupole (Q3). Thus, MRM is highly suitable for selective and sensitive detection of targeted metabolites on a single cell level.

Although the LSC-MS was successfully applied in many applications [8, 15–18], it is difficult to distinguish isomers due to the lack of a separation step. However, using differential mobility spectrometry (DMS) attempts to answer this important limitation. DMS separates ions in the gas phase according to the subtle differences in their physical structures [19]. DMS works by applying separation voltages (SVs) which cause ions in the DMS cell to move radially toward the opposing electrode due to the differential in mobility between the high and low electric fields. To neutralize the radial motion of the ions, a compensation voltage (COV) is applied which focuses the ions on the center of the cell/tube. Following is the general protocol of MRM and DMS analysis using a triple quadrupole instrument (QTRAP 5500) coupled to a DMS module (SelexION). Set the mass spectrometer settings to the following settings (the rest is set at default values):

1. Curtain gas (CUR) (*see* **Note 5**): 5.0; Ion source gas 1 & 2 (GS1 & GS2): zero; Ion spray voltage (IS): +1800 for positive mode and −1500 for negative mode.
2. Set the desired precursor and product ion pairs and start automatic compound optimization in Analyst software, it will automatically tune instrument parameters such as declustering potential (DP), collision energy (CE), and collision cell exit potential (CXP) that gives the highest signal for the chosen compounds.
3. Set up an MRM experiment by writing the precursor and product ions in Q1 and Q3 and adjust DP, CE, and CXP to the chosen parameters in the previous step.

4. For DMS analysis, adjust the parameters as follows. Modifier: None; MoC: Low; DMS offset: 5.0; DMS resolution enhancement (DR): 0. SVs and COV should be optimized manually because those are specific to each target compounds (*see* **Note 6**).
5. Place the nanospray tip on the nanospray ion source and start the MS analysis.

3.5 Data Processing and Analysis

This section will focus mainly on data generated from untargeted analysis of samples using an orbitrap equipment, since targeted analysis data processing and interpretation is relatively simple using the default software accompanied with the triple quadrupole instruments. Unlike data generated from LC-MS experiments, data from the LSC-MS is two dimensional (*m/z* and intensities), thus it is imperative to establish good data filtering and processing protocols to minimize error and false positive results.

1. Extract peaks from Thermo.raw proprietary file format (*see* **Note 7**). Open.raw file in Xcalibur Qual Browser software, set the mass range to the one used during the experiment and the extraction parameters, right click the peak table, and select export to clipboard (Exact Mass) (Fig. 5a–c). Paste the peaks table in an Excel spread sheet, we usually delete relative intensities and only keep the *m/z*, absolute intensity and noise columns. Save the Excel file as a tab-delimited text file.

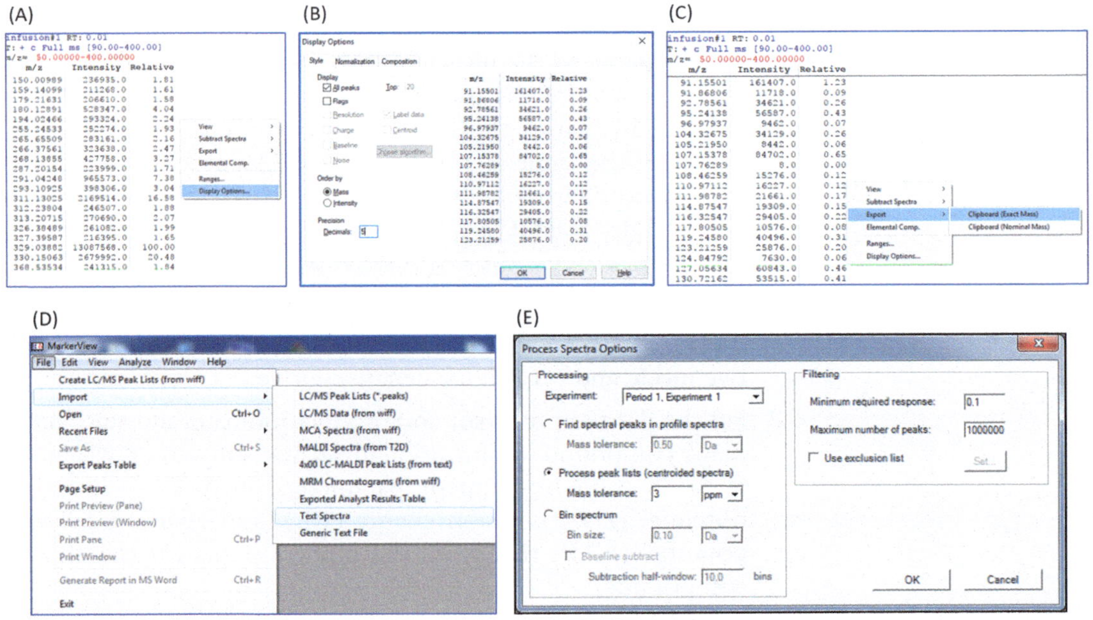

Fig. 5 Processing of data obtained from an orbitrap instrument. (**a–c**) Peak extraction from raw file obtained from orbitrap mass spectrometer. (**d–e**) Peak alignment using MarkerView software

2. Data filtration can be done manually on Excel or using custom scripts in Python, R, etc. First, noise filtering: peaks with signal to noise ratio (S/N) ≤ 3 should be removed. Second, replicate filtering (if there are sample replicates): peaks not present in 70% of the replicates are removed. Finally, blank filtering: peaks present in blank (ionization solution only without sample) should be removed from sample peaks.
3. For aligning peaks using MarkerView software, import the text files and set the alignment threshold to 5 ppm for LTQ XL and Velos Pro instruments, or to 3 ppm for Q-Exactive instrument (Fig. 5d–e) (*see* **Note 2**). The software automatically aligns all the peaks and displays the data in a matrix format with columns representing samples and rows representing m/z values. If normalization is necessary, highlight the internal standard peak and select (Analyze -> Normalize using selected peak), this will normalize peaks across all samples according to the intensity of the selected internal standard peak.
4. A simple t-test can be performed between two samples to establish which peaks/compounds are specific to each sample; peaks with p-value <0.05 can be selected and searched through databases for potential hits.
5. For more than two samples, dimensional reduction techniques such as principle component analysis (PCA) are recommended (*see* **Note 8**). To perform PCA in MarkerView software, simply select (Analyze -> Perform PCA) then select Logarithmic weighing and Pareto scaling (if PCA will be done by another software, make sure to perform logarithmic conversion of intensities of the peaks first prior to PCA). The distant dots from the center represent peaks that contribute greatly to variation between each group of samples.
6. Perform peak annotation and molecular identification using several databases such as Kyoto Encyclopedia of Genes and Genomes (KEGG). KEGG also shows the possible metabolic pathway for the molecules of interests. The following basic rules must be followed before identifying a possible peak by its exact mass alone. First, mass shift error between the detected peak m/z and theoretical m/z should be less than 3 ppm. Second, detected m/z should have the same isotopic pattern as the standard of the potential candidate molecule. Finally, detected molecule m/z should follow the nitrogen rule (*see* **Note 9**).
7. Even with these precautions, false positives are still a possibility, therefore it is recommended to use MS/MS for verification of molecules of interest. MetFrag database can be used to match measured MS/MS data with possible candidates.

4 Notes

1. Contrary to orbitrap instruments' nanospray source, AB Sciex nanospray ion source cannot be used for offline LSC-MS experiments without customization. However, it is a simple process to modify the NanoSpray III source to be used in this experiment (Fig. 6) by modifying a plastic holder to have a groove to hold the nanospray and connecting the nanospray to the voltage supply of the source by a thin metal filament. Moreover, the simplest way to achieve the modification is to fix the capillary between a plastic holder and a screw, and connecting the screw to the source power supply by metal clippers. Finally, care should be taken since touching the ion source metal connections during analysis can cause electrocution, so all adjustments and connections should be tested with a multimeter offline prior to analysis.
2. After external calibration, the mass shift should be less than 5 ppm in LTQ XL and Velos Pro instruments, and less than

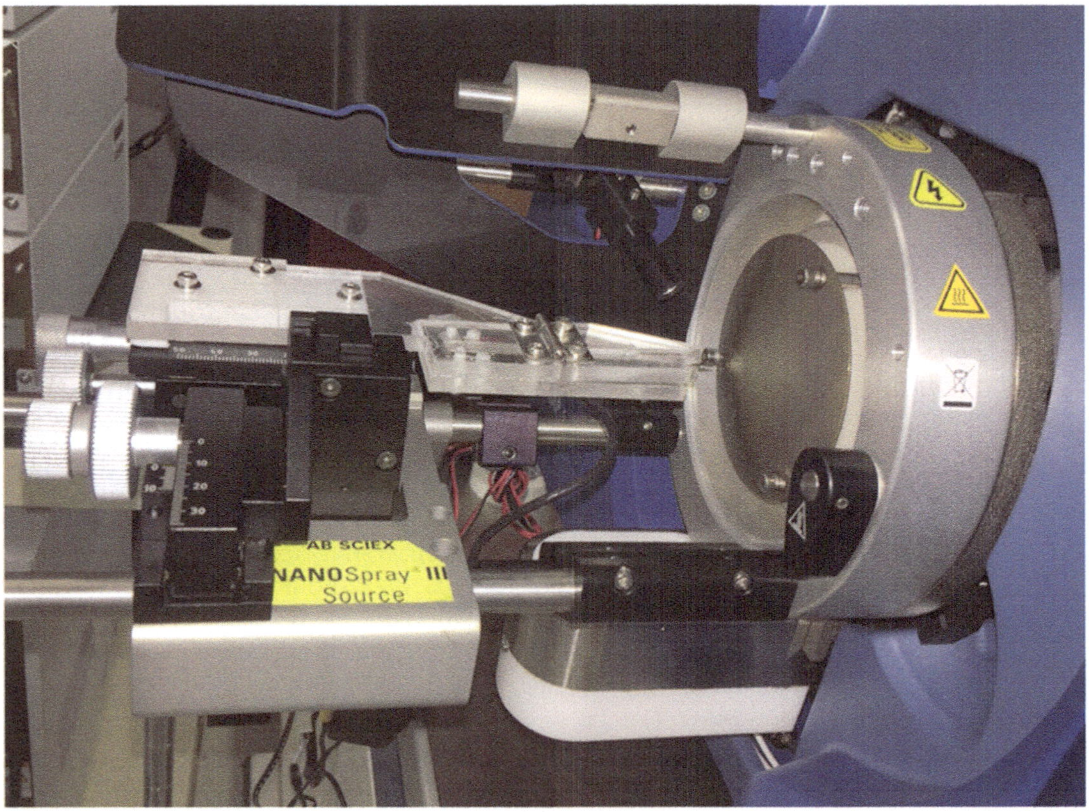

Fig. 6 Nanospray ion source for triple quadrupole instrument. NanoSpray III source (AB Sciex) needs to be modified for the LSC-MS experiment. A plastic holder has a groove to hold the nanospray tip and is connected to the voltage supply of the source by a thin metal filament

3 ppm in Q-Exactive instruments. However, internal calibration can be performed by using "lock-mass" in Tune software, thus reducing the mass shift to less than 1 ppm.

3. In DMS, sometimes adding volatile modifiers to the DMS cell enhances the separation performance. However, with our modified nanospray source for AB Sciex instrument, it is not possible to introduce a modifier using the default machine settings. This can be circumvented by introducing the modifier to the DMS module by adding it to the sharp end of the tip itself.
4. The LSC-MS uses no separation technique (HPLC or capillary electrophoresis) therefore, having the highest possible resolution and mass accuracy is paramount to the successful identification of coeluting metabolites with almost identical mass to charge values. Despite the increased scan time and slight loss of sensitivity.
5. Curtain gas flows vertically across the inlet of the mass spectrometer to protect the mass spectrometer from foreign objects. However, it adversely affects the quality of the spray while running LSC-MS experiments, so running it at its lowest setting is recommended. Another method would be to remove the top plate completely (thus effectively setting the curtain gas to zero) but care should be taken since this exposes the mass spectrometer to potential risks.
6. As a possible application, a method to separate two isomers from a single tea leaf cell (−)-gallocatechin gallate (GCg) and (−)-epigallocatechin gallate (EGCg) had been described by Fujii et al. [6].
7. Since data generated from the LSC-MS is two dimensional only (without retention time), and most proprietary software that can open Thermo.raw files deals with LC-MS data (with retention time), its best to export the data to a more manageable format (text peak table). This allows you to perform data analyses with free public software for PCA and *t*-test (e.g., R, GNU GPL) or *t*-test only by Excel, and annotation using public databases.
8. In PCA, a summary of the data can be seen in the scores plot (between PC1 and PC2, which accounts for the highest variation between samples). Each dot corresponds to a single sample; this is useful since it can show possible clusters of samples as well as possible technical outliers. The loadings plot however, shows all the m/z peaks as dots corresponding to each m/z. The central dots correspond to peaks that are common to most samples, while the distant dots from the center represent peaks that contribute greatly to variation between each group of samples.

9. The molecule which has an odd nominal mass has an odd number of nitrogen atoms. The molecule which has an even nominal mass has an even number of nitrogen atoms.

Acknowledgments

We thank Dr. Patrick J. Rudewicz, Novartis Institutes for BioMedical Research, for his support with writing this chapter. This work was conducted under the RIKEN Pioneering Projects of "Single Cell Science," "Integrated Lipidology," and "Biology of Symbiosis" and fund for the Development of Advanced Measurement and Analysis System (SENTAN), by the Japan Agency for Medical Research and Development (AMED).

References

1. Misra BB, Assmann SM, Chen S (2014) Plant single-cell and single-cell-type metabolomics. Trends Plant Sci 19:637–646
2. Schauer N, Fernie AR (2006) Plant metabolomics: towards biological function and mechanism. Trends Plant Sci 11:508–516
3. Hall R, Beale M, Fiehn O et al (2002) Plant metabolomics: the missing link in functional genomics strategies. Plant Cell 14:1437–1440
4. Jorge TF, Rodrigues JA, Caldana C et al (2016) Mass spectrometry-based plant metabolomics: Metabolite responses to abiotic stress. Mass Spectrom Rev 35:620–649
5. Fujii T, Matsuda S, Tejedor ML et al (2015) Direct metabolomics for plant cells by live single-cell mass spectrometry. Nat Protoc 10:1445–1456
6. Altschuler SJ, Wu LF (2010) Cellular heterogeneity: do differences make a difference? Cell 141:559–563
7. Oikawa A, Saito K (2012) Metabolite analyses of single cells. Plant J 70:30–38
8. Lorenzo Tejedor M, Mizuno H, Tsuyama N et al (2012) In situ molecular analysis of plant tissues by live single-cell mass spectrometry. Anal Chem 84:5221–5228
9. Schripsema J (2010) Application of NMR in plant metabolomics: techniques, problems and prospects. Phytochem Anal 21:14–21
10. De Vos RC, Moco S, Lommen A et al (2007) Untargeted large-scale plant metabolomics using liquid chromatography coupled to mass spectrometry. Nat Protoc 2:778–791
11. Ernst M, Silva DB, Silva RR et al (2014) Mass spectrometry in plant metabolomics strategies: from analytical platforms to data acquisition and processing. Nat Prod Rep 31:784–806
12. Dan Y, Zhang W, Xue R et al (2015) Characterization of gold nanoparticle uptake by tomato plants using enzymatic extraction followed by single-particle inductively coupled plasma-mass spectrometry analysis. Environ Sci Technol 49:3007–3014
13. Ito J, Herter T, Baidoo EE et al (2014) Analysis of plant nucleotide sugars by hydrophilic interaction liquid chromatography and tandem mass spectrometry. Anal Biochem 448:14–22
14. Masujima T (2009) Live single-cell mass spectrometry. Anal Sci 25:953–960
15. Shimizu T, Miyakawa S, Esaki T et al (2015) Live Single-cell plant hormone analysis by video-mass spectrometry. Plant Cell Physiol 56:1287–1296
16. Esaki T, Masujima T (2015) Fluorescence probing live single-cell mass spectrometry for direct analysis of organelle metabolism. Anal Sci 31:1211–1213
17. Hiyama E, Ali A, Amer S et al (2015) Direct lipido-metabolomics of single floating cells for analysis of circulating tumor cells by live single-cell mass spectrometry. Anal Sci 31 (12):1215–1217
18. Ali A, Abouleila Y, Amer S et al (2016) Quantitative live single-cell mass spectrometry with spatial evaluation by three-dimensional holographic and tomographic laser microscopy. Anal Sci 32:125–127
19. Schneider BB, Nazarov EG, Londry F et al (2016) Differential mobility spectrometry/mass spectrometry history, theory, design optimization, simulations, and applications. Mass Spectrom Rev 35:687–737

Part III

Data Analysis & Statistics Chapters

Chapter 20

Statistical and Multivariate Analysis of MS-Based Plant Metabolomics Data

Carsten Jaeger and Jan Lisec

Abstract

Raw data from metabolomics experiments are initially subjected to peak identification and signal deconvolution to generate raw data matrices $m \times n$, where m are samples and n are metabolites. We describe here simple statistical procedures on such multivariate data matrices, all provided as functions in the programming environment R, useful to normalize data, detect biomarkers, and perform sample classification.

Key words Untargeted metabolomics, Bioinformatics tools, Multivariate statistics, Quality control, Data normalization, R programing language, R software packages

1 Introduction

Modern metabolomics technologies reveal plant metabolism in unprecedented detail as both analytical instrument, i.e., mass spectrometry (MS) and nuclear magnetic resonance (NMR) and bioinformatics tools available in data analysis continue to advance at a high pace. Dealing with untargeted data in practice, however, remains challenging even to experts in the field, not least due to the various technical issues encountered in the untargeted approach. We here present an interactive statistical procedure based on the free, open source environment R that we routinely use in our lab when analyzing new experimental data. In a detailed step-by-step guide, we demonstrate (1) how to load and prepare the data in R, (2) how to check for possible batch effects and correct them, (3) how to detect biomarkers, and (4) how to perform a basic classification to predict class association of yet unknown samples.

While learning R can be challenging, especially to novices in programming, we believe that the flexibility that is gained over using commercial point-and-click solutions is well worth the effort. As R scripts, once programmed, can be reused for all subsequent related problems we are convinced that new users will quickly

Carla António (ed.), *Plant Metabolomics: Methods and Protocols*, Methods in Molecular Biology, vol. 1778, https://doi.org/10.1007/978-1-4939-7819-9_20,

experience substantial productivity gains. When becoming acquainted with R, we highly recommend taking a learning-by-doing approach using "real life" data from the very beginning. We take this approach in the present chapter and hope to contribute to easier and more widespread adoption of R for the benefit of truly transparent and reproducible research.

2 Materials

2.1 Software and Hardware Requirements

The statistical procedures described below require an installation of R, freely available from https://cran.r-project.org/ for all major operating systems. We highly recommend to also installing RStudio, a convenient development environment for R (https://www.rstudio.com). Within R or RStudio, proceed by installing *MetabolomicsBasics*, the package that accompanies this chapter and provides the discussed functions and datasets. Type **install.packages ("MetabolomicsBasics")** into the R console or use the respective RStudio menu. Hardware requirements for R and *MetabolomicsBasics* mainly depend on the amount of data to be analyzed. To reproduce the discussed examples or process similarly sized custom datasets, any recent desktop or notebook computer with at least 3 GB RAM will be sufficient.

2.2 Data Preparation (Sample Table, Metabolite Table)

We usually organize experimental data in three different text files: data table, sample table and metabolite table. We will call these files *raw*, *sam*, and *met*. *raw* contains measured values (peak heights or areas), with metabolites organized in rows and samples organized in columns. *sam* defines sample names and one or more group levels that the samples are associated with (e.g., genotype, treatment, time point), while *met* defines metabolite information like names, retention times, spectra or data base IDs. Consequently, the row numbers of *sam* and *met* should match the dimensions of *raw*. The raw, sam, and met datasets provided by *MetabolomicsBasics* give examples that will be discussed in more detail below.

3 Methods

3.1 Data Preparation

We start a new R session and begin by loading the package and the sample datasets. Type the following into the R console:

1. **library(MetabolomicsBasics)** # attaches package and dependencies.
2. **data(raw)** # load peak intensities (112 metabolites × 120 samples).

3. *raw* contains intensities for 112 metabolites and 120 samples measured by GC-MS. The samples were maize root tips that were analyzed regarding metabolic patterns in heterosis [1].
4. **data(sam)** # load sample names and metadata.
5. *sam* contains sample identifiers, sample genotypes, analytical batch and measurement order (columns *ID*, *GT*, *Batch*, and *Order*). In addition, we use *sam* to predefine colors and symbols to be used later in a variety of plots (columns *cols*, *pchs*, with the latter for "plot characters").
6. **data(met)** # load metabolite names.
7. Column *MP* in *sam* gives the median peak height that we use in normalization as an approximation of sample dry weight. Columns *Class* and *Origin* assign each sample to be a hybrid or parental genotype (H or P) and of American or European origin (Am or EU). In general it is helpful to have as much as possible of secondary data annotating samples at hand to be used as cofactors in plotting or statistical functions.
8. **flt** <- **sam[,"Class"]=="P"** # index of non-hybrid samples.
9. For the purpose of simplicity, we set up a filter variable *flt* which may be used to reduce the original experiment to 36 samples, removing all hybrid strains.
10. To import your own data, use:

    raw <- **read.csv("your_intensity_data.csv");**
    sam <- **read.csv("your_sample_data.csv")**
    met <- **read.csv("your_metabolite_names.csv")**
11. For comma-separated value (csv) files, or the related function read.delim() or read.xlsx() from the xlsx package for common alternative data formats.
12. Consult the respective help page (accessible for example by typing **?read.csv**) to ensure correct data import.
13. Pay particular attention to the "sep" and "decim" arguments, determining field and decimal separators, respectively.
14. We recommend using the same column names as in the examples to apply the following procedures easily.

3.2 Data Transformation

Peak intensity data provided in raw are already $\log_{10}$-transformed as log-transformation helps to improve normality [2] which is an important assumption for many statistical tests usually performed on omix data. If not yet done, consider this step for you own data as well by using: **raw** <- **log10(raw)**.

3.3 Checking for Outliers

Outliers often arise from analytical issues, e.g., coelution of contaminating peaks, and/or issues during peak detection. In case of replicate measurements of biological groups (here we used six replicates per genotype), which is highly recommended in metabolomics experiments, one may use assumptions on the statistical distribution of replicate values to detect outliers within the intensity matrix, by:

tmp <- apply(raw, 2, CheckForOutliers, group=sam$GT, method="dist").

We compute the multiple of standard deviations each value is off the group mean. This allows for example to apply the five-sigma rule to discard dubious values by:

raw[tmp>5] <- NA # set five-sigma values to not available (NA).

If measurement values would be Gaussian distributed, 99.99994% of all values should be found within a band of five standard deviations around the mean.

3.4 A First Look at the Data: Principal Component Analysis (PCA)

One of our first steps when inspecting a new dataset is usually to employ a PCA, which is a widely used and relatively simple non-parametric tool in data analysis [3]. By reducing dimensionality, it helps to quickly identify the major effects (or variance components) in a complex dataset. We set up a PCA model and produce a basic scores plot:

1. **pca_res <- pcaMethods::pca(raw, method="rnipals", scale="pareto")**.
2. **PlotMetabolitePCA(pca_res=pca_res, sam=sam)**.
3. **pca_res <- pcaMethods::pca(raw[flt,], method="rnipals", scale="pareto")**.
4. **sam[,c("cols","pchs")] <- AdjustSymbols(cols=as.numeric (sam$GT), pchs=sam$GT)**.
5. **PlotMetabolitePCA(pca_res=pca_res, sam=sam[flt,], medsd=T, g=sam$GT[flt], text.col"Order", comm=c("in greyscale","only parental lines"))**.

As visible in Fig. 1a, PCA indicates genotype class and origin to be major sources of variance (Fig. 1). Figure 2b indicates some metabolic differences between individual genotypes (we will elaborate on this further below) (Fig. 2). You might have noticed that PlotMetabolitePCA() uses the colors and plot symbols defined in *sam*, allowing for flexible changes in the plot layout. We included a function AdjustSymbols() to automate such adjustments. *See* the respective help file for details (**?AdjustSymbols**).

3.5 Identifying and Removing Technical Artifacts

Technical artifacts often occur due to unintended (and often unnoticed) changes during sample preparation and analysis, e.g., due to use of different batches of chemicals, vials, GC liners, etc.

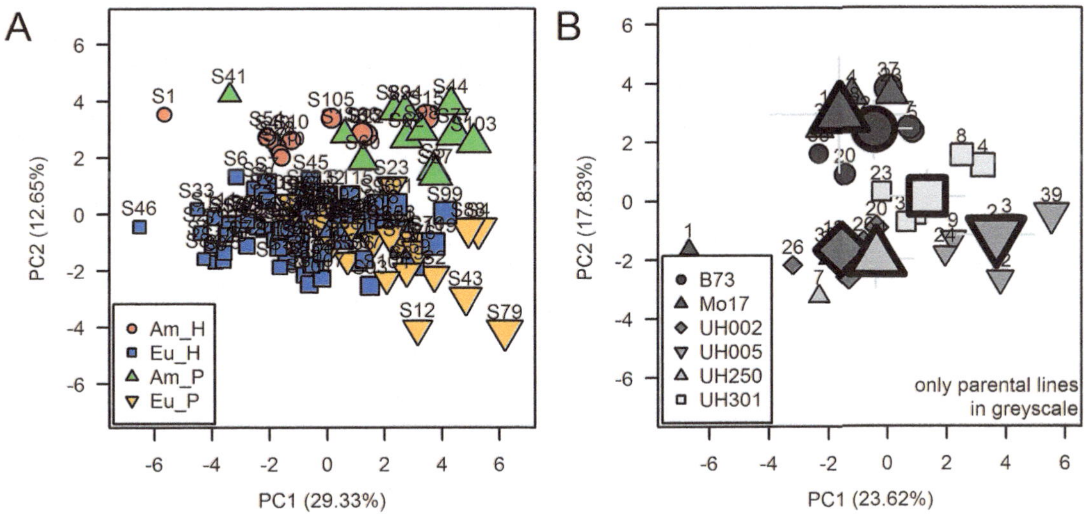

Fig. 1 (**a**) PCA scores plot of all samples. Symbols were selected to visualize major variance components in the data being class (H or P, separated on PC1) and origin (Am or EU, separated on PC2). Data points are annotated with their ID. (**b**) A PCA can be performed on subsets of the data and with different color schemes. Here, only parental genotypes were included and annotated with the Order number from the batch. Please note that group medians and standard deviations were added using the "medsd" argument. Such labeling and coloring schemes help to quickly identify major trends in the data and identify outliers (i.e., S1 one replicate of Mo17, which was the first sample in the measurement batch)

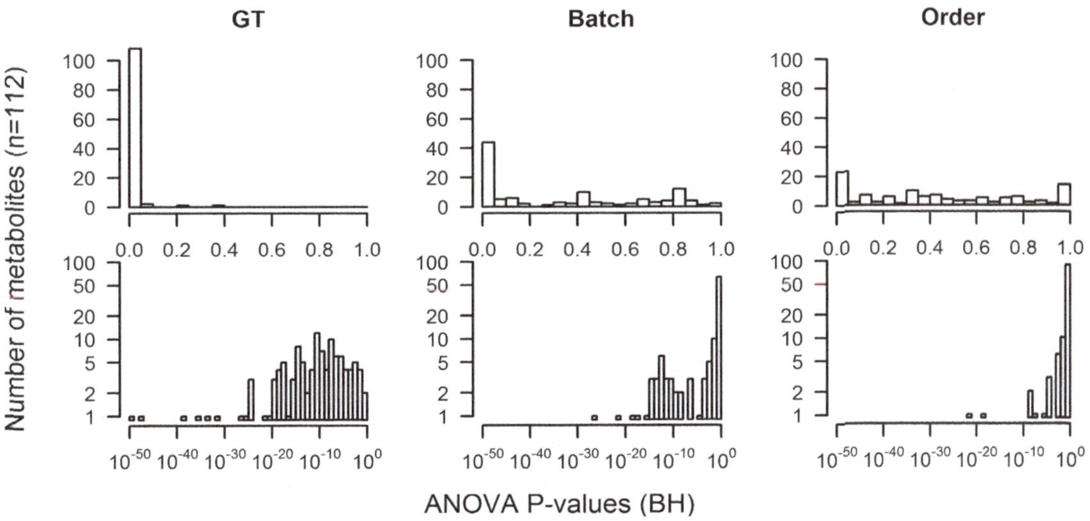

Fig. 2 *p*-Value histograms. *p* values of metabolite-wise ANOVA models are summarized as histograms for each independent variable, i.e., "GT," "Batch," and "Order." The lower row of histograms uses logarithmic scales to better differentiate small *p* values. The total number pf metabolites ($n = 112$) and the multiple testing correction method (*BH* Benjamini–Hochberg) are annotated in the axis labels

Identifying and correcting such "batch effects" is important to avoid misinterpretation in later analyses. Analysis of variance (ANOVA) is a suitable tool for this purpose because it allows dissecting the variances associated with each independent variable. We test if factor *Batch*, as defined in sam, explains a significant portion of the total variance by setting up an ANOVA model and calculating *p* values for each metabolite:

1. **fmod <- "GT+Batch+Order"** # define the full model.
2. **pvals <- MetaboliteANOVA(dat=raw, sam=sam, model=fmod, method="BH")** # Benjamini-Hochberg-adjusted p values for each metabolite.
3. **pk <- which(pvals[,"GT"]<0.01 & pvals[,"Batch"]<0.01 & pvals[,"Order"]<0.01)** # metabolites significant by GT and also affected by Batch & Order.
4. **pvals[pk,]**

	GT	Batch	Order
M60	2.79e−08	8.36e−10	1.38e−05
M61	5.14e−12	8.99e−12	4.73e−05
M74	4.74e−08	1.42e−04	4.92e−08
M80	9.46e−04	1.22e−05	3.34e−05

We see that four metabolites are significantly different between samples for all factors from our model (*p* adjusted <0.01). We produce a histogram to inspect this in a more general way:

PlotPValueHist(out=pvals) # plot a p value histogram.

sum(pvals[,"Batch"]<0.05).

[1] 44.

We find that in total 44 metabolites are affected by *Batch* (*p* adjusted <0.05). To remove the contribution of *Batch* to the total variance, we define a second ANOVA model and recalculate *p* values as follows:

kmod <- "GT" # reduced model (Batch+Order removed).

rawc <- RemoveFactorsByANOVA(y=raw, sam=sam, fmod=fmod, kmod=kmod) # calculate intensity values corrected for Batch+Order.

MetaboliteANOVA(dat = rawc, sam = sam, model = fmod) [pk,].

	GT	Batch	Order
M60	6.68e−09	1	1
M61	1.13e−11	1	1
M74	2.91e−07	1	1
M80	4.74e−04	1	1

We see that the *Batch* (and also *Order*) effects have been completely eliminated from the response variables. Note that the procedure is applicable to both categorical variables (such as *Batch*) and continuous variables (such as *Order*) and that it allows to correct for technical biases while acknowledging biological differences as cofactors. This is in contrast to the often used division by a factor for normalization. One should be aware of the issue of overfitting; however, it arises with small group sizes (*see* **Note 1**). Please also note that we use multiple testing-corrected p values (*see* **Note 2**).

3.6 Visualization of Differential Features (Biomarkers)

We next investigate metabolic features that are significantly influenced by genotype (*GT*). Reusing the ANOVA results from above, we visualize the most significant metabolite with regard to *GT* (Fig. 3):

1. **par(mfrow=c(1,3)).**
2. **MBoxplot(pk=pk[1], pca.dat=raw[flt,], sam=sam[flt,], met=met, g=sam$GT[flt], plot_sample_n=TRUE, an=c ("ID","Name"))** # raw data (log10-transformed).

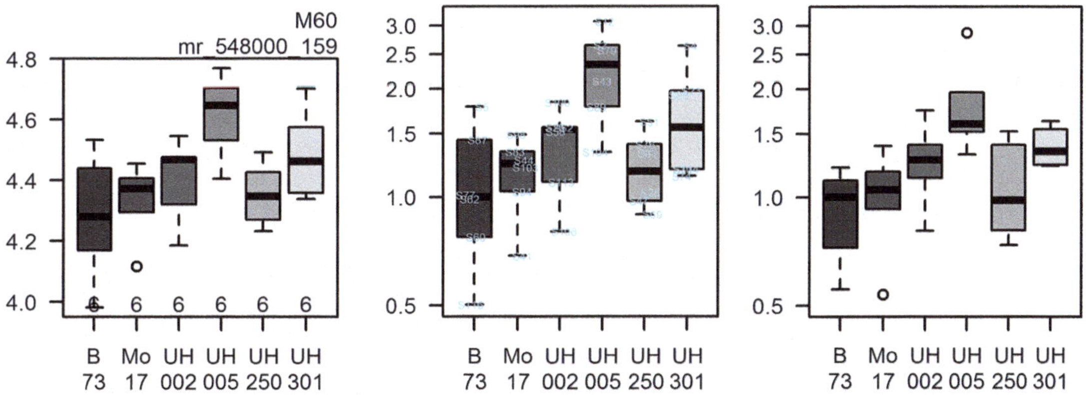

Fig. 3 Visualization of a differential metabolite, using different options available in `MBoxplot()`. *Left plot*: raw data ($\log_{10}$-transformed) with annotated number of replicates per group. *Middle plot*: original data, relative to "B73" as reference with annotated sample IDs. *Right plot*: data corrected for *Batch* and *Order*, relative to "B73" as reference. Note how correction results in smaller variances when comparing *right* and *middle* plots

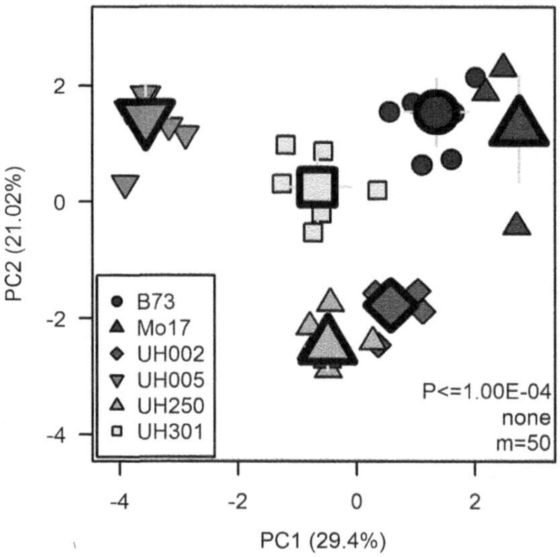

Fig. 4 A PCA restricted to metabolites significantly differentiating samples according to a specified factor. *p*-value threshold, multiple testing correction, method and number of remaining metabolites are annotated at the bottom of the figure

3. **MBoxplot(pk=pk[1], pca.dat=10^raw[flt,], sam=sam [flt,], met=met, g=sam$GT[flt], plot_rel_axis="B73", ylim=c(0.5,3))** # original data, "B73" as reference.
4. **MBoxplot(pk=pk[1], pca.dat=10^rawc[flt,], sam=sam [flt,], met=met, g=sam$GT[flt], plot_rel_axis="B73", ylim=c(0.5,3))** # corrected data, "B73" as reference.

Our usual procedure is to produce such plots for all metabolites and to save them into a common PDF for later reference.

3.7 Principal Component Analysis Revisited

We finally demonstrate a variant of PCA that is particularly useful to differentiate biological groups. Our function RestrictedPCA() combines calculation of metabolite-wise ANOVA models as above and performs a PCA only on metabolites significant for a specified factor (here: *GT*, Fig. 4). When compared to Fig. 1b, one clearly sees much improved differentiation of the different genotypes:

RestrictedPCA(dat=raw, sam=sam, use.sam=which(flt), sign.col="GT", group.col="GT", fmod="GT+Batch+Order", P=10^-4, medsd=T).

3.8 Missing Value Imputation

Some statistical methods such as the classification method introduced below, require complete matrices, i.e., matrices without missing values (NAs). We demonstrate here how to impute missing values using the "nipals" algorithm which is also used for fault-tolerant PCA (*see* **Note 3**). As we do not have missing values in the data matrix we set all outliers to NA before substituting these:

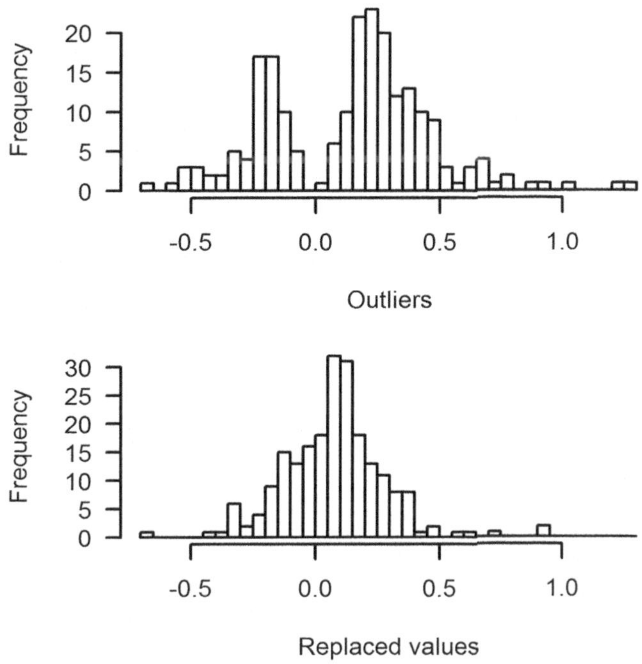

Fig. 5 Histogram of absolute distances between group means and detected outliers. As expected these deviate from zero as this is the outlier detection criteria. After non supervised replacement distances are distributed around zero, indicating that missing value substitution performed as expected

1. **idx <- apply(raw, 2, CheckForOutliers, group=sam$GT, n_sd=5, method="logical")** # find outliers using five-sigma.
2. **raw_repl <- raw**.
3. **raw_repl[idx] <- NA**.
4. **raw_repl <- ReplaceMissingValues(x=raw_repl)**.

After replacing outliers with NA, and substituting these against real values, we can test if the substitution introduced rather noise or meaningful data.

To this end we compare the distance of each substitute value from its group mean before (outlier) and after (replaced) substitution (Fig. 5).

3.9 Classification

We finally present convenience functions to perform sample classification using Decision Trees (DT) or Support Vector Machines (SVM) as classification methods. We know from the initial PCA that genotype origin (Am or Eu) is a major component of variance in the metabolite data. To obtain the optimal DT classifying maize roots with respect to their origin we would use:

1. **class_res <- ClassificationCV(d=raw, g=sam$Origin, method="rpart")**.
2. **fancyRpartPlot(class_res)**.

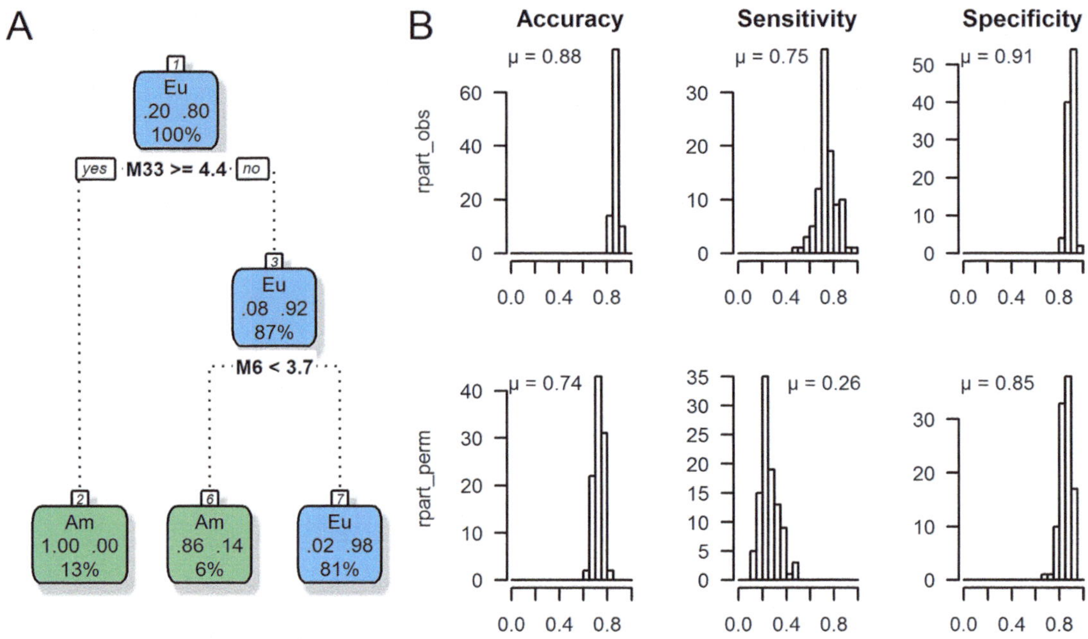

Fig. 6 Classification according to genotype origin using Decision Trees. (**a**) The optimal solution achieving 98% accuracy by splitting the data evaluating only two metabolites. (**b**) Robust evaluation of the classification method, pointing at problems regarding imbalance and overfitting. Imbalance in class distribution (only 24 Am but 96 Eu samples) often leads to high accuracies. Overfitting may be indicated because the mean accuracy of cross validated observed data (0.88, *upper left panel*), is not strongly different from the accuracy achieved on permuted data (0.74, *lower left panel*)

It includes only two metabolites (M33 and M6), which if evaluated according to defined split points (4.4 and 3.7) allow to group all samples (top node of the tree) into three final nodes (bottom of the tree), resulting in an classification accuracy of 98%, i.e., assigning 117 of 120 samples to the correct class (Fig. 6).

Many classification algorithms are prone to overfitting (*see* **Note 4**). Therefore, the approach should be performed using cross validation and permuted datasets:

class_res <- ClassificationWrapper(d=raw, g=sam$Origin, method="rpart").

ClassificationHistogram(class_res).

4 Notes

1. *Overfitting in normalization*: ANOVA-based normalization allows to investigate technical biases (e.g., run order or sample batch but also differences in sample amount or a common internal standard) while keeping biological relevant information (genotype, treatment etc.) as cofactors. If replicate number of biological groups is small however compared to the factors

incorporated in the model, artifacts may be introduced in ANOVA-based normalization. A good strategy is therefore to compare results before and following normalization, e.g., plotting raw and normalized boxplots side by side. An overview of different normalization methods for omics data can be found in [4].

2. *Multiple testing*: As in any omics-based analysis we usually perform statistical test on multiple traits in parallel, often on low numbers or biological replicates. Testing group mean difference between two genotypes in a *t*-test for example on 100 metabolites will lead to five *p*-values smaller than 0.05 by chance, i.e., even if none of the metabolites is different. Significance thresholds (often alpha = 0.05), therefore, have to be adjusted to this problem. In R several methods are implemented for this purpose in the function p.adjust(). We recommend using Benjamini–Hochberg [5] in general, because Bonferroni often is too strict [6].
3. *Missing values*: Missing values are common in metabolomics analysis, especially if performed nontargeted. We recommend to generally accept NA values wherever possible because avoiding false positive (FP, reporting some significant result which is not there) results is of higher importance than avoiding false negative results (FN, failing to significantly detect a difference which is truly present). If missing value substitution is necessary, choice of a method should be guided by minimizing the possibility to find FPs in the follow up method. For example, a *t*-test will fail if one group contains only NAs. Substituting NAs by zero will lead to highly significant *p*-values even if the comparison group shows levels just above detection limit. Here, a better approach would be to substitute against values just below detection limit including noise reflecting the reproducibility of the analytical system.
4. *Overfitting in classification*: Classification models always have to be evaluated robustly which involves cross validation and permutations [7]. The most common issue with classification models is overfitting, which occurs when models contain too many explanatory variables as compared to the number of data points. Such models usually only describe random noise in the data and fail to predict true class membership of the samples, which becomes obvious when the model is applied to a second, independent dataset. Therefore, a common validation procedure is internal cross validation, which involves splitting the dataset into a training and a test set, often in a ratio of 2/3 to 1/3 or similar. The training set is then used to develop the model and the test set serves to determine its prediction accuracy. Only models that are transferrable from the training to the test set should be reported. We recommend performing this approach repeatedly (e.g., 1000 times) on different

combinations of test and training samples. Applying the same procedure on permuted data, i.e., where the true class information is randomly distributed, helps to access overfitting. For two equally distributed classes prediction accuracy on permuted data should be on average 50% within a small band. Significance of classification robustness can be calculated using a t-test between the accuracy distributions obtained from observed and permuted data.

Acknowledgments

J. Lisec gratefully acknowledges Lothar Willmitzer for the time and freedom to learn about metabolomics and Eleanor Eife-Horn for the continuous support and commitment.

References

1. Lisec J, Römisch-Margl L, Nikoloski Z et al (2011) Corn hybrids display lower metabolite variability and complex metabolite inheritance patterns. Plant J 68:326–336
2. Steinfath M, Groth D, Lisec J, Selbig J (2008) Metabolite profile analysis: from raw data to regression and classification. Physiol Plant 132:150–161
3. Wold S, Esbensen K, Geladi P (1987) Principal component analysis. Chemometrics Intelligent Lab Syst 2:37–52
4. Li B, Tang J, Yang Q et al (2016) Performance evaluation and online realization of data-driven normalization methods used in LC/MS based untargeted metabolomics analysis. Sci Rep 6:38881
5. Benjamini Y, Hochberg Y (1995) Controlling the false discovery rate: a practical and powerful approach to multiple testing. J R Statis Soc B 57:289–300
6. Franceschi P, Giordan M, Wehrens R (2013) Multiple comparisons in mass-spectrometry-based-omics technologies. Trends Anal Chem 50:11–21
7. Arlot S, Celisse A (2010) A survey of cross-validation procedures for model selection. Statist Surv 4:40–79

Chapter 21

Integration of Plant Metabolomics Data with Metabolic Networks: Progresses and Challenges

Nadine Töpfer, Samuel M. D. Seaver, and Asaph Aharoni

Abstract

In the last decade, plant genome-scale modeling has developed rapidly and modeling efforts have advanced from representing metabolic behavior of plant heterotrophic cell suspensions to studying the complex interplay of cell types, tissues, and organs. A crucial driving force for such developments is the availability and integration of "omics" data (e.g., transcriptomics, proteomics, and metabolomics) which enable the reconstruction, extraction, and application of context-specific metabolic networks. In this chapter, we demonstrate a workflow to integrate gas chromatography coupled to mass spectrometry (GC-MS)-based metabolomics data of tomato fruit pericarp (flesh) tissue, at five developmental stages, with a genome-scale reconstruction of tomato metabolism. This method allows for the extraction of context-specific networks reflecting changing activities of metabolic pathways throughout fruit development and maturation.

Key words Flux balance analysis, Plant metabolic network, Constraint-based optimization, Data integration, Metabolomics, Python programming language

1 Introduction

1.1 Constraint-Based Modeling and Data Integration

Since the advent of genome sequencing computational modeling of metabolism has shifted from detailed, kinetic modeling of particular pathways of interest to holistic, genome-scale modeling of metabolic networks. Flux balance analysis [1] is the method of choice for many large-scale modeling approaches due to its simplicity and the need for only few input variables; the method uses linear optimization to determine flux distributions solely based on the knowledge of the structure of the metabolic network under consideration, physicochemical constraints (e.g., maximum uptake rates or thermodynamic reversibility) and a predefined cellular objective such as the maximization of growth (Fig. 1). Kinetic parameters such as maximum velocities, v_{max} or metabolite

Author contributed equally with all other contributors. Nadine Töpfer and Samuel M. D. Seaver

Carla António (ed.), *Plant Metabolomics: Methods and Protocols*, Methods in Molecular Biology, vol. 1778,
https://doi.org/10.1007/978-1-4939-7819-9_21,

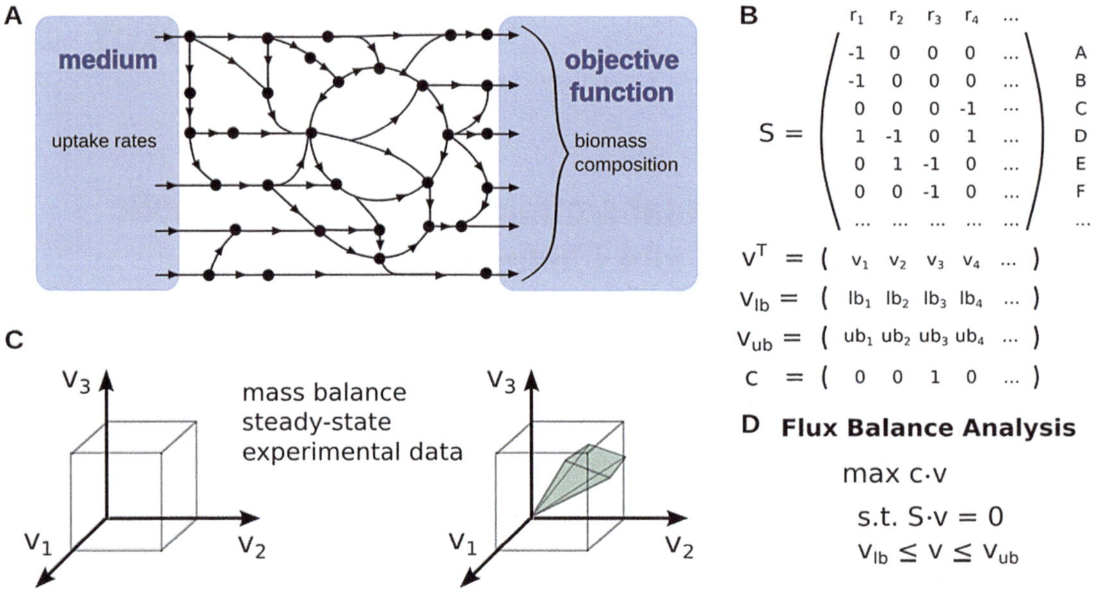

Fig. 1 Constraint-based modeling. (**a**) Schematic representation of a genome-scale metabolic reconstruction. Inorganic compounds are taken up from the medium and converted to precursors for biomass. Often the biomass function, i.e., the measured ratios of the biomass composition, coincides with the objective function of the Flux Balance Analysis formulation. (**b**) Representation of a stoichiometric matrix *S*; the rows denote the metabolites and the columns the reactions of the system under consideration. Entries in the matrix indicate at which ratios metabolites are being produced or consumed in the respective reaction. The flux vector v is constrained by lower and upper boundaries, v_{lb} and v_{ub}, respectively, and the vector c denotes which reaction of the metabolic network is the objective function to be optimized (maximized or minimized). (**c**) Applying mass balance constraints as represented by the stoichiometric matrix, a (quasi-)steady-state assumption and, if available, integration of experimental data, e.g., measured rates of uptake or excretion/accumulation, narrows down the feasible solutions space. Each cone face represents a single optimal solution to the optimization problem. (**d**) Generic Flux Balance Analysis formulation. The equations formalize the optimization of the objective function while fulfilling constraints on the stoichiometric matrix and the allowed flux values. The solution is a single, usually, nonunique flux distribution and a unique value for the objective

concentrations, as required in kinetic modeling approaches, are not needed but might be integrated as additional constraints [2–4].

The first genome-scale metabolic network reconstructions aimed at modeling plant metabolism at a single cell level [5, 6] and were solely based on information extracted from metabolic pathway databases such as [7–9] and biochemical literature. In recent years modelers' focus has broadened to the representation of specific cell types [6, 10], tissues [11, 12], and organs [13]. This specificity can be achieved by manually adding or removing certain pathways based on textbook knowledge [14]. Yet this is limited to relatively few well characterized cases. To represent context-specific networks that reflect the metabolic network active in a particular tissue, growth stage, or environmental condition the integration of large-scale transcript, protein and metabolite data has proven useful for making inferences about reaction activities [11, 12, 15].

Transcript and protein data-integrative approaches typically enforce data-driven upper and lower boundaries on the reactions velocities. In contrast, the integration of metabolomics data only indirectly applies constraints on reaction fluxes by enforcing those reactions to be active that are necessary for the synthesis of a measured metabolite. To date, metabolomics data have been used alone [16, 17] or in conjunction with transcript and protein data [2, 3, 18–20]. In this chapter we will briefly review (1) metabolite identification methods, (2) FBA approaches that integrate metabolomics data and (3) exemplify a pipeline to (a) match GC-MS metabolite data with the ModelSEED biochemical database [21] and (b) integrate the mapped metabolomics data with a genome-scale metabolic reconstruction of tomato from the PlantSEED project [22].

1.2 Mass Spectra Libraries and Annotation

The process of identifying metabolites by deconvoluting a complex set of mass spectral peaks results in a set of metabolites annotated with metabolic names or identifiers. For the purpose of integrating this data with a metabolic reconstruction in order to model and simulate metabolism, the metabolites themselves must be identified at least by name. It would be convenient to all metabolic modelers that the resulting annotation is generated in a consistent manner, or annotated with a universally recognized identifier such as CAS [23] or KEGG [7, 8], so that any integration of the data can be applied consistently to any metabolic reconstruction. However, due to the various approaches by which the peaks are generated, deconvoluted, and annotated, via means of the various hardware, experiments, samplings, software, and libraries of standards, the variation in the results from sample to sample, and from laboratory to laboratory, makes metabolomic approaches almost impossible to standardize. The diverse means of annotating mass spectra extends to the use of metabolic names and identifiers for annotation, and it follows, as many laboratories develop their own internal protocols, that there is a proliferation of libraries of mass spectrometry standards, within which different metabolic names and identifiers may be used for each spectra. Several plant metabolomics laboratories in the US and around the world are listed in Table 1, many notably situated at land-grant universities and colleges because of their historical association with crops, and whether they have public releases of the library of standards they use (Table 1).

1.3 Integration of Metabolite Annotation with Metabolic Reconstruction

The integration of metabolomics data with a metabolic reconstruction requires that each metabolite is matched using the annotated name or identifier. This section describes the likely process by which this integration may occur (Fig. 2).

Many metabolomics datasets may only have a single annotated name, and it follows that the metabolites in the metabolic reconstruction should have the same name in order for the match to occur.

Table 1
MS library/deconvolution references

Plant metabolomics laboratories	Public releases of standards library
N. Dudareva (Purdue)	https://link.springer.com/protocol/10.1007%2F978-1-60327-563-7_17
L. Sumner (Missouri)	https://www.ncbi.nlm.nih.gov/pubmed/26107650
A. Hegeman (Minnesota)	http://psb.stanford.edu/psb-online/proceedings/psb07/abstracts/2007_p157.html
GMD@Golm	https://academic.oup.com/bioinformatics/article-lookup/doi/10.1093/bioinformatics/bti236 http://onlinelibrary.wiley.com/doi/10.1002/9783527669882.ch18/summary
A. Fernie (MPI)	https://www.ncbi.nlm.nih.gov/pubmed/15733837
B. Nikolau (Iowa State)	https://www.ncbi.nlm.nih.gov/pubmed/24828308 https://www.ncbi.nlm.nih.gov/pubmed/25342293 https://www.ncbi.nlm.nih.gov/pubmed/22449044
O. Fiehn (UC Davis)	https://www.ncbi.nlm.nih.gov/pubmed/27824832 https://www.ncbi.nlm.nih.gov/pubmed/28436590
A. Aharoni (Weizmann Institute of Science)	https://www.ncbi.nlm.nih.gov/pubmed/27571918
Handbook of Plant Metabolomics (2013)	http://onlinelibrary.wiley.com/book/10.1002/9783527669882

Many metabolites can have multiple names, which would improve the likelihood of a match to occur, but, reconstructions do not usually store multiple names for each metabolite. Indeed, this is not the case for a reconstruction generated using PlantSEED. It follows on then that the best source of names to use for performing the matches will be the ones provided by the underlying biochemical database, which, in this case, is the ModelSEED biochemical database [21], wherein multiple databases and metabolic reconstructions were combined, and the names of the metabolites were collected and curated. The curated names are collected into a single file available via their repository (https://github.com/ModelSEED/ModelSEEDDatabase) and links the names to all of the ModelSEED identifiers used in every metabolic reconstruction built via ModelSEED or PlantSEED [22]. Having each of the ModelSEED and PlantSEED metabolic reconstructions built from the same underlying biochemical database means that each metabolomics dataset can be consistently and reliably integrated using the same names provided by the biochemical database.

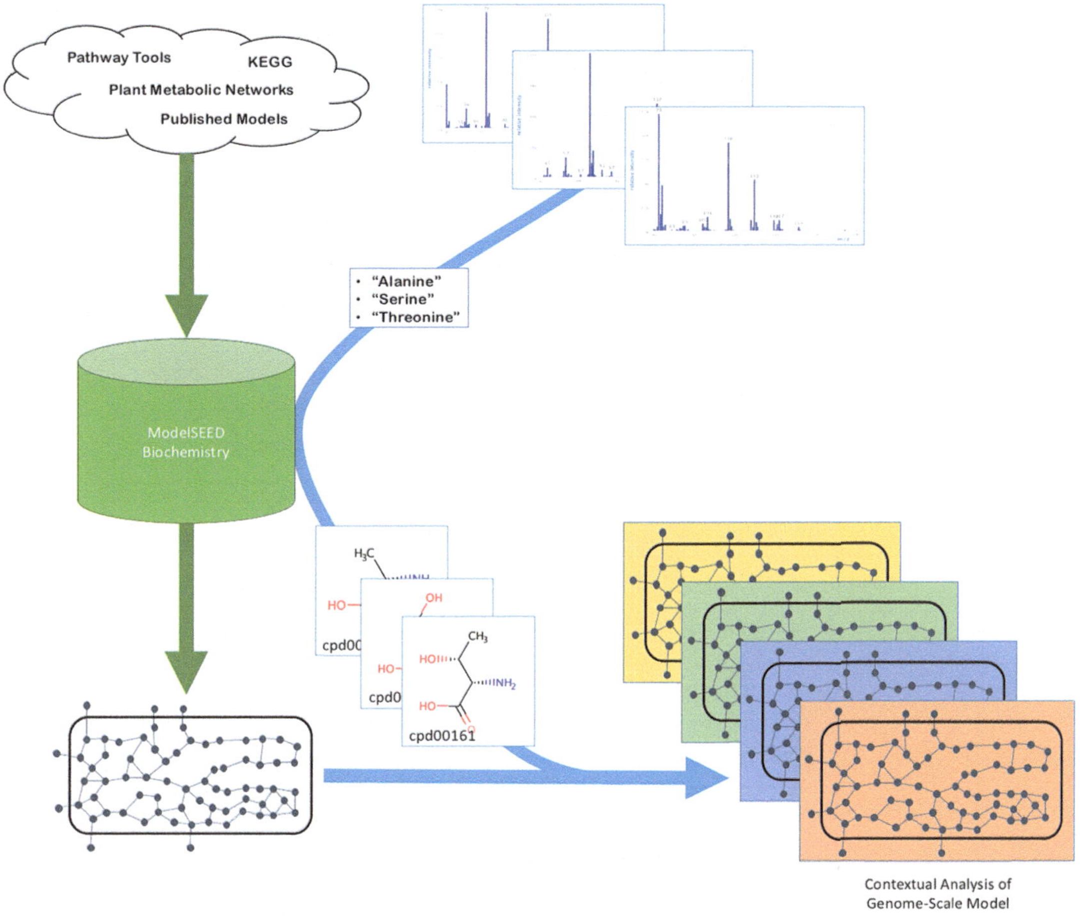

Fig. 2 Integration of metabolomics data with a PlantSEED metabolic reconstruction. The ModelSEED biochemical database was built from the integration of various online resources as depicted by the cloud [22]. In turn, the ModelSEED biochemical database acts as the central hub from which all ModelSEED and PlantSEED metabolic reconstructions are built, and within which all names, identifiers, and aliases for metabolites are found. A common result from the process of annotating mass spectra is a list of compound names, which can be matched with metabolites within the ModelSEED biochemical database. The result of the matching process is a list of ModelSEED compound identifiers that are found in the PlantSEED metabolic reconstruction, and in turn, allows the integration of metabolomics data with the PlantSEED metabolic reconstruction to produce Flux Balance Analysis simulations that correspond with the experimental context from which the metabolomics datasets were sampled

1.4 Approaches to Using Metabolomics with FBA

Metabolite data integrative approaches can be classified according to several criteria, e.g., whether they aid the network reconstruction, the extraction or flux prediction, as reviewed in [24]. Here we nonexhaustively summarize approaches based on the type of metabolite data, i.e., relative and absolute quantification (Table 2).

1.4.1 Integration of Absolute Metabolite Data

A set of approaches makes use of absolute metabolite levels or estimated concentration ranges to infer reaction reversibility based on Gibbs Free Energies [25–28] or integrates kinetic

Table 2
Overview of FBA approaches that integrate metabolomics data

	Absolute metabolite data	Relative metabolite data
Model reconstruction		MBA [18]
Network extraction		GIM3E [19]
Flux prediction based on differential metabolite levels	InDisMinimizer [3] "Modelling central metabolic fluxes" [4] TREM-flux [20] uFBA [17]	iReMet-flux [16]
Flux prediction based on thermodynamics	"Gibbs free energy based FBA" [25] "Algorithm for systematic reaction direction assignment" [26] TMFA [27] "Pathway thermodynamics" [28]	
Flux prediction based on kinetics	IOMA [2] DFBA [29]	

constraints with stoichiometric modeling [2, 29]. Other studies calculate metabolites' rates of change from time-series absolute metabolite data and include this information as constraints to the optimization problem by replacing the steady-state constraint by lower and upper fluxes boundaries marking the maximum deviation from the steady-state [3, 4, 17, 20].

1.4.2 Integration of Relative Metabolite Data

As the vast majority of metabolomics studies yields relative metabolite data, several approaches make use of these information for metabolic reconstruction [18] or reaction/pathway activation [19]. A recent study integrates relative metabolomics data to predict differential fluxes between wild-type and mutant plants [16]. Here, we demonstrate a pipeline that maps GC-MS metabolite data of tomato fruit flesh tissue at five stages of fruit development [30] to the ModelSEED biochemical database and subsequently applies this mapping to impose flux constraints on a genome-scale metabolic reconstruction of tomato. The integrated data are mainly primary metabolite measurements that were sampled at five fruit developmental stages (immature green (IG, 25 Days Post Anthesis (DPA)), mature green (MG, 42 DPA), breaker (BR, 44 DPA), orange (OR, 46 DPA), red (RD, 48 DPA)).

2 Materials

To share the script used in this pipeline, we created a public github repository at https://github.com/samseaver/Plant_Metabolomics_Chapter and users interested in integrating metabolomics data with a ModelSEED or PlantSEED metabolic reconstruction can download and run the scripts. Both the metabolomics data and the metabolic reconstruction used for this demonstration are made available in the same repository.

1. A computer running python programming language 2.7.12 or above (python.org).
2. Additional python packages: numpy, cobra, and matplotlib, python-libsbml (*see* **Note 1**).
3. Git (optional).
4. A linear optimization solver, e.g., cplex (IBM) or GUROBI (Gurobi Optimization) or glpk (GNU project) (*see* **Note 2**).
5. PlantSEED metabolic reconstruction (*see* **Note 3**; can be downloaded from http://modelseed.org).
6. Processed metabolite data set to be integrated with the metabolic reconstruction (*see* **Note 4**).
7. Internet connection to access the ModelSEED biochemical database and to download the example scripts and data.
8. A python editor (e.g., spyder python, pydev, emacs, vim, or any other favorite editor).

3 Methods

We wrote a python script to perform the data integration. The overall goal of this script is (1) to use the names in the metabolomics dataset as an input, and (2) to print out the list of ModelSEED identifiers that are used to represent metabolites in the metabolic reconstruction. More specifically, this script will interact with the ModelSEED biochemical database, and attempt to match the names. The script will print out a list of metabolite matches that have been made, and a list of metabolites that have not been matched. Most unmatched metabolites will have unusual or generic names (or even be listed as "unknown," as is often the case). We recommend the user to review the list of names in the dataset, and edit the names to increase the matching probability when rerunning the script.

For the sake of this demonstration, in the second part, we use a simplified version of the approach presented in [19] in which for each measured metabolite above a certain threshold (in this example we consider all metabolites that are above 50% of the average over all conditions as "present," the rest as "not present") a sink reaction j is included in the metabolic reconstruction and a minimum turnover flux ($v_{min} = 0.1$) is enforced in the problem formulation. If the problem formulation is feasible, the constraints are permanently added to the context-specific metabolic reconstruction. If a metabolite is annotated with several compartments, the constraints are formulated in a way that the sum of all sink fluxes in all compartments fulfills the minimum turnover flux:

$$\max c\tilde{n}v \tag{1}$$

$$S_{\text{sink}}\tilde{n}v = 0 \tag{2}$$

$$v_{\text{lb}} \geq v \geq v_{\text{ub}} \tag{3}$$

$$v_j \geq v_{\text{min}}, \forall j \in \text{sink reactions}, \tag{4}$$

where c is a vector of weights, indicating how much each reaction contributes to the objective function, v is the vector is fluxes, S_{sink} is the stoichiometric matrix including the sink reactions, and v_{lb} and v_{ub} are the lower and upper flux boundaries, respectively. Equations 1–3 represent the classic FBA problem with the maximization of biomass as the objective function. Equation 4 represents the constraints on the turnover sink reactions for the measured metabolites. Our workflow is demonstrated in the next section.

3.1 Demo

To demonstrate our workflow, we take the following steps:

1. Use git to clone the repository, and copy the data and code available within. Open a console:

   run: git clone https://github.com/samseaver/Plant_Metabolomics_Chapter

   Alternatively, download the files directly from the webpage.
2. The metabolomics data is made available in both spreadsheets and tab-separated values (TSV) flat-files. Any spreadsheet software can export data as a TSV flat-file (*see* **Note 5**). The data is readily available as *"GCMS_data.txt"* in the newly cloned folder.
3. Navigate to the new Plant_Metabolomics_Chapter folder in a console, and run the python script named ModelSEED_Metabolomics_Matching.py:

   run: python ModelSEED_Metabolomics_Matching.py GCMS_data.txt
4. The script generates two files named "*Matched_Metabolomics_-Data.txt*" and "*Unmatched_Metabolomics_Data.txt*". Examine the latter file for any compounds that were not matched (*see* **Note 6**). In the sample metabolomics data, we find that there are 20 compounds that were not matched. We see that several of them are unknown, and we ignore these, for example, "EIQTMS_N12C_SD1_1228.8_7.24" in "GCMS_data.txt". Several other compounds are not matched because of typos or

unusual names; for example, "fumaric acid" is misspelled as "fumeric acid."

5. Edit the metabolomics dataset so that the unusual names are replaced with more commonly recognized names, and rerun the script on the resulting saved files.

   run: python ModelSEED_Metabolomics_Matching.py GCMS_data_edited.txt

   The script now finds 10 of the previously unmatched metabolites and matches a total number of 63 metabolites. The matched metabolites and data are saved in the file "*Matched_-Metabolomics_Data.txt*" that allows to integrate the data directly with the PlantSEED metabolic reconstruction.

6. The python file *"Plant_Metabolomics_Modeling.py"* contains all the code necessary to perform the integration of metabolite data. The file can be run from the console (this step can take several minutes to run depending on the solver), and the final output can be further analyzed:

   run: python Plant_Metabolomics_Modeling.py

3.2 Step-by-Step Metabolomics Data Integration

In this section, we will go through the integration procedure step by step. As the source code is too long to be discussed here in detail, it is split into small sections, each fulfilling a specific function, separated by comments, and print statements in the code. We recommend the user to follow the comments in the code to see where each output corresponds with each section:

0. *Importing modules and defining functions.*
1. *Reading PlantSEED tomato model.*
2. *Printing basic model properties*: This section will print the size of the network. We see that the reconstruction has 1082 reactions, 1138 metabolites and 743 genes.
3. *Printing exchange reactions*: This section will list all exchanges and the lower and upper flux boundaries.
4. *Printing biomass composition*: This section will print a list of all biomass components and their ratio in the biomass function (*see* **Note 7**).
5. *Running initial FBA*: This section runs an initial unconstrained FBA simulation for the given exchange flux boundaries and prints a summary of the results showing the influxes and outfluxes of metabolites that are being taken up or produced by the reconstruction and the value of the objective function (indicating the generation of biomass). For the given constraints this value is 90.9.

6. *Changing rates of exchange and rerunning FBA*: This section will block the consumption of urea, light, and biomass and the production of sucrose and recalculates the objective function value for the new constraints. We see that this value has now dropped to 70.4.
7. *Reading matched metabolomics data*: This section reads the data file that we generated in the metabolomics integration **steps 3–5**.
8. *Integrating metabolomics data and extracting active subnetworks*: This section integrates the data with the metabolic reconstruction according to Eqs. 1–4 for the five developmental stages, and extracts the active subnetworks. We discuss the context of these subnetworks in Subheading 3.2.
9. *Analyzing results*: Finally, this section takes the active subnetworks and uses the results to generate plots for exchange fluxes, biomass production and pFBA solutions, in SVG format, named "Exchange_Fluxes.svg", "Biomass.svg", and "pFBA_Solutions.svg", respectively. These plots are shown in Fig. 3.
10. Running this step also lists the reactions whose fluxes exhibit the greatest change between the immature and the mature green stage. We discuss these further in Subheading 3.3.

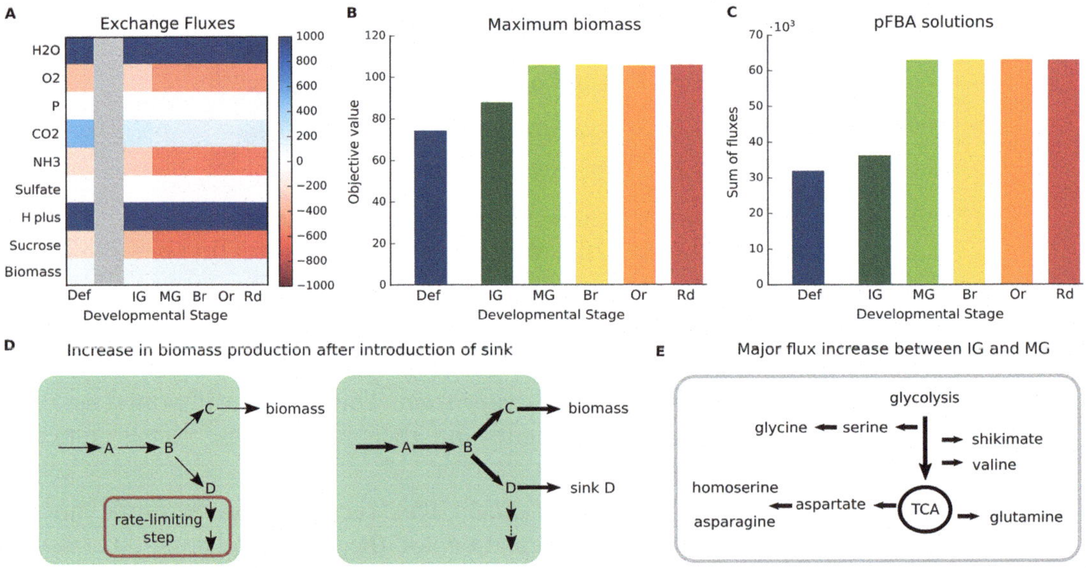

Fig. 3 Overview of FBA simulations results for the five fruit developmental stages. (**a**) Heatmap representation of exchange fluxes. Blue colors indicate that the respective metabolite is being produced while red colors indicate consumption. (**b**) Maximum biomass productions and (**c**) pFBA solutions for the default network (Def) and the five fruit developmental stages; immature green (IG), mature green (MG), breaker (Br), Orange (Or), and Red (Rd), respectively. (**d**) The illustration shows how adding sink reactions can attenuate the effect of rate-limiting steps in downstream pathways and lead to increased biomass production. (**e**) Schematic representation of pathways with most changing fluxes between the IG and MG stages of fruit development

3.3 Genome-Scale Reconstruction of Tomato Metabolism

Figure 3 summarizes main characteristics of the condition-specific networks. While the "default" active network shows fluxes through 309 reactions and produces 325 metabolites, after data integration the number of active reactions and produced metabolites increases to 339–365 reactions and 344–358 metabolites (Fig. 3). We extract the smallest active network for the immature green state and the largest active network for the red state (339/365 reactions and 344/358 metabolites, respectively).

Figure 3a shows the exchange fluxes for the default network and the condition specific networks. All context-specific networks show similar exchange patterns with the immature green stage being the most distinct. Furthermore, both, the value of the maximum biomass production and the objective value of the pFBA [31] (the minimum sum of fluxes that are necessary to support the biomass objective value and imposed constraints from the metabolite data integration), are higher for the condition specific networks than the default solutions and increase over time (Fig. 3b and c). While the latter is intuitive, as more reactions are being activated due to constraints from the data integration, the increased capability to produce biomass seems rather counterintuitive. To explain this behavior we can recall that constraints imposed by the metabolite data integration force certain fluxes to be directed toward sink reactions. These sink fluxes can allow "stoichiometric by-products" to leave the system that would otherwise constrain the biomass production, e.g., because they cannot be further converted due to rate-limiting steps in the downstream pathways. An illustrative example is given in Fig. 3d. Figure 3c shows that the strongest increase in total fluxes in the system occurs between the immature and the mature green stage. The time between these two measurements is also the longest (i.e., 17 days).

Which reactions contribute most to the overall increase in fluxes between these two time points? Investigating the reactions whose fluxes show the strongest increase between immature and mature green fruit, reveals that glycolysis and several pathways involved in the synthesis of amino acids, especially those involved in serine and downstream metabolite synthesis are being up-regulated. These results are in line with the experimental data that shows increased levels of valine, serine, homoserine, aspartate, asparagine, and glutamine. A graphical overview is given in Fig. 3e.

3.4 Final Considerations

Genome-scale metabolic modeling has emerged as a state-of-the-art tool to simulate plants' metabolism at a large scale. The integration of "omics" data aids constraining the feasible solution space and the extraction of context-specific subnetworks. In this chapter we presented a python-based pipeline to map a tomato GC-MS metabolite dataset with the PlantSEED metabolites database. We further demonstrated an approach to use these metabolite data to activate reactions required for the synthesis of the measured metabolites in the PlantSEED tomato metabolic network reconstruction. In summary, genome-scale metabolic reconstructions are a suitable tool to contextualize "omics" data.

4 Notes

1. We recommend using pip to install the packages.
2. GLPK is a freely available open source solver for linear and mixed Integer programming. CPLEX and Gurobi are commercial optimization solvers that have 1-year free academic licenses. It should be noted that depending on the solver and the solvers default settings the optimization results might slightly differ, especially the number of active metabolites and reactions. Nevertheless, this does not affect the overall network behavior.
3. All SEED models, including ModelSEED and PlantSEED, are built from the same biochemical database and can be used with this pipeline. Other network reconstructions from other platforms use different identifiers and cannot readily be used.
4. The here used metabolite data integration approach relies on binarized data, i.e., we need to define a threshold for calling a metabolite ON or OFF. We chose a threshold of 50% of the average level of a metabolite over all measured conditions. Only if the measured level if above the threshold, we enforce a turnover flux. This procedure is somewhat arbitrary, and parameters should be adjusted to the used datasets and model system. Despite this drawback, the approach can give us clues about pathway (de)activation and flux-rerouting when comparing different samples.
5. In Excel it can be saved as "TSV," in LibreCalc it must be saved as "CSV" with "field delimiter {TAB} and subsequently renamed to *<file>.txt*"
6. We typically have to make a range of arbitrary decisions when using metabolomics data. If there are multiple compounds that are stereochemically related (i.e., L-alanine and D-alanine, or alpha-glucose and beta-glucose) it is recommended to combine them, or in case of a metabolite name that consists of multiple names (i.e., Phenylalanine/Chorismate in "GCMS_data.txt") one of them should be selected and used. Finally, since a PlantSEED reconstruction consists of primary metabolism, if a secondary metabolite is detected, but not found in the PlantSEED reconstruction, the user may consider replacing it with a precursor that is found in the reconstruction so that its data points can be integrated.

7. The biomass composition used for this example is a simplified core biomass equation. It is generic, in that it contains all the primary components of biomass, and a few cofactors that are common to all plants. Whilst the coefficients used in the biomass are not specific to tomato fruit, it is only important, for the sake of analysis, that the components are present and actively biosynthesized.

Acknowledgments

N. Töpfer and A. Aharoni thank the Adelis Foundation, the Leona M. and Harry B. Helmsley Charitable Trust; the Jeanne and Joseph Nissim Foundation for Life Sciences; and especially the Tom and Sondra Rykoff Family Foundation Research for supporting the A. Aharoni lab activity. A. Aharoni is the incumbent of the Peter J. Cohn Professorial Chair. S.M.D. Seaver was supported by National Science Foundation Grant Number IOS-1444202 and by the Office of Science, Office of Biological and Environmental Research, of the US Department of Energy under contract number DE-ACO2-06CH11357, as part of the DOE Systems Biology Knowledgebase.

References

1. Orth JD, Thiele I, Palsson BØ (2010) What is flux balance analysis? Nat Biotechnol 28:245–248
2. Yizhak K, Benyamini T, Liebermeister W et al (2010) Integrating quantitative proteomics and metabolomics with a genome-scale metabolic network model. Bioinformatics 26: i255–i260
3. Recht L, Töpfer N, Batushansky A et al (2014) Metabolite profiling and integrative modeling reveal metabolic constraints for carbon partitioning under nitrogen starvation in the green algae Haematococcus pluvialis. J Biol Chem 289:30387–30403
4. Colombié S, Nazaret C, Bénard C et al (2015) Modelling central metabolic fluxes by constraint-based optimization reveals metabolic reprogramming of developing Solanum lycopersicum (tomato) fruit. Plant J 81:24–39
5. Poolman MG, Miguet L, Sweetlove LJ et al (2009) A genome-scale metabolic model of Arabidopsis and some of its properties. Plant Physiol 151:1570–1581
6. de Oliveira Dal'Molin CG, Quek L-E, Palfreyman RW et al (2010) AraGEM, a genome-scale reconstruction of the primary metabolic network in Arabidopsis. Plant Physiol 152:579–589
7. Kanehisa M, Sato Y, Kawashima M et al (2016) KEGG as a reference resource for gene and protein annotation. Nucleic Acids Res 44: D457–D462
8. Kanehisa M, Goto S, Sato Y et al (2014) Data, information, knowledge and principle: back to metabolism in KEGG. Nucleic Acids Res 42: D199–D205
9. Caspi R, Billington R, Ferrer L et al (2016) The MetaCyc database of metabolic pathways and enzymes and the BioCyc collection of pathway/genome databases. Nucleic Acids Res 44:D471–D480
10. Simons M, Saha R, Amiour N et al (2014) Assessing the metabolic impact of nitrogen availability using a compartmentalized maize leaf genome-scale model. Plant Physiol 166:1659–1674
11. Mintz-Oron S, Meir S, Malitsky S et al (2012) Reconstruction of Arabidopsis metabolic network models accounting for subcellular compartmentalization and tissue-specificity. Proc Natl Acad Sci U S A 109:339–344
12. Seaver SMD, Bradbury LMT, Frelin O et al (2015) Improved evidence-based genome-

scale metabolic models for maize leaf, embryo, and endosperm. Front Plant Sci 6:142

13. de Oliveira Dal'Molin CG, Quek L-E, Saa PA et al (2015) A multi-tissue genome-scale metabolic modeling framework for the analysis of whole plant systems. Front Plant Sci 6:4
14. de Oliveira Dal'Molin CG, Quek L-E, Palfreyman RW et al (2010) C4GEM, a genome-scale metabolic model to study C4 plant metabolism. Plant Physiol 154:1871–1885
15. Töpfer N, Caldana C, Grimbs S et al (2013) Integration of genome-scale modeling and transcript profiling reveals metabolic pathways underlying light and temperature acclimation in Arabidopsis. Plant Cell 25:1197–1211
16. Sajitz-Hermstein M, Töpfer N, Kleessen S et al (2016) iReMet-flux: constraint-based approach for integrating relative metabolite levels into a stoichiometric metabolic models. Bioinformatics 32:i755–i762
17. Bordbar A, Yurkovich JT, Paglia G et al (2017) Elucidating dynamic metabolic physiology through network integration of quantitative time-course metabolomics. Sci Rep 7:46249
18. Jerby L, Shlomi T, Ruppin E (2010) Computational reconstruction of tissue-specific metabolic models: application to human liver metabolism. Mol Syst Biol 6:401
19. Schmidt BJ, Ebrahim A, Metz TO et al (2013) GIM3E: condition-specific models of cellular metabolism developed from metabolomics and expression data. Bioinformatics 29:2900–2908
20. Kleessen S, Irgang S, Klie S et al (2015) Integration of transcriptomics and metabolomics data specifies the metabolic response of Chlamydomonas to rapamycin treatment. Plant J 81:822–835
21. Henry CS, DeJongh M, Best AA et al (2010) High-throughput generation, optimization and analysis of genome-scale metabolic models. Nat Biotechnol 28:977–982
22. Seaver SMD, Gerdes S, Frelin O et al (2014) High-throughput comparison, functional annotation, and metabolic modeling of plant genomes using the PlantSEED resource. Proc Natl Acad Sci U S A 111:9645–9650
23. American Chemical Society Chemical Substances—CAS REGISTRY, https://www.cas.org/content/chemical-substances
24. Töpfer N, Kleessen S, Nikoloski Z (2015) Integration of metabolomics data into metabolic networks. Front Plant Sci 6:49
25. Hoppe A, Hoffmann S, Holzhütter H-G (2007) Including metabolite concentrations into flux balance analysis: thermodynamic realizability as a constraint on flux distributions in metabolic networks. BMC Syst Biol 1:23
26. Kümmel A, Panke S, Heinemann M (2006) Systematic assignment of thermodynamic constraints in metabolic network models. BMC Bioinformatics 7:512
27. Henry CS, Broadbelt LJ, Hatzimanikatis V (2007) Thermodynamics-based metabolic flux analysis. Biophys J 92:1792–1805
28. Noor E, Bar-Even A, Flamholz A et al (2014) Pathway thermodynamics highlights kinetic obstacles in central metabolism. PLoS Comput Biol 10:e1003483
29. Mahadevan R, Edwards JS, Doyle FJ 3rd (2002) Dynamic flux balance analysis of diauxic growth in Escherichia coli. Biophys J 83:1331–1340
30. Adato A, Mandel T, Mintz-Oron S et al (2009) Fruit-surface flavonoid accumulation in tomato is controlled by a SlMYB12-regulated transcriptional network. PLoS Genet 5: e1000777
31. Lewis NE, Hixson KK, Conrad TM et al (2010) Omic data from evolved E. coli are consistent with computed optimal growth from genome-scale models. Mol Syst Biol 6:390

Chapter 22

Understanding the Functionality of a Biological System as a Whole: Comparative Data Analysis

Isabel Orf

Abstract

As part of a systems biology approach, metabolomics often aim at broadening our understanding of the functionality of biological systems as a whole. Observations from stand-alone experiments may reveal interesting changes in metabolites of a specific pathway or metabolite class. However, bringing these observations into context with more general biological processes requires the integration and comparison of different datasets. This chapter aims at introducing and explaining methods of comparative data analysis for plant metabolomics using the statistical software framework R.

Key words Systems biology, Comparative analysis, Data integration, Correlation, Clustering, Summary statistics, R programing language, R software packages

1 Introduction

Comparative analysis—that is not restricted to plant metabolomics, but can be applied to data from any kind of research field—focuses on the explanation of differences and/or the explanation of similarities between datasets. Depending on their aim, comparative data analyses may be divided into three different subtypes (adapted from [1, 2] on comparative analysis in social sciences): (1) "Individualizing comparisons" that try to grasp the specific features of each dataset, (2) "universalizing comparisons" that aim to understand each dataset as an observation of a general phenomenon, and (3) "variation-finding comparisons" that seek to examine systematic differences between datasets to define a principle of variation. Although the analytical methods for each subtype of comparison might be highly similar or even the same, the aim of the comparison can have a strong influence on the interpretation of the data. Ideally, a comparative analysis tries to examine all aspects of the different subtypes of comparisons to prevent bias. Another crucial point in comparative analysis is the definition of what counts as a difference or similarity, as most mathematical measures have

Carla António (ed.), *Plant Metabolomics: Methods and Protocols*, Methods in Molecular Biology, vol. 1778,
https://doi.org/10.1007/978-1-4939-7819-9_22, © Springer Science+Business Media, LLC, part of Springer Nature 2018

continuous output variables. It is therefore important to carefully report, e.g., p-values for hypothesis tests.

Comparative data analysis in plant metabolomics aims to make comparisons across datasets derived from different experiments. This includes comparisons across experiments measured using different analytical methods, performed in different seasons, applying different stresses or treatments, including different tissues, or even different organisms. It may focus on similarities and differences in: (1) properties of variables, e.g., absolute values of metabolites or metabolite variance within an experiment; (2) relationships between variables, e.g., metabolite ratios or metabolite interaction; (3) behavior of variables, e.g., changes in metabolite levels in a time series experiment or between treatments.

The strength of comparative data analysis is its ability to extrapolate observations to general biological concepts; a major weakness is its requirement of equivalent data. Equivalent data in the case of metabolomics means that the metabolites measured/identified in the datasets either need to be the same, or only the overlap between the datasets can be used for the comparison. In addition, the data needs to have comparable units. Therefore, normalization of the datasets before starting the comparative analysis is crucial.

This chapter covers a selection of methods to compare variable properties, relationships between variables and behavior of variables. The methods described in this chapter are optimized to find similarities between datasets rather than differences. Some remarks on data preparation/normalization steps required to perform comparative data analysis are provided as well.

2 Materials

If not detailed differently, all analysis described in this chapter are performed using the statistical software framework R (http://www.r-project.org/). Specific R software packages are mentioned throughout Subheading 3.

3 Methods

1. *Missing data imputation*: Use only such metabolites for the comparative analysis that are present in more than 70% of all samples in all datasets (*see* **Note 1**). For metabolites that are missing in only a few samples, impute the missing data in each dataset. Use for example the "missForest" algorithm from the "missForest" package, which is based on the random forest algorithm [3].
2. *Z-score normalization, combining datasets*: If you do not have quantitative metabolite data (which is true in most of the

cases) and for most statistical tests your datasets need to be normalized. In case the datasets were generated using the same machine and including a quality control sample, normalize each dataset to the quality control. Then combine the datasets. You can now perform metabolite-wise *Z*-score normalization using the "scale" function on the combined dataset so metabolites are centered on 0 with a standard deviation of 1 (*see* **Note 2**). In case the datasets cannot be normalized to a quality control sample median-centre each metabolite in each dataset. Then combine the datasets. Use sample-wise *Z*-score normalization to obtain comparable units for each sample. Using this technique, you lose some of the information on each metabolite, but allow comparisons across very distinct datasets. The combination of all datasets into one matrix simplifies many comparative analyses and is therefore recommended.

3. *Distance between samples*: Distances between samples (Fig. 1) can be calculated according to correlation distance. The Pearson correlation coefficient (r) is not only a measure of the strength of the relationship between two variables, or in this case samples, but also of the direction of the relationship (*see* **Note 3**). First, calculate the average linkage of r using the "cor" function. The correlation distance is $1 - \text{cor}(x,y)$. For a visual output of the distance between samples transform the distance into a distance matrix using the "as.dist" function and cluster the resulting matrix using the "hclust" function. Objects of the type produced by "hclust" can be plotted using the ordinary "plot" function, the result will be a distance dendrogram. A distance matrix and the corresponding dendrogram give you a measure of how similar samples are to each other. Cutting the dendrogram at a certain distance (arbitrarily chosen cut-off) will allow you to group samples into different clusters. A more unsupervised method to group samples is *k*-means clustering [5, 6] or partitioning around medoids (PAM) clustering (*see* **Note 4**) [7]. For the latter, use the "pam" function in the package "cluster." The number of clusters (k) can again be chosen arbitrarily, or using the silhouette coefficient [8] provided by the same function (*see* **Note 5**). Other methods to determine the optimal number of clusters that are not detailed further include calculating the "within groups"--sum of squared error, the Calinski–Harabasz index [9], the Bayesian information criterion [10], affinity propagation clustering [11], and gap statistics [12] (*see* also [13] for internal clustering validation measures).
4. *Distance between metabolites*: Distances between *Z*-scored metabolites (Fig. 2) can be calculated in exactly the same way as distances between samples. To gain a simultaneous overview of both, you can create a heatmap including the distance

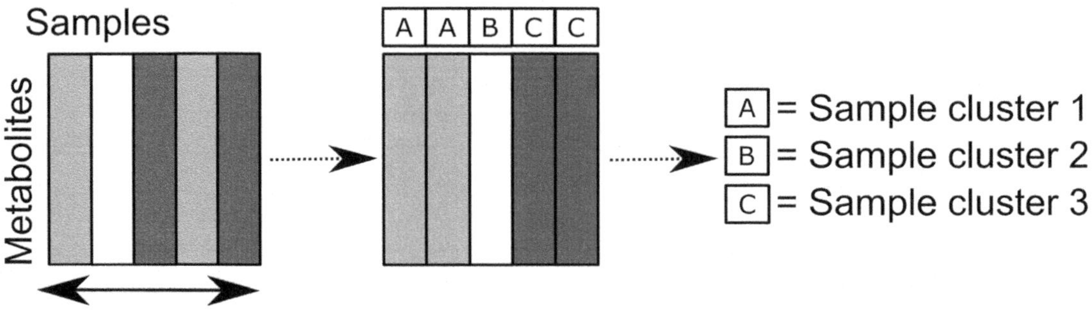

Fig. 1 Distance between samples. Data used for comparative analysis is best described as a matrix, with *n* columns representing the samples from different experiments and *x* rows representing the measured metabolites within the samples. Using a clustering algorithm like *k*-means or PAM, samples can be grouped together based on the distance between them, adapted from [4]

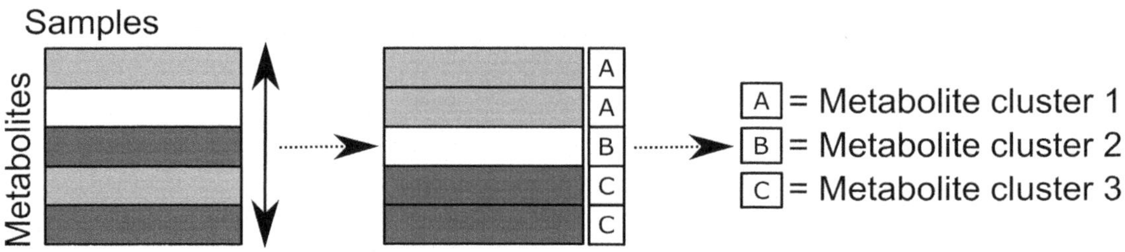

Fig. 2 Distance between metabolites. To identify clusters of metabolites within the data matrix, the same clustering algorithms as described for samples (Fig. 1) can be used, adapted from [4]

dendrograms of both matrix dimensions. Use the function "heatmap.2" from the "gplots" package [14], or any other heatmap function for this purpose.

5. *Summary statistics*: The term "summary statistics" refers to simple observations describing data including mean, median, minimum and maximum value, standard deviation, and coefficient of variation (CV). These statistics can first of all be used to describe overall differences between datasets; second, they can be used to compare different conditions across datasets (Fig. 3). Assume that you want to compare two different abiotic stresses, for example heat and cold stress, and how similar they render the metabolic state of plant leaves. You have several metabolite datasets for each of the stresses. You can now group the samples into sample groups across experiments and calculate and compare summary statistics for every metabolite. Comparing the CV (*see* **Note 6**) or variability of metabolites in different conditions is especially interesting as it can be regarded as a measure of regulatory mechanisms having an impact on the respective metabolite or the plasticity of this metabolite. Write your own simple function, or use the "cv" function in the package "raster" to calculate the CV [15]. To

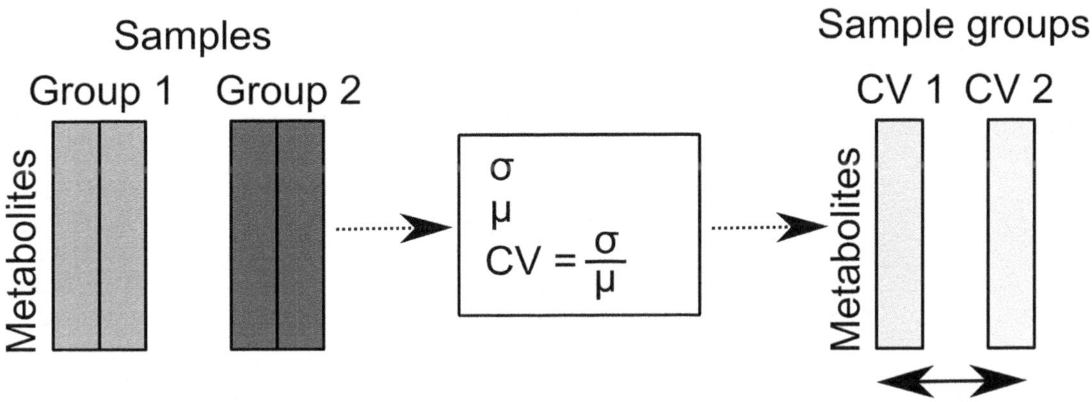

Fig. 3 Comparison of samples based on summary statistics. Summary statistics, (for example: variance σ, mean μ, and CV) are calculated within sample groups. Groups are then compared to each other based on these statistical properties, adapted from [4]

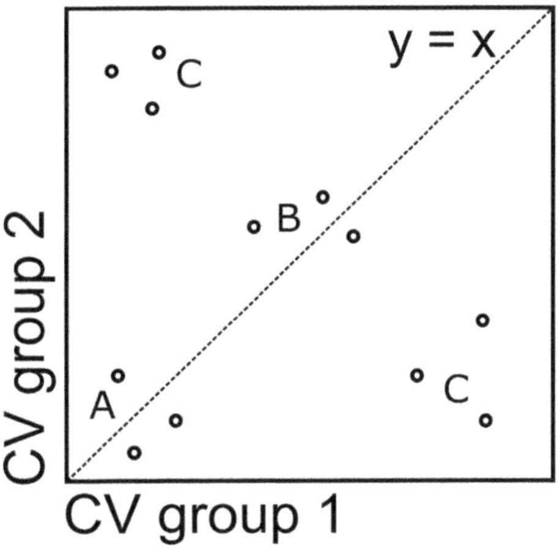

Fig. 4 Graphical assessment of the similarity of CV values. CV values are calculated for each metabolite in two sample groups and used as *x*- and *y*-values in a scatter plot. The segmented line represents the $y = x$ line, dots represent metabolites. A = Metabolites with comparable, low variability in the two sample groups. B = Metabolites with comparable, high variability in the two sample groups. C = Metabolites with different variability in the two sample groups

test for the equality of CV values perform either the "Asymptotic test for the equality of coefficients of variation from k populations" or the "Modified signed-likelihood ratio test (SLRT) for equality of CVs," respectively [16, 17]. Both tests are available in the "cvequality" package [18]. It is also possible to assess the similarity of CV values graphically (Fig. 4). Coming back to the example, you would have each one CV value per metabolite for the heat stress and the cold stress conditions. You can now plot this data as a simple scatter plot with the heat stress CVs as x-values and the cold stress CVs as y-values. Metabolites that have a comparable CV in both conditions are found close to the $y = x$ line (segmented line in Fig. 4). The closer a metabolite is to the origin of the graph (metabolites marked with A in Fig. 4), the less variability is found in this metabolite in the compared conditions. Respectively, the further away a metabolite is from the origin, the more variability is found in this metabolite within the compared conditions (metabolites marked with B in Fig. 4). Metabolites that do not have a comparable CV in the compared conditions are found distant from the $y = x$ line (metabolites marked with C in Fig. 4).

6. *Transformation of metabolite levels*: The ratios between metabolites can give implications on the kinetics of the enzyme reactions linking the metabolites to each other. Assuming constant reaction rates, comparable metabolite ratios imply consistent balancing of metabolites and likely cofactors in the compared samples, conditions, or experiments. As samples in a metabolomics experiment are basically vectors of metabolite values, calculating the ratios for all pairs of metabolites will result in a square ratio matrix for each sample (Fig. 5). To compare conditions, calculate the average values of all metabolites in every condition before the transformation of metabolite levels. Besides comparing ratios for specific metabolite pairs to each other, you can compare the ratio matrices as a whole, using the Rv coefficient [19]. As a multivariate generalization of the squared Pearson correlation coefficient, the Rv coefficient takes values between 0 and 1. Use the "coeffRV" function in the package "FactoMineR" to calculate the Rv coefficient for two matrices [20].

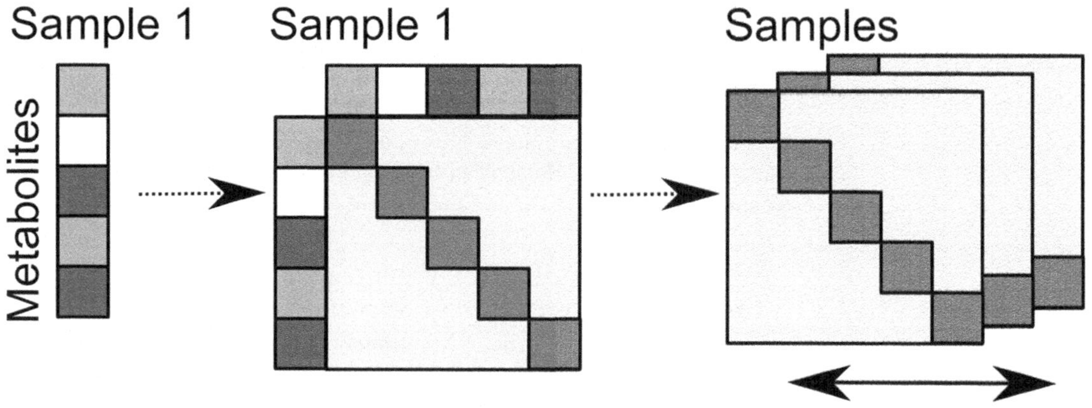

Fig. 5 Comparison of samples after transformation of metabolite levels. To compare relationships between certain metabolites across datasets, the metabolite levels of each sample can be transformed into a matrix of metabolite ratios. The resulting sample ratio-matrices can be compared to each other based on distance-measures, adapted from [4]

4 Notes

1. It is recommended to calculate the percentage of missing data for each dataset before starting the comparative analysis. This percentage gives you a quality measure for your datasets. You may exclude datasets with too high percentage of missing values from your analysis.
2. The result of *Z*-score normalization is that the features (values of metabolites over samples or metabolites within each sample) will be rescaled to have the properties of a standard normal distribution with $\mu = 0$ and $\sigma = 1$ (μ is the mean (average) and σ is the standard deviation from the mean). The *Z*-scores of the features are calculated as follows: $z = (x - \mu)/\sigma$.
3. The Pearson correlation coefficient is the covariance of the compared variables cov(X,Y) divided by the sample standard deviation σ. Be aware that the Pearson correlation coefficient is a measure of similarity. You can also choose to use direct distance measures to examine the relationship between samples. As different results can appear using different distance measures, it is important to choose the distance measure carefully. Euclidean distance, for example, is an appropriate measure when variables are continuous and numerical, and absolute distances should be reflected. As every variable is taken into account, highly correlated variables (redundant variables) have a high relative weight. Mahalanobis distance as a second example takes redundancies into account and removes them [21].
4. Like the *k*-means algorithm the PAM algorithm breaks the dataset into groups and tries to minimize the distance between the data points belonging to each group and a central point of this cluster, respectively. In contrast to *k*-means clustering, the central point in each cluster is an actual data point in PAM clustering. In addition, the PAM algorithm minimizes a sum of dissimilarities rather than a sum of squared Euclidean distances as in the *k*-means algorithm.
5. The silhouette coefficient is a measure of how distant the data points of one cluster are to the points of the other clusters. It ranges between −1 and 1, where positive values indicate a big distance to other clusters (good match of data points to their designated cluster), values close to 0 indicate that data points are close to the decision boundary between clusters, and negative values indicate a potential misassignment (poor match of the data points with their designated cluster). Calculating the silhouette coefficients for a range of different numbers of *k* (e.g., from 1 to 10) will allow you to choose the number of *k* with the highest silhouette coefficients for all of the clusters. Note that the highest silhouette coefficients are often found for

$k = 2$, however, this is mostly due to a very unequal splitting of the data points (most data points are in one big cluster, a small outgroup in a second cluster). Try to opt for equal numbers of data points per cluster.

6. The CV allows the direct comparison of variation in samples or metabolites with different units or means. It is defined as the ratio of the sample standard deviation σ and the mean μ ($CV = \sigma/\mu$). Note that the CV should only be calculated for data on a ratio scale (nonnegative values). For data with $\mu = 0$ the CV cannot be computed, when the mean is close to 0 the CV can be inaccurate. For log normalized data CV values are generally uninformative. Use the rawdata of the experiments you want to compare to calculate the CV.

Acknowledgments

This work was supported by the Minerva Fellowship Program of the Max Planck Society.

References

1. Tilly C (1984) Big structures, large processes, huge comparisons. Russell Sage Foundation, New York
2. Pickvance CG (2001) Four varieties of comparative analysis. J Housing Built Environ 16:7–28
3. Stekhoven DJ, Bühlmann P (2012) MissForest-non-parametric missing value imputation for mixed-type data. Bioinformatics 28:112–118
4. Orf I, Timm S, Bauwe H et al (2016) Can cyanobacteria serve as a model of plant photorespiration? A comparative meta-analysis of metabolite profiles. J Exp Bot 67:2941–2952
5. Steinhaus H (1956) Sur la division des corp materiels en parties. Bull Acad Polon Sci 1:801–804
6. MacQueen J (1967) Some methods for classification and analysis of multivariate observations. In: Proceedings of the fifth Berkeley symposium on mathematical statistics and probability, vol 1: Statistics. University of California Press, Berkeley, CA
7. Reynolds AP, Richards G, de la Iglesia B, Rayward-Smith VJ (2006) Clustering rules: a comparison of partitioning and hierarchical clustering algorithms. J Math Model Algorithm 5:475–504
8. Rousseeuw PJ (1987) Silhouettes: a graphical aid to the interpretation and validation of cluster analysis. J Comput Appl Math 20(C):53–65
9. Caliński T, Harabasz J (1974) A dendrite method for cluster analysis. Commun Stat 3:1–27
10. Kass RE, Wasserman L (1995) A reference Bayesian test for nested hypotheses and its relationship to the Schwarz criterion. J Am Stat Assoc 90:928–934
11. Frey BJ, Dueck D (2007) Clustering by passing messages between data points. Science 315:972–976
12. Tibshirani R, Walther G, Hastie T (2001) Estimating the number of clusters in a data set via the gap statistic. J R Statist Soc B 63:411–423
13. Liu Y, Li Z, Xiong H et al (2010) Understanding of internal clustering validation measures. In: Proceedings of the 2010 I.E. international conference on data mining. IEEE Computer Society, pp 911–916
14. Warnes G, Bolker B, Bonebakker L, Gentleman R, Liaw WHA, Lumley T, Maechler M, Magnusson A, Moeller S, Schwartz M, Venables B (2016) gplots: various R programming tools for plotting data. R package version 3.0.1
15. Hijmans RJ, van Etten J (2016) raster: geographic data analysis and modeling. R package version 2.5–8
16. Feltz CJ, Miller GE (1996) An asymptotic test for the equality of coefficients of variation from k populations. Stat Med 15:646–658

17. Krishnamoorthy K, Lee M (2014) Improved tests for the equality of normal coefficients of variation. Comput Stat 29:215–232
18. Marwick B (2016) cvequality: tests for the equality of coefficients of variation from multiple groups. R package version 0.1.1
19. Robert P, Escoufier Y (1976) A unifying tool for linear multivariate statistical methods: The RV-coefficient. J R Statist Soc C 25:257–265
20. Husson F, Josse J, Lê S, Mazet J (2017) FactoMineR: multivariate exploratory data analysis and data mining. R package version 1.38
21. Mahalanobis PC (1936) On the generalised distance in statistics. In: Proceedings National Institute of Science, India

Chapter 23

Regression-Based Modeling of Complex Plant Traits Based on Metabolomics Data

Francisco de Abreu e Lima, Lydia Leifels, and Zoran Nikoloski

Abstract

Bridging metabolomics with plant phenotypic responses is challenging. Multivariate analyses account for the existing dependencies among metabolites, and regression models in particular capture such dependencies in search for association with a given trait. However, special care should be undertaken with metabolomics data. Here we propose a modeling workflow that considers all caveats imposed by such large data sets.

Key words Metabolomics, Plants, Trait, Regression, Prediction, Modeling, R programing language, R software packages

1 Introduction

In the last decade, high-throughput metabolic profiling methods, such as gas chromatography–mass spectrometry (GC-MS) and liquid chromatography–mass spectrometry (LC-MS), capillary electrophoresis–mass spectrometry (CE-MS), and nuclear magnetic resonance (NMR) have become increasingly affordable for applications in plant research [1]. Metabolites are synthesized, used, or degraded by a network of interconnected biochemical reactions whose activity is affected by the genetic make-up, the environment [2] and the cellular context (e.g., tissues, cell types) [3]. Metabolism then comprises the entirety of biochemically reactions through which metabolites are interconverted. Since these reactions are affected by epigenetic, posttranscriptional, and posttranslational regulation, the levels of metabolite can be regarded as the end product of diverse molecular interactions taking place *in planta*. Therefore, it has been stipulated that metabolomics data, capturing the levels of measured metabolites in a given sample, are more revealing of plant phenotypic variation compared to other omics data [4].

Carla António (ed.), *Plant Metabolomics: Methods and Protocols*, Methods in Molecular Biology, vol. 1778,
https://doi.org/10.1007/978-1-4939-7819-9_23,

Consequently, metabolomics data have been used to address a wide variety of problems from plant sciences: from food nutritional improvement [5] to identification of biomarkers for stress tolerance [6] and to reconstruction and completion of metabolic pathways [7]. However, the exciting opportunity of using metabolomics to quantitatively predict complex traits (e.g., molecular and agronomic phenotypes) is still very novel [8–10].

Here we aim to give an introduction to regression methods for prediction and biomarker discovery using plant metabolomics data. The covered approaches can also be employed with other "omics" data resulting from different high-throughput technologies (e.g., transcriptomics). We focus on data preprocessing tasks, cross-validation, and model fitting using methods that can handle high-dimensional data. Finally, we discuss tools for interpreting and comparing models. We hope our contribution will serve both applied and fundamental research on the molecular basis of different plant traits.

2 Materials

All methods described in this chapter are performed using R language and environment for statistical computing and graphics (http://www.r-project.org/), if not detailed differently.

3 Methods

In this section, we will refer to metabolite levels as predictors (the p independent variables), an agronomic trait as a response and plants as samples (the n observations). In essence, predictive modeling consists of three sequential steps: data processing, cross-validation and model fitting to establish a relationship between responses and predictors. First, the individual predictors should be complete (i.e., missing values should be imputed), transformed and standardized on the grounds we elaborate next. Second, models should be cross-validated to avoid overfitting and find the optimal tuning parameters that may render the model generalizable to untested samples (Fig. 1). Third, the final model should be determined using the cross-validated tuning parameters and its components should be carefully examined. Residuals diagnostics, coefficient estimates, and permutations are particularly informative and help setting the stage for interpretation. This will be the structure we will use for outlining the following steps.

3.1 Data Processing and Normalization

1. *Treatment of missing values*: Assuming the integrity of all samples, metabolites should be inspected for the proportion of missing values. Only metabolites carrying a moderate amount of missing values should be kept for posterior imputation.

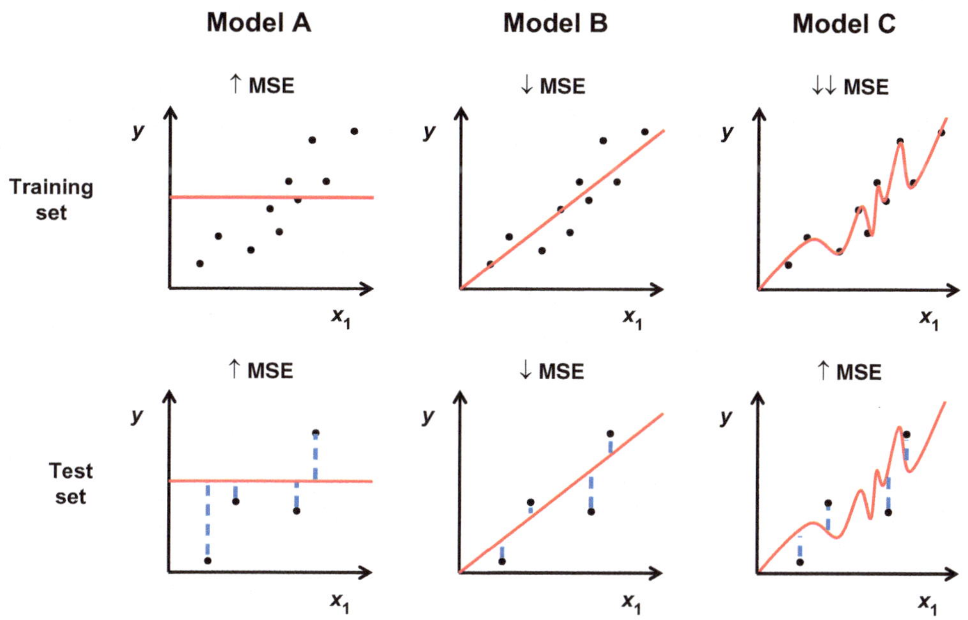

Fig. 1 Model complexity underlies underfitting and overfitting. Here we compare three hypothetical models trained and tested with the same train-test sets of a hypothetical population. Model A represents a strict model resulting from the usage of too few preditors (or high regularization), therefore prone to underfit. The mean square error (MSE) is relatively high in both the training and test sets. Model B represents an adequate model, with the right amount of regularization or predictors. The MSE is relatively low in the training set, and most importantly the lowest in the test set. As opposed to model A, model C represents a flexible model resulting from the usage of too many predictors (or insufficient regularization), therefore prone to overfit. For simplification we compare the response y to the first predictor, x_1. Red and dashed blue lines represent the model fit and the residuals, respectively

Imputation of largely undetected metabolite levels can severely affect the results and ultimately model interpretation. Thus, we encourage the application of stringent cutoffs (e.g., 20%), depending on the sample size. Methods for imputing missing values include median imputation, *K*-nearest neighbor [11] (KNN), Bayesian principal component analysis [12] (bPCA), and random forests [13] (a predictive model per se).

2. *Ensure the levels of each metabolite are normally distributed*: Transformations such as Box-Cox, logarithm, and square root safeguard unbiased coefficient estimates and prediction. In our experience, GC-MS-based metabolite quantification yields highly right-skewed distributions. In that case, a log-transformation (e.g., ln, $\log_2$, $\log_{10}$) will approximately render normally distributed levels.
3. *Identify and filter redundant metabolites*: Seldom will a metabolomics set include metabolites whose levels are independent. It is easy to mathematically demonstrate that there are infinite solutions for determining the coefficient estimates of two

identical metabolites in an ordinary least squares. This is the issue of collinearity, and although the impact is highly variable for different models, we recommend discarding redundant metabolites. One possible criteria would be, for every pair of metabolites correlated above a given threshold (e.g., Pearson correlation coefficient $r = 0.9$), identifying and discarding the one having the highest average correlation with all other metabolites.

4. *Standardize metabolite levels*: Considering from all previous steps that the metabolites constitute complete observations, are normally distributed and uncorrelated, it still is necessary to bring their levels to a common range (standardization). Assuming that no standardization is performed, metabolites will be treated unevenly which in turn will compromise interpretation and coefficient estimation. Nonparametric models (e.g., decision tree-based) are scale invariant and thus do not require standardization. Nevertheless, to center and scale is a cautious general principle. Autoscaling (mean-centering and unit variance) is probably one of the most popular standardization methods. The advantages and caveats of other methods are discussed by van den Berg and colleagues [14].

3.2 Cross-Validation

1. *Choose the model*: Models suited for high-dimensional problems include the partial least-squares (PLS), the least absolute shrinkage and selection operator (LASSO), the ridge, the elastic net (ENET, a combination of LASSO and ridge), support vector regression (SVR) and random forests (RF). PLS, LASSO, ridge and ENET are all linear in their parameters, in contrast to the nonlinear models SVR and RF that explore basis functions and predictor interactions respectively (Table 1, *see* **Note 1**).
2. *Define a range of values for the tuning parameters*: A proper range for fine-tuning might be established based on prior knowledge (e.g., $p \gg n$, which will be the case in most metabolomics experiments) or from a first run using a broad range. When models with more than one tuning parameter are to be cross-validated (e.g., ENET), the values of the different parameters should be combined in a grid design.
3. *Choose the metric of goodness-of-fit (i.e., predictability)*: In regression problems, R^2 and the mean squared error (MSE) are two popular metrics. In most cases, R^2 is equivalent to the squared Pearson correlation coefficient between the observed and the predicted response. One of many differences is that underfit models (e.g., only fit with an intercept) will prompt an error in the calculation of R^2 since the variance in the predicted response is null. Here, the MSE would pose no problem. In terms of interpretability, R^2 might be favored over MSE. Nonetheless, the root of the mean squared error (RMSE), which represents the average error in the original units, is informative as well.

Table 1
High-dimensional regression methods

Model	Corr	Scale inv	Parameters	Type	Runtime
PLS	×		n comps	Linear	+
LASSO	×		α	Linear	++
Ridge	×		λ	Linear	++
ENET	×		α, λ	Linear	+++
SVR	×		C, σ	Nonlinear	+++
RF	×	×	n trees, m_{try}	Nonlinear	+++

All methods shown manage high-dimensional and correlated data (Corr). However, with the exception of RF, all require standardization (Scale inv). With respect to the tuning parameters, PLS is controlled by the number of latent variables (n comps); the LASSO by the L_1 penalty weight α the ridge by the L_2 penalty weight λ; the ENET by both α and λ; the SVR by the cost C and σ which determines the bandwidth of a kernel, if applicable; the RF by the number of trees (n trees) and the number of predictors sampled in each tree (m_{try})—tree depth and bagging can also be parameterized. With respect to runtime, PLS is generally the fastest compared to more heavily parameterized models such as the ENET, SVR and RF

4. *Choose the resampling method*: There are numerous cross-validation routines available, tailored for different sample sizes and purposes. Bootstrap, n-fold and leave-one-out (LOO) cross-validation are probably the most popular. Again, different methods suit different purposes. For example, n-fold cross-validation with a small number of folds (e.g., 3 or 5) is robust against outlying samples but yields highly variable mean predictability estimates. Conversely, LOO is sensitive to extreme observations but more precise with respect to mean predictability estimates. Repeated n-fold cross-validation simultaneously suits small sample sizes and provides precise estimates of the goodness-of-fit. A bootstrap consists in sampling with replacement and therefore is also adequate for small samples.
5. *Perform the cross-validation and inspect the results*: The cross-validated R^2 (resp. MSE) profiles should display maxima (resp. minima) or plateauing (Fig. 2). When this is not the case, it is advisable to redefine the range of tuning parameter values and repeat the procedure (*see* **Note 2**).

3.3 Model Fitting

1. *Refit the model on the entire dataset with the optimal tuning parameters selected from Subheading 3.2*: Use the same goodness-of-fit metric optimized at the cross-validation stage. Examine the distribution and variance of the resulting residuals. Residuals exhibiting heterogeneous variance (e.g., a "funnel" structure in the residual-response plot) give rise to the problem of heteroscedasticity, whereby the assumption of normally distributed residuals, $\varepsilon \sim N(0, \sigma^2)$, is violated. In such cases, the parameterization of the residual variance component with linear mixed-effects models should be considered.

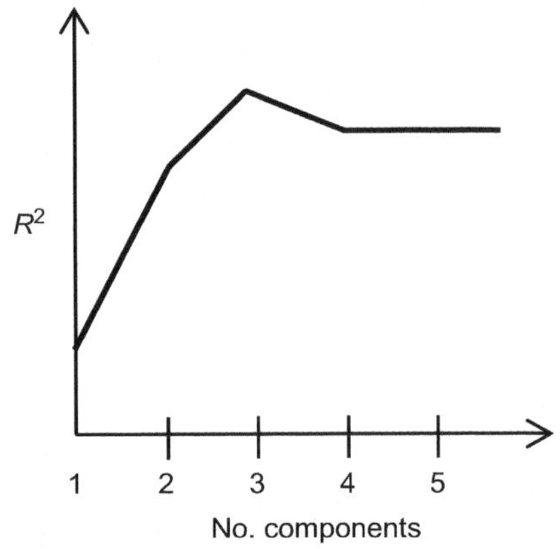

Fig. 2 Cross-validation profile. In this theoretical example, we cross-validate the number of components of a PLS model using the R^2. In the range 1–5, we see that the PLS model with three components has the optimal compromise between underfit (values to the left) and overfit (values to the right) models

2. *Investigate the results*: Informative summaries include observed-predicted plot and boxplots displaying the distribution of predictability, in case the practitioner wants to compare models. With respect to its significance, the comparison of the model results against the results obtained from permutation of the observations (or the response) should demonstrate that metabolites are indeed predictive of the trait under analysis. The variable importance in the projection (ViP) determines which predictors are more important in the prediction and might provide biological insights into the problem and delineate potential biomarkers. In most parametric models, ViP is equivalent to the coefficient estimates—the greater the value in absolute terms, the greater the importance of that particular predictor. In PLS it additionally considers the loadings of the latent variables. In RF, it averages the importance over all trees that include that particular predictor (determined in each tree from permuting predictors one by one). Finally, reverse engineering from constrained metabolite levels allows for determining the combination of metabolites that might associate with optimal levels of the agronomic trait (*see* **Note 3**).
3. *Supporting practical tutorial*: A practical tutorial that uses data from de Abreu e Lima et al. [8] to predict biomass in maize hybrids in the field using the metabolic profiles of primary roots from the hybrid seedlings can be found in www.github.com/monogenea/metabolomicsRegression. This tutorial follows many of the tools presented here using R software framework. For further reading we suggest the comprehensive work of Hastie and colleagues [15] and the more focused contribution from Kuhn and Johnson [16].

4 Notes

1. Linear models can still be given polynomial (e.g., x_1^2, x_2^2, ...) and interaction terms (e.g., $x_1 \cdot x_2$, $x_1 \cdot x_3$, ...). In the case of existing nonlinear relationships with the response, this substantially helps and might actually represent an enhancement compared to nonlinear models. The increase in p can still be dealt by the methods discussed above.
2. The "one SE" rule picks the simplest tuning parameters that provide an average cross-validation metric within one standard error (SE) of the highest average estimate derived from more complex models. This rule favors more parsimonious models.
3. If prediction is the sole objective, comparing different models is key. There is no all-around best model ("no free lunch" theorem).

References

1. Lisec J, Schauer N, Kopka J et al (2006) Gas chromatography mass spectrometry-based metabolite profiling in plants. Nat Protoc 1:387–396
2. Omranian N, Kleessen S, Tohge T et al (2015) Differential metabolic and coexpression networks of plant metabolism. Trends Plant Sci 20:266–268
3. Robaina Estévez S, Nikoloski Z (2015) Context-specific metabolic model extraction based on regularized least squares optimization. PLoS One 10:e0131875
4. Fiehn O (2002) Metabolomics-the link between genotypes and phenotypes. Plant Mol Biol 48:155–171
5. Tieman D, Zhu G, Resende MFR et al (2017) A chemical genetic roadmap to improved tomato flavor. Science 355:391–394
6. Gauthier L, Atanasova-Penichon V, Chéreau S, Richard-Forget F (2015) Metabolomics to decipher the chemical defense of cereals against Fusarium graminearum and deoxynivalenol accumulation. Int J Mol Sci 16:24839
7. Wen W, Li K, Alseek S et al (2015) Genetic determinants of the network of primary metabolism and their relationships to plant performance in a maize recombinant inbred line population. Plant Cell 27:1839–1856
8. de Abreu e Lima F, Westhues M, Cuadros-Inostroza Á et al (2017) Metabolic robustness in young roots underpins a predictive model of maize hybrid performance in the field. Plant J 90:319–329
9. Dan Z, Hu J, Zhou W et al (2016) Metabolic prediction of important agronomic traits in hybrid rice (*Oryza sativa* L.) Sci Rep 6:21732
10. Xu S, Xu Y, Gong L, Zhang Q (2016) Metabolomic prediction of yield in hybrid rice. Plant J 88:219–227
11. Hastie T, Tibshirani R, Narasimhan B, Chu G (2001) Impute: imputation for microarray data. Bioinformatics 17:520–525
12. Stacklies W, Redestig H, Scholz M et al (2007) pcaMethods – a bioconductor package providing PCA methods for incomplete data. Bioinformatics 23:1164–1167
13. Stekhoven DJ, Bühlmann P (2012) Missforest-non-parametric missing value imputation for mixed-type data. Bioinformatics 28:112–118
14. van den Berg RA, Hoefsloot HCJ, Westerhuis JA et al (2006) Centering, scaling, and transformations: improving the biological information content of metabolomics data. BMC Genomics 7:142–142
15. Hastie T, Tibshirani R, Friedman J (2009) The elements of statistical learning: data mining, inference, and prediction, 2nd edn. Springer, New York
16. Kuhn M, Johnson K (2013) Applied predictive modeling. Springer, New York

Chapter 24

Mathematical Modeling Approaches in Plant Metabolomics

Lisa Fürtauer, Jakob Weiszmann, Wolfram Weckwerth, and Thomas Nägele

Abstract

The experimental analysis of a plant metabolome typically results in a comprehensive and multidimensional data set. To interpret metabolomics data in the context of biochemical regulation and environmental fluctuation, various approaches of mathematical modeling have been developed and have proven useful. In this chapter, a general introduction to mathematical modeling is presented and discussed in context of plant metabolism. A particular focus is laid on the suitability of mathematical approaches to functionally integrate plant metabolomics data in a metabolic network and combine it with other biochemical or physiological parameters.

Key words Metabolomics, Plant biochemistry, Mathematical modeling, ODE, Jacobian matrix, Multivariate statistics, Time series analysis, Granger causality

1 Introduction

Due to the need for theoretical methods to combine the comprehensive output of plant systems biology studies, approaches of multivariate statistics, pattern recognition, and mathematical modeling have become more and more attractive to plant biologists over the recent decades. Ultimately, many modeling approaches in plant biology aim at linking genotype-specific information to an observed phenotype, e.g., by modeling plant metabolic pathways [1], growth and development [2] or by functionally integrating multilevel high-throughput results [3–5]. However, deriving predictive mathematical models of plant metabolism is challenging due to several reasons. First, plants are sessile organisms having to cope with a strongly fluctuating environment. As a consequence, plant metabolism has a complex, interlaced and nested structure with numerous redundant pathways being responsive to specific stress conditions. Hence, to unambiguously dissect the contribution of a

Lisa Fürtauer and Jakob Weiszmann contributed equally to this work.

Carla António (ed.), *Plant Metabolomics: Methods and Protocols*, Methods in Molecular Biology, vol. 1778, https://doi.org/10.1007/978-1-4939-7819-9_24, © Springer Science+Business Media, LLC, part of Springer Nature 2018

metabolic pathway to plant stress response numerous experimental setups and mutants are necessary. Second, plant cells are highly compartmentalized, and it has become evident that subcellular resolution of plant metabolism is necessary to fully understand stress-induced metabolic reprogramming [6–8]. Finally, it is the pure amount and complexity of multidimensional high-throughput data, such as that of plant metabolomic, proteomic, or transcriptomic studies, preventing an intuitive recognition of regulatory principles.

Conclusively, mathematical methods are preliminary for a functional and causal interpretation of experimental high-throughput studies, e.g., plant metabolomics studies, by enabling the efficient analysis of multidimensional data. Hence, the following paragraphs intend to provide the reader with basic principles of selected mathematical procedures for the analysis of plant metabolomics data.

2 Representation of Plant Metabolism by Differential Equations

Differential equations are commonly applied in natural sciences for the quantitative description of dynamic systems being also utilized for the description of natural laws, e.g., Newton laws. A differential equation describes the relationship between functions and related derivative functions, where a derivative function describes a rate of change. There are two main types of differential equations, (1) ordinary differential equations (ODEs), and (2) partial differential equations (PDEs). Ordinary differential equations are such equations where the searched function depends only on one variable and just *ordinary* derivatives appear. On the contrary, partial differential equations comprise *partial* derivatives, and the searched function depends on several variables. PDEs and related solution strategies are nowadays still in the focus of mathematical research and will not be discussed in this chapter.

In a biological context, a dynamic system can be defined by a set of ordinary differential equations (Eq. 1):

$$\begin{aligned}\frac{\mathrm{d}x_i(t)}{\mathrm{d}t} &= \text{influx} - \text{efflux} = \text{synthesis} - \text{degradation}\\ &\pm \text{transport} \pm \text{chemical interconversion} \pm \text{shuttling}\\ &\pm \text{chemical modification} \pm \ldots\end{aligned} \tag{1}$$

The state x_i of the system, which may represent a metabolite concentration in the plant metabolome, depends on other state variables, i.e., can be rewritten by $x_i(t) = \{x_1(t), x_2(t), x_3(t), \ldots\}$, finally resulting in a *coupled* system of ODEs. To solve the system, the entire set of differential equations has to be integrated simultaneously. In context of plant metabolism, this integration step represents the summation of all metabolic fluxes and reactions affecting a metabolite concentration x_i over a time period, predominantly being dependent on enzymatic reactions [9]. The mathematical integration of a metabolite's ODE finally results in the actual metabolite concentration at time point t_0, $x_i(t_0)$. Mathematically, ODEs are defined as follows (Eq. 2):

$$\begin{aligned} \dot{x}(t) &= f(t, x(t)) \qquad \text{for all } t \in \mathbb{R} \\ \dot{x}(t) &= f(t, x) \qquad \text{for all } t \in \mathbb{R} \end{aligned} \tag{2}$$

Here, $x(t)$ represents the searched variable within function f, and $\dot{x}(t)$ is the derivative function of $x(t)$.

In a system of (ordinary) differential equations, the state of the system, i.e., the value of $x(t)$, is set in context of its time-dependent derivative function $\dot{x}(t)$. The power of ODEs in mathematical modeling stems from the sufficiency to describe the relationship of a system's state and its derivative within a given time interval, and, based on this temporally local definition, to be able to predict the behavior of the system beyond this time interval. Solving ODEs means to find the searched function $x(t)$ $t \in \mathbb{R}$ (i.e., the solution) for which the differential equation is fulfilled at any time point. In context of computational biology, analytical solutions of ODEs are hardly definable and, hence, most studies rely on numerical approximation performed by optimization algorithms. While, therefore, optimization theory and its application plays a crucial role in computational biology, the present chapter does not focus on this comprehensive topic and we recommend the interested reader to refer to more specialized literature [10–12].

2.1 Solving a Single Linear Ordinary Differential Equation

To consider general solution strategies for ODEs, we start with a simple linear example:

$$\dot{x}(t) = \frac{\mathrm{d}x(t)}{\mathrm{d}t} = k\, x(t) \tag{3}$$

Separation of variables, here $x(t)$ and t, yields Eq. 4:

$$\frac{\mathrm{d}x(t)}{x(t)} = k\mathrm{d}t \tag{4}$$

To solve this ODE, both sides of the equation are integrated (Eq. 5):

$$\int \frac{1}{x(t)}\, \mathrm{d}x(t) = \int k\mathrm{d}t \tag{5}$$

This finally yields Eq. 6:

$$\ln |x(t)| = kt + c_1 \tag{6}$$

Applying the rule that for $a > 0$, $e^{\ln a} = a$ (where e is Euler's number, $e = 2.718281\ldots$), we can rewrite Eq. 6 by Eq. 7:

$$|x(t)| = e^{kt+c_1} \tag{7}$$

Rewriting the absolute value of $x(t)$ applying the substitution $c = e^{c_1}$, two cases can be distinguished:

$$|x(t)| = ce^{kt} \begin{Bmatrix} \text{I.for } x(t) > 0 \rightarrow x(t) = ce^{kt} \\ \text{II.for } x(t) < 0 \rightarrow x(t) = -ce^{kt} \end{Bmatrix} \tag{8}$$

If $x(t)$, in the context of plant metabolism, represents a (time-dependent) metabolite concentration, the second case defining a negative value for $x(t)$ is (biologically) irrelevant. In contrast, the first case implies that, if c and k are known, $x(t)$ can be calculated at any time point. Additionally, we can solve this equation for the integration constant c by inserting the initial (known) condition at time point t_0 (Eq. 9):

$$c = x(t_0)e^{-kt_0} \tag{9}$$

Plugging Eq. 9 in Eq. 8I yields Eq. 10:

$$x(t) = x(t_0)e^{k(t-t_0)} \tag{10}$$

Hence, based on a given concentration $x(t_0)$ at a time point t_0 and the parameter k, metabolite concentrations at any time point t_1, t_2, t_3, etc. can be calculated.

2.2 ODEs in Mathematical Modeling of Plant Metabolism

Mathematical models of plant metabolism have been developed and reviewed for decades [13–19], and numerous research questions have benefitted from the application of ODEs to quantitatively describe, analyze, and predict complex and nonintuitive system dynamics in plant biochemistry and physiology.

Pettersson and Ryde-Pettersson developed a kinetic model of the Calvin cycle [20]. This model takes into consideration that Calvin cycle activity is affected by metabolite concentrations outside the chloroplast, e.g. the concentration of external orthophosphate which was experimentally confirmed with isolated chloroplasts. Additionally, the dependency of photosynthetic rate on CO_2 and O_2 concentrations and the rate of starch production was predicted and experimentally confirmed. Besides, this model was used and extended to study properties like feedback regulation or Rubisco control in the Calvin cycle [21].

Another example for the application of ODEs to study plant metabolic pathways is a mathematical model of the plastidial 2-C-methyl-d-erythritol 4-phosphate (MEP) pathway. The model was developed to simulate and analyze diurnal pathway regulation in *Arabidopsis thaliana* [22]. Comprising several mutant lines, the authors validated their model assumptions according to substrate availability, circadian regulation and feedback regulation. Finally, the authors unraveled a feedback regulation mechanism stabilizing MEP pathway fluxes against fluctuations in substrate concentration.

While there are numerous further examples for a successful application of ODE-based mathematical models, unraveling non-intuitive and interlaced regulatory networks, the two selected studies already provide evidence for the suitability of ODE models to analyze pathway regulation and enzyme activity. Conclusively, mathematical modeling of plant metabolism potentially provides insights into complex network regulation which might be overlooked without such approaches.

2.3 How to Develop a Model of Carbohydrate Metabolism

The development of a biochemical ODE model typically starts with a graphical representation of metabolites and (enzymatic) reactions which will be analyzed. To provide an example, the central carbohydrate metabolism of leaf mesophyll cells is graphically summarized in Fig. 1.

Frequently, model construction comprises several critical steps of structural and/or kinetic simplification and, hence, has to be considered very carefully. In the provided example, cellular carbohydrate metabolism has been simplified by ignoring aspects like subcellular compartmentation, which, under in vivo conditions, would split sucrose cleavage, catalyzed by different invertase isoforms, into cytosolic and vacuolar reactions. Consequently, model simulations will not reveal information about subcellular partitioning of biochemistry which has to be considered when interpreting model output. A graphical model can be developed by hand with pen and paper, in a graphical program or in cellular simulator software, e.g., CellDesigner™ [23], which allows for a direct combination of graphical representation, ODEs, and simulations.

In the given example, the translation of the graphical into an ODE model results in the following equation system (Table 1).

In the next step, all reaction rates have to be defined by kinetic equations. For example, the ODE of G6P depends on the reactions *PGI* and *GlcK* (influx) and *PGM* (efflux). To quantify the influx and efflux, kinetic laws are defined to resemble in vivo kinetics as closely as possible with the available enzyme kinetic parameters, e.g., assuming mass action kinetics (Eqs. 11 and 12) or Michaelis–Menten kinetics which might also comprise inhibition terms (Eq. 13, *here:* competitive inhibition of GlcK by G6P).

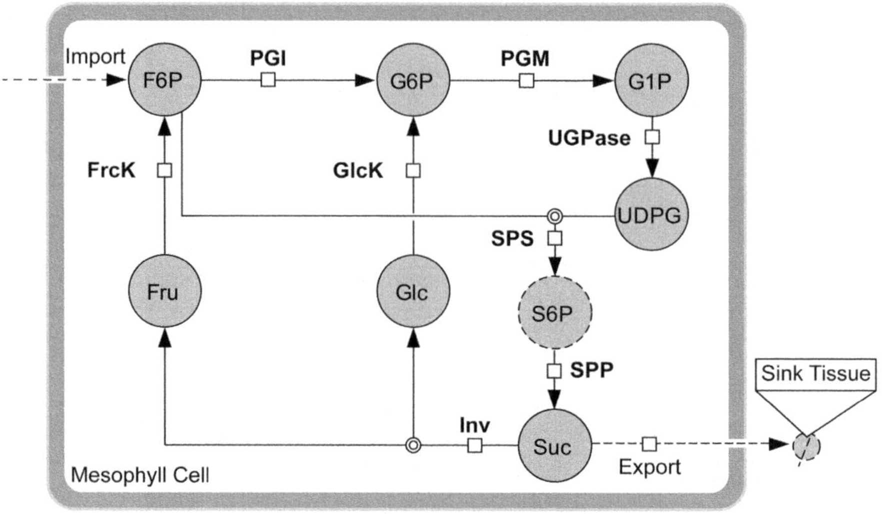

Fig. 1 A schematic overview of the central carbohydrate metabolism of plant leaf cells. Metabolites are described in circles, while enzymatic reactions are indicated by arrows. *F6P* fructose-6-phosphate, *G6P* glucose-6-phosphate, *G1P* glucose-1-phosphate, *UDPG* uridinediphosphate glucose, *S6P* sucrose-6-phosphate, *Suc* sucrose, *Glc* glucose, *Fru* fructose, *PGI* phosphoglucoisomerase, *PGM* phosphoglucomutase, *SPS* sucrose phosphate synthase, *SPP* sucrose phosphate phosphatase, *Inv* invertase, *FrcK* fructokinase, *GlcK* glucokinase

Table 1
Summary of ODEs to describe the graphical model shown in Fig. 1. Reactions of SPS and SPP have been summarized by the SPS reaction and S6P is not exlicit part of the ODE structure

$\frac{\mathrm{d}\,[\mathrm{F6P}]}{\mathrm{dt}} = \mathrm{Import} + \mathrm{FrcK} - \mathrm{PGI} - \mathrm{SPS}$	$\frac{\mathrm{d}\,[\mathrm{Suc}]}{\mathrm{dt}} = \mathrm{SPS} - \mathrm{Inv} - \mathrm{Export}$
$\frac{d\,[\mathrm{G6P}]}{dt} = \mathrm{PGI} + \mathrm{GlcK} - \mathrm{PGM}$	$\frac{\mathrm{d}\,[\mathrm{Glc}]}{\mathrm{dt}} = \mathrm{Inv} - \mathrm{GlcK}$
$\frac{d\,[\mathrm{G1P}]}{dt} = \mathrm{PGM} - \mathrm{UGPase}$	$\frac{\mathrm{d}\,[\mathrm{Frc}]}{\mathrm{dt}} = \mathrm{Inv} - \mathrm{FrcK}$
$\frac{\mathrm{d}\,[\mathrm{UDPG}]}{\mathrm{dt}} = \mathrm{UGPase} - \mathrm{SPS}$	$\frac{\mathrm{d}\,[\mathrm{Sink}]}{\mathrm{dt}} = \mathrm{Export}$

$$\mathrm{PGI} = [\mathrm{F6P}] * k_{\mathrm{PGI}} \tag{11}$$

$$\mathrm{PGM} = [\mathrm{G6P}] * k_{\mathrm{PGM}} \tag{12}$$

$$\mathrm{GlcK} = \frac{v_{\max} * [\mathrm{Glc}]}{[\mathrm{Glc}] + K_{\mathrm{M}} * \left(1 + \frac{[\mathrm{G6P}]}{K_{\mathrm{i,G6P}}}\right)} \tag{13}$$

Finally, the system of ODEs has to be solved, i.e., integrated, revealing the time course of the metabolite concentrations. As explained before, analytical solutions of such ODE systems fulfilling all experimental preliminaries, i.e., kinetic parameters and metabolite concentrations, can hardly be assigned. Instead, numerical solutions are determined applying local and/or global optimization algorithms, searching for locally or globally optimal solutions. Optimal solutions are searched within a predefined range of kinetic parameters being derived from experiments and/or database entries. A common optimality criterion is the so-called cost function which, for example, quantifies the squared error between simulated and experimentally determined metabolite concentrations. Optimization algorithms and solvers are implemented in modeling software like CellDesigner™ [23, 24] or COPASI [24], or are available for other numerical software environments [25].

A limiting step of kinetic (ODE) modeling is the availability of sufficient kinetic information being necessary to develop and solve ODEs within biochemically and physiologically meaningful boundaries. Although enzyme databases, e.g., BRENDA (http://www.brenda-enzymes.org/) [26], contribute comprehensive kinetic information across numerous organisms, pathways and enzyme isoforms, in many studies additional experimental assays are necessary to reveal diurnal or stress-induced dynamics in enzyme activities (v_{max}), protein abundance, substrate affinity (K_M), and inhibition/activation of enzymes by metabolic compounds (K_i). Particularly, large-scale modeling approaches comprising hundreds, or even thousands, of metabolites and proteins are inherently limited by experimental data. To (partly) overcome these limitations, alternative approaches have been developed, frequently comprising steps of system linearization, or parameter normalization [27–30]. A mathematical term playing a central role in these approaches is the so-called Jacobian matrix being introduced in the following section.

3 Steady States and the Biochemical Jacobian Matrix

Due to the rapid development of genome sequencing technologies, numerous whole genome sequences have become available for a diverse set of organisms, comprising more than 180 plant reference sequences [31]. Further, protocols have been developed using genome sequence information to deduct metabolic networks attempting to resemble in vivo reaction networks [32, 33]. To combine these genome-scale networks with results from transcriptomics, proteomics and metabolomics studies, mathematical methods are needed being able to functionally, i.e., causally, connect metabolite concentrations and/or enzyme activities with biochemical network information. ODE-based kinetic modeling of such large-scale networks is currently precluded by the lack of kinetic information being necessary to numerically integrate ODEs within a physiologically feasible range. Further, finding a global optimum for kinetic parameters and metabolite concentrations of large metabolic systems is challenging due to the nonlinear dynamics being particularly characteristic of plant–environment interactions. Hence, instead of solving the complete nonlinear system, deriving solutions at a steady state of the system is a way to deal with these limitations.

The steady state assumption is expressed by the term $\frac{d[\boldsymbol{M}]}{dt} = \mathbf{0}$ stating that no change in metabolite concentrations $[\mathbf{M}]$ can be observed over a change of time. Geometrically spoken, $\frac{d[\boldsymbol{M}]}{dt}$ represent the slopes of tangents of all functions $\boldsymbol{M}(t)$, i.e., of the time course of metabolite concentrations. At steady state conditions, slopes of these tangents are zero. Hence, steady state conditions imply that reaction rates of the system are constant, and consuming and synthesizing reactions of each metabolite are in equilibrium. While these steady state conditions are hardly met by any plant biological study focusing on metabolic reprogramming, e.g., due to stress response or diurnal rhythms, two additional requirements can be defined which still justify the biochemical interpretation:

1. Instead of a strict steady state, a *quasi*-steady state might be considered by changing the preliminaries from ($\frac{d[\boldsymbol{M}]}{dt} = \mathbf{0}$; **no** change) to ($\frac{d[\boldsymbol{M}]}{dt} \approx \mathbf{0}$; **no significant** change).
2. The interpretation of results is only valid within a very small (or infinitesimal small) time period.

Mathematically, $\frac{d[\boldsymbol{M}]}{dt}$ can be described via the stoichiometric matrix $\boldsymbol{S}\,(m \times n;\ m$ number of metabolites, n number of reactions), where the entries represent the participation of a metabolite in a reaction, multiplied with the vector $\boldsymbol{v}$ containing the reaction rates. Resulting from this, a system of linear algebraic equations can be defined (Eq. 14).

$$\frac{d\boldsymbol{M}}{dt} = \boldsymbol{S} * \boldsymbol{v} = \boldsymbol{f}(\boldsymbol{M}, \boldsymbol{p}, t) = 0 \qquad (14)$$

Here, metabolic functions, $\boldsymbol{f}(\boldsymbol{M}, \boldsymbol{p}, t)$, summarize the right side of the ODEs depending on metabolite concentrations $\boldsymbol{M}$, parameters $\boldsymbol{p}$ (e.g., kinetic parameters) and time t.

Explicitly applying this equation to the previously defined model (Fig. 1), the reaction network can be written as a steady state equation matrix system (Eq. 15).

$$\frac{d}{dt}\begin{bmatrix} \text{F6P} \\ \text{G6P} \\ \text{G1P} \\ \text{UDPG} \\ \text{Suc} \\ \text{Glc} \\ \text{Frc} \\ \text{Sink} \end{bmatrix} = \begin{bmatrix} 1 & 1 & -1 & -1 & 0 & 0 & 0 & 0 & 0 \\ 0 & 0 & 1 & 0 & 1 & -1 & 0 & 0 & 0 \\ 0 & 0 & 0 & 0 & 0 & 1 & -1 & 0 & 0 \\ 0 & 0 & 0 & -1 & 0 & 0 & 1 & 0 & 0 \\ 0 & 0 & 0 & 1 & 0 & 0 & 0 & -1 & -1 \\ 0 & 0 & 0 & 0 & -1 & 0 & 0 & 1 & 0 \\ 0 & -1 & 0 & 0 & 0 & 0 & 0 & 1 & 0 \\ 0 & 0 & 0 & 0 & 0 & 0 & 0 & 0 & 1 \end{bmatrix}$$

$$* \begin{bmatrix} \text{Import} \\ \text{FrcK} \\ \text{PGI} \\ \text{SPS} \\ \text{GlcK} \\ \text{PGM} \\ \text{UGPase} \\ \text{INV} \\ \text{Export} \end{bmatrix} = \begin{bmatrix} 0 \\ 0 \\ 0 \\ 0 \\ 0 \\ 0 \\ 0 \\ 0 \end{bmatrix} \tag{15}$$

To mathematically analyze system behavior at a *quasi*-steady state condition, metabolic functions can be linearized at time point t_0 of the steady state. In the context of metabolic networks, this linearization leads to tangents of metabolic functions at $(t_0, f(\boldsymbol{M}, \boldsymbol{p}, t_0))$.

To characterize (regulatory) effects of metabolites on metabolic functions, partial derivatives are built quantifying changes in metabolic functions with respect to changes in metabolite concentrations. Summarizing all partial derivatives in a matrix results in the so-called Jacobian matrix $\boldsymbol{J}$ (Eq. 16).

$$J = \begin{pmatrix} \frac{\delta f_1}{\delta M_1} & \frac{\delta f_1}{\delta M_2} & \cdots & \frac{\delta f_1}{\delta M_m} \\ \frac{\delta f_2}{\delta M_1} & \frac{\delta f_2}{\delta M_2} & \cdots & \frac{\delta f_2}{\delta M_m} \\ \vdots & \vdots & \ddots & \vdots \\ \frac{\delta f_m}{\delta M_1} & \frac{\delta f_m}{\delta M_2} & \cdots & \frac{\delta f_m}{\delta M_m} \end{pmatrix} \tag{16}$$

As described before, metabolic functions reflect all internal and external factors affecting the dynamics of a metabolite. Therefore, the Jacobian entries report on the response of metabolic functions to (slight) perturbations of the metabolic (steady) state. Hence,

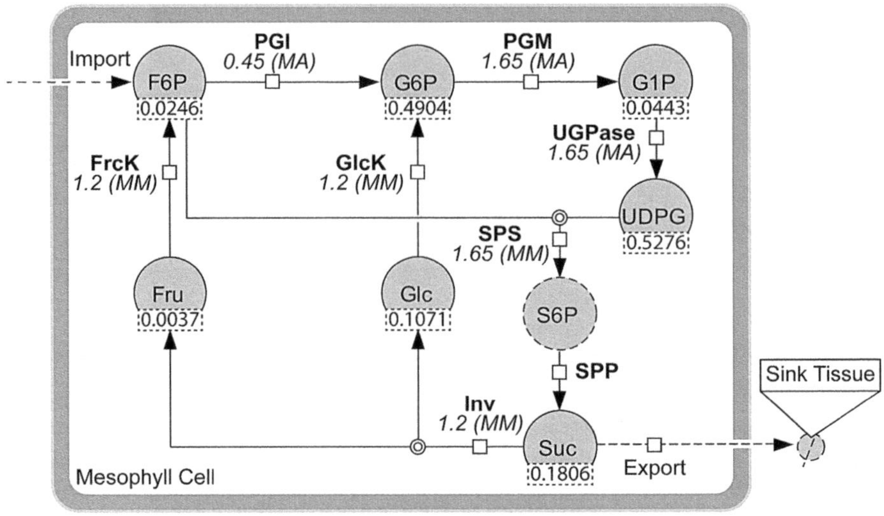

		δ[M]						
		F6P	G6P	G1P	UDPG	Suc	Fru	Glc
δ(*f*)	F6P	-44.87	0	0	-1.59	0	277.33	0
	G6P	18.46	-4.04	0	0	0	0	9.05
	G1P	0	4.04	-37.33	0	0	0	0
	UDPG	-26.41	0	37.33	-1.59	0	0	0
	Suc	26.41	0	0	1.59	-8.08	0	0
	Fru	0	0	0	0	5.62	-277.33	0
	Glc	0	0	0	0	5.62	0	-9.05

Fig. 2 A steady state model of the central carbohydrate metabolism. The graphical model on the left side contains information about an arbitrarily chosen steady state comprising metabolite concentrations (below circles, in mM) and fluxes (below reaction names, in mM h^{-1}). Reactions follow either mass-action (MA) or Michaelis–Menten (MM) kinetics. The table summarizes entries of the Jacobian matrix of this steady state. Entries correspond to the definition in Eq. 16. For example, the entry in the first column and the first row represents the value of ($\delta f_{F6P}/\delta[F6P]$). Abbreviations refer to Fig. 1

high absolute values of Jacobian entries indicate a strong effect of concentration changes on metabolic functions.

In the following paragraph, examples are depicted to visualize the effect of different metabolic constitutions on Jacobian entries. The previously introduced network of central carbohydrate metabolism will serve as an example, in which all kinetic parameters and metabolite concentrations are arbitrarily chosen.

The first scenario describes a metabolic state of the system in which every reaction is unidirectional and no activating or inhibitory influences are included in the network (Fig. 2).

The corresponding Jacobian matrix is nonzero in all entries where a metabolite represents either a reaction substrate or a reaction product.

For example, the metabolic function of F6P (f_{F6P}; derivatives δf_{F6P} first row in table of Fig. 2) depends on changes in the concentration of Fru being a substrate for the formation of F6P by the activity of FrcK. Further, f_{F6P} depends on changes in the concentration of UDPG because F6P and UDPG are substrates for the SPS reaction.

In the given example, highest absolute values of Jacobian entries describe the dependency of f_{F6P} and f_{Fru} on fructose dynamics. This indicates a strong effect of (slight) dynamics in fructose concentration on the rate of fructose phosphorylation, affecting the metabolic function f_{F6P} to increase (F6P: reaction product) and affecting f_{Fru} to decrease (Fru: reaction substrate). Another example of such a strong product-substrate dependency can be observed for the reaction catalyzed by UGPase where [G1P] dynamics decrease f_{G1P} and increase f_{UDPG}. Hence, both reactions have in common that reaction substrates are affecting reaction rates stronger than observed for other reactions, for example compared to the effect of Glc on GlcK reaction or G6P on the PGM reaction. This is due to different ratios of substrate concentrations to kinetic parameters and resulting fluxes: while the ratio of the FrcK flux to Fru concentration is ~300, the ratio of UGPase flux to G1P concentration is ~37 and PGM flux related to G6P concentration results in a ratio of (only) ~3. Further, these three reactions differ in the kinetic law which was assumed for flux simulation. While FrcK was assumed to follow Michaelis–Menten kinetics, UGPase- and PGM-driven reactions were described by mass-action kinetics. While Michaelis–Menten kinetics are characterized by a saturation level above which higher substrate concentrations do not affect the reaction rate anymore, mass-action law (for monosubstrate reactions) states a linear dependence of reaction rate on substrate concentration. The consequence for dynamics of Jacobian entries is exemplarily shown for both kinds of reaction kinetics (Fig. 3). While the mass action-related Jacobian entry $\delta(f_{UDPG})/\delta[G1P]$ is constant over G1P concentrations (Fig. 3a, b), the Michaelis–Menten related entry $\delta(f_{F6P})/\delta[Fru]$ is asymptotically decreasing over increasing Fru concentrations (Fig. 3c, d).

While the above discussed metabolic network (Fig. 2) does not contain any regulatory effectors, a modified network structure contains feedback inhibition by reaction products of invertase, fructokinase, and glucokinase reactions (Fig. 4).

Inhibitions are directly reflected in the Jacobian entries. Invertase inhibition by Fru and Glc results in positive entries of $\delta(f_{Suc})/\delta[Fru]$ and $\delta(f_{Suc})/\delta[Glc]$ due to their potential reduction of sucrose consumption. Analogously, inhibition of FrcK by F6P and GlcK by G6P are reflected by positive entries $\delta(f_{Fru})/\delta[F6P]$ and $\delta(f_{Glc})/\delta[G6P]$. In contrast, product inhibition of invertase by Fru and Glc results in negative entries for $\delta(f_{Fru})/\delta[Glc]$ and $\delta(f_{Glc})/\delta[Fru]$ because both inhibitors (negatively) affect each other's production by sucrose cleavage.

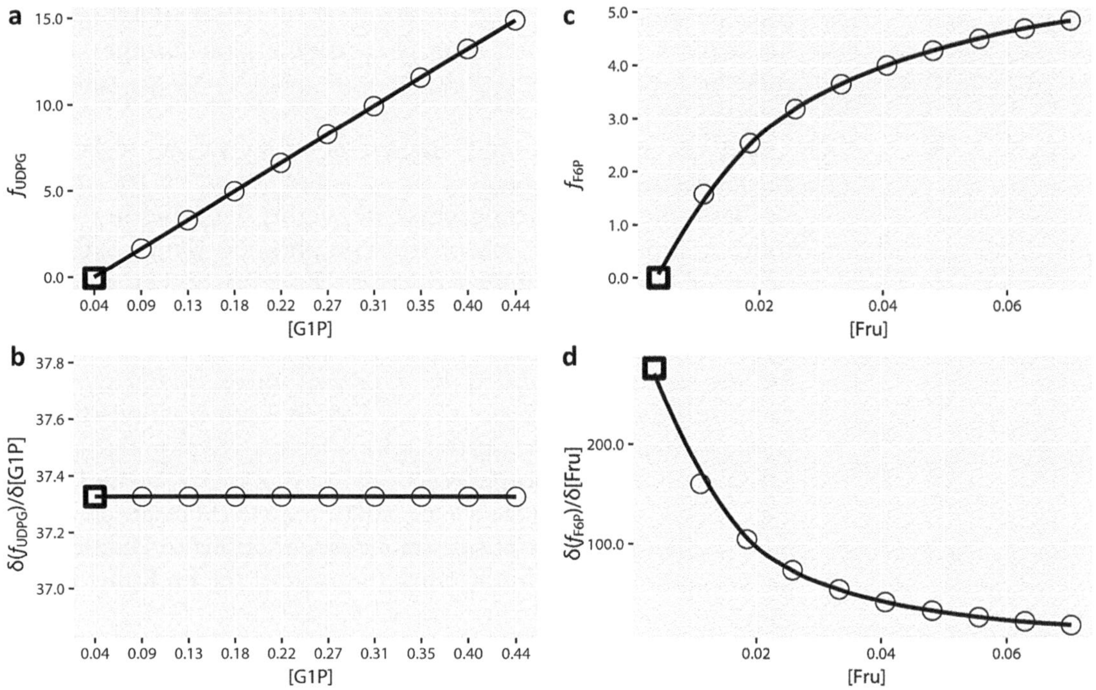

Fig. 3 Metabolic functions in relation to Jacobian entries for mass-action kinetics (**a**, **b**) and Michaelis–Menten kinetics (**c**, **d**). The metabolic function of UDPG (**a**) and its Jacobian entry (**b**) are related to the substrate concentration for its biosynthesis, [G1P]. The metabolic function of F6P (**c**) and its Jacobian entry (**b**) are related to the substrate concentration for its biosynthesis, [Fru]. Squares indicate the concentration of G1P and Fru at the simulated steady state (Fig. 2)

In summary, a high value of Jacobian entries of a metabolic system does not necessarily reflect a high flux through an enzymatic reaction, yet rather delivers information about the influence of changes in a metabolite concentration on a metabolic function. Considering the relation between substrates and products, this influence is especially large if the enzymatic reaction is far away from saturation, so a small change in substrate concentration will have a large effect on the product's metabolic function. This is of particular importance for biochemical systems, which usually do not follow linear kinetics, and, hence, a biochemical Jacobian matrix allows insight in the system's capabilities to cope with intra-cellular or environmental fluctuations.

The analysis of metabolic networks at (*quasi*) steady state conditions significantly simplifies mathematical modeling and optimization procedures. Further, only a fraction of experimental data is needed when compared to time dynamic and nonlinear systems. However, even under these simplified conditions many kinetic parameters have to be determined, and particularly for large-scale networks this still limits the application of mathematical modeling to metabolomics data. To overcome such limitations, approaches

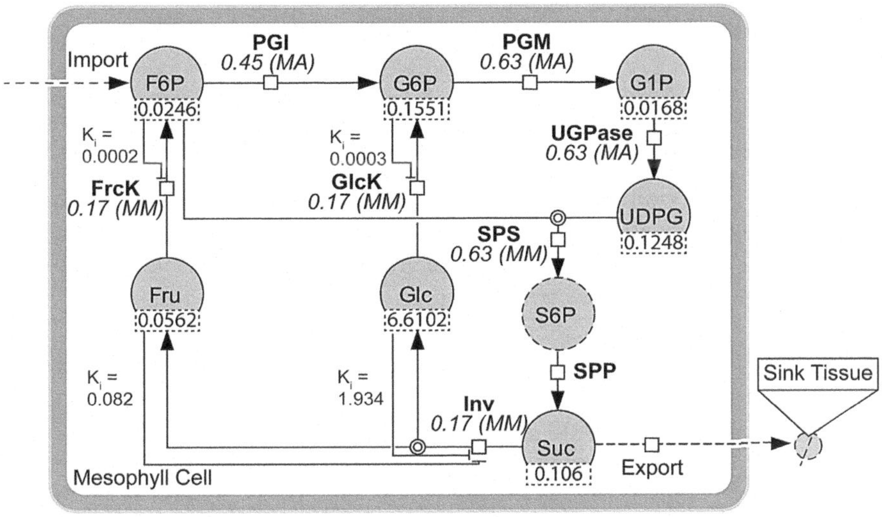

		δ[M]						
		F6P	G6P	G1P	UDPG	Suc	Fru	Glc
δ(*f*)	F6P	-35.22	0	0	-4.09	0	2.99	0
	G6P	18.46	-5.11	0	0	0	0	0.03
	G1P	0	4.04	-37.33	0	0	0	0
	UDPG	-10.01	0	37.33	-4.09	0	0	0
	Suc	10.01	0	0	4.09	-3.31	1.12	0.02
	Fru	6.75	0	0	0	0.86	-4.11	-0.02
	Glc	0	1.07	0	0	0.86	-1.12	-0.05

Fig. 4 A steady state model of the central carbohydrate metabolism comprising feedback inhibition reactions of Inv, FrcK, and Glck by their products. The graphical model on the left side contains information about an arbitrarily chosen steady state comprising metabolite concentrations (below circles, in mM), fluxes (below reaction names, in mM h^{-1}) and inhibitory constants, K_i (in mM). Reactions follow either mass-action (MA) or Michaelis–Menten (MM) kinetics. The table summarizes entries of the Jacobian matrix of this steady state. Entries correspond to the definition in Eq. 16. For example, the entry in first column and first row represents the value of ($\delta f_{F6P}/\delta[F6P]$). Grey colored entries are related to feedback inhibition. Abbreviations refer to Fig. 1

have been developed reducing the amount of kinetic information by normalization procedures [27] or by inference of biochemical Jacobian matrices from experimental covariance data [9, 34–39]. Such approaches may then be used for the comparison of two observed steady states under different conditions to get insight in the regulatory state of a biochemical system. Finally, it still has to be investigated how far *quasi* steady state-derived information about Jacobian matrices might be extrapolated to gain (still) valid information about dynamic properties of the metabolic system. These properties might be experimentally analyzed within metabolomics time-course experiments being dealt with in the next section.

4 Time Series Analysis of Multidimensional Metabolomics Data

The previous paragraphs of this chapter introduced strategies to deal with the challenge of functionally connecting experimental plant metabolomics data with biochemical network information. While solving ODE systems using enzyme kinetic parameters and absolute metabolite concentrations is desirable as it yields a causal model to simulate and predict reprogramming of plant metabolism due to environmental cues, particularly for large-scale models it is highly challenging, if possible at all, to derive a biochemically feasible solution due to the lack of knowledge about kinetic parameters. On the contrary, approximation of large-scale systems by linearization methods, e.g., resulting in the Jacobian matrix J, is limited by its steady state assumption hampering the extrapolation and interpretation of solutions for J in a dynamic and nonlinear context. Hence, to complement the presented toolbox of mathematical procedures, methods are needed coping with the identification and quantification of dynamic, and often nonlinear, signatures in metabolomics data sets. Methods of mathematical time series analysis are able to decipher such signatures in reiterated measurements over a time period which might comprise diurnal rhythms, stress-induced dynamics, or developmental reprogramming of metabolism. Mathematical procedures of time series analysis are diverse and numerous textbooks are available providing detailed information about algorithms and mathematical theory [40, 41]. In addition, statistical software tools are providing a comprehensive compilation of ready-to-use algorithms with a detailed documentation. For example, the freely available language and environment for statistical computing and graphics, R (https://www.r-project.org/) provides a collection of algorithms, procedures, and documentation summarized by the so-called *CRAN Task View: Time Series Analysis* (https://CRAN.R-project.org/view=TimeSeries; Date: 06/2017). While many of these algorithms might have been developed in contexts other than metabolomics, e.g., finance and econometrics, they prove to be very useful in the research field of metabolomics (e.g., the Granger causality analysis; *see* Subheading 5). Moreover, tutorial reviews and web-based tools are available particularly focusing on the analysis of dynamic metabolomics data [42–44]. Finally, commercial software, being frequently used for mathematical modeling, provides a comprehensive library of methodologies in the fields of statistical time series analysis and/or pattern recognition, e.g., the numerical programming environment MATLAB® (https://www.mathworks.com). In addition to commercially available toolboxes, there are numerous freely available graphical user interfaces, based on the MATLAB®software, supporting the application of multivariate statistics, mathematical modeling, and time series analysis [45–48].

5 Granger Causality Analysis of Metabolomics Data: Procedure and Limitations

Previously, we have applied the MATLAB®-based multivariate statistics toolbox COVAIN [37] to analyze stress-induced dynamics between primary and secondary metabolism in *Arabidopsis thaliana* [36]. In this study, Granger causations were calculated describing a time series correlation between experimental data collected on a GC-MS and LC-MS platform. Selecting the most significant correlations between both data sets revealed an abiotic stress-induced initiation of cyanidine glucoside biosynthesis being accompanied by a deregulation of the shikimic acid pathway.

Granger causality analysis was developed several decades ago to investigate causal relations by econometric models [49]. It considers the time series of two variables, here X and Y, which, in a plant metabolomics context, might represent concentrations of two different metabolites (Eqs. 17 and 18)

$$X(t) = \sum_{i=1}^{n} C_{X,i}X(t-i) + \sum_{i=1}^{n} C_{XY,i}Y(t-i) + R_X(t) \quad (17)$$

$$Y(t) = \sum_{i=1}^{n} C_{YX,i}X(t-i) + \sum_{i=1}^{n} C_{Y,i}Y(t-i) + R_Y(t) \quad (18)$$

$X(t)$ and $Y(t)$ are metabolite concentrations at time point t. In Eq. 17, $C_{X,\ i}$ and $C_{XY,\ i}$ represent the regression coefficients between $X(t)$ and $X(t-i)$ and between $X(t)$ and $Y(t-i)$, respectively. In Eq. 18, $C_{YX,\ i}$ is the regression coefficient between $Y(t)$ and $X(t-i)$ while $C_{Y,\ i}$ represents the regression coefficient between $Y(t)$ and $Y(t-i)$. $R_X(t)$ and $R_Y(t)$ are residual errors of the regression. The variable n describes the maximum time lag between the variables X and Y, i.e., metabolite concentrations. Hence, significant cross-coefficients $C_{XY,\ i}$ and $C_{YX,\ i}$ indicate a putatively causal relationship between the two variables X and Y within the recorded time series and across a time lag n. In general, the regression coefficients might be positive or negative indicating how X influences Y (and/or vice versa). In context of metabolite levels derived from a metabolomics experiment, positive cross-coefficients indicate a parallel trend of the metabolites X and Y, shifted by the lime lag n. This might either be represented by a decrease or an increase of both metabolite concentrations. If the cross-coefficients are negative, this indicates an opposite trend of X and Y, i.e., a decrease of the concentration of X and an increase of the concentration of Y, or vice versa. The maximum time lag n corresponds to the maximum time period which is considered for building the regression coefficients. If adequately chosen, n resembles the minimal time frame which it takes to yield an observable output of the potentially causal interaction of X and Y.

In a biochemical context this time lag might be related to signaling processes or enzymatic reactions rates interconverting X into Y. Although only discussed for two variables, this procedure is also applicable to multiple variables (e.g., m variables) using a model of mth order [49].

Limitations of applying Granger causality, and also other strategies of time series analysis, to metabolomics time course data are mainly attributable to (1) comparability of experimental data from different platforms, and (2) interpretability of putatively causal relationships in a biochemical context.

To be able to compare data from different platforms, e.g., relative metabolite levels derived from GC-MS and LC-MS measurements, it is crucial to normalize and scale the data before applying the time series analysis. This is, of course, not only necessary for Granger causality analysis but for various statistics and modeling approaches. Commonly, chromatographic peak areas are normalized to internal or external standards, sample protein amount, sample fresh weight, or sample dry weight. If calibration solutions of substances are available for data normalization, this enables the absolute quantification of metabolite concentrations which allows a direct comparison of metabolomics data among different analytical platforms. If calibration solutions are not available, for example in (untargeted) metabolite profiling approaches, it is crucial to scale normalized data before performing statistical analysis. A frequently used scaling method is *autoscaling* (also known as *zero mean-unit variance* or *z-scores*) which subtracts the mean of each variable, i.e., metabolite, from each single value and divides by its standard deviation. Finally, metabolites have the same standard deviation of one making them equally important. While this scaling method is advantageous for making data comparable between different experimental platforms, measurement errors are strongly weighted, and hence might enhance data scattering due to technical limitations. Alternatively, experimental results might be normalized, centered, scaled or transformed in various ways for meaningful and convincing data analysis. However, a careful choice of these mathematical operations is crucial to reduce the chance of misinterpretation. For this, excellent reviews and methodological summaries are available supporting the experimentalist to choose an adequate method [50].

6 Conclusions

While many statistical approaches yield information on correlations between metabolites, this does not necessarily resemble realistic, i.e., biochemically observable, interactions. Hence, it is crucial to combine the statistical output with biochemical information in order to enhance the reliability and predictive power of the output. In the context of metabolomic studies, this biochemical information can typically be derived from a genome-scale metabolic network which connects cellular metabolites by biochemical reactions which are predicted by the genome sequence of an organism [9, 33]. Numerous genome-scale metabolic network models are available, in particular for plant model organisms, e.g., Arabidopsis [51], maize [52], and rice [53]. With the list of enzymatically catalyzed reactions in each of those models it is possible to select the biochemically interpretable correlations between two metabolites derived from statistical time series analysis. In a previously published approach we automatized this connection of biochemical network information and time-continuous metabolomics data analysis within a MATLAB®-based graphical user interface, FEMTO (Functional Evaluation of Metabolic Time series Observations) [54]. FEMTO unravels time points of interests, where perturbed dependencies in metabolic functions indicate a deregulated biochemical interaction. A regularly updated version of FEMTO, including a user manual and a test data set, is freely available at http://mosys.univie.ac.at/resources/software/.

Acknowledgments

This work was supported by the Austrian Science Fund (FWF), Project I 2071, and the Vienna Metabolomics Center ViMe at the University of Vienna.

References

1. Morgan JA, Rhodes D (2002) Mathematical modeling of plant metabolic pathways. Metab Eng 4:80–89
2. Prusinkiewicz P (2004) Modeling plant growth and development. Curr Opin Plant Biol 7:79–83
3. Fiehn O (2002) Metabolomics-the link between genotypes and phenotypes. Plant Mol Biol 48:155–171
4. Weckwerth W (2011) Unpredictability of metabolism-the key role of metabolomics science in combination with next-generation genome sequencing. Anal Bioanal Chem 400:1967–1978
5. Liberman LM, Sozzani R, Benfey PN (2012) Integrative systems biology: an attempt to describe a simple weed. Curr Opin Plant Biol 15:162–167
6. Hoermiller II, Naegele T, Augustin H et al (2017) Subcellular reprogramming of metabolism during cold acclimation in *Arabidopsis thaliana*. Plant Cell Environ 40:602–610
7. Hurry V (2017) Metabolic reprogramming in response to cold stress is like real estate, it's all about location. Plant Cell Environ 40:599–601
8. Fürtauer L, Weckwerth W, Nägele T (2016) A benchtop fractionation procedure for subcellular analysis of the plant metabolome. Front Plant Sci 7:1912
9. Nägele T (2014) Linking metabolomics data to underlying metabolic regulation. Front Mol Biosci 1:22
10. Wang Y, Zhang X-S, Chen L (2010) Optimization meets systems biology. BMC Syst Biol 4: S1
11. Banga JR (2008) Optimization in computational systems biology. BMC Syst Biol 2:47

12. Reali F, Priami C, Marchetti L (2017) Optimization algorithms for computational systems biology. Front Appl Math Stat 3:6
13. Loomis RS, Rabbinge R, Ng E (1979) Explanatory models in crop physiology. Annu Rev Plant Physiol 30:339–367
14. Rios-Estepa R, Lange BM (2007) Experimental and mathematical approaches to modeling plant metabolic networks. Phytochemistry 68:2351–2374
15. Klipp E, Liebermeister W (2006) Mathematical modeling of intracellular signaling pathways. BMC Neurosci 7(Suppl 1):S10
16. Chew YH, Smith RW, Jones HJ et al (2014) Mathematical models light up plant signaling. Plant Cell 26:5–20
17. Rohwer JM (2012) Kinetic modelling of plant metabolic pathways. J Exp Bot 63:2275–2292
18. Gombert AK, Nielsen J (2000) Mathematical modelling of metabolism. Curr Opin Biotechnol 11:180–186
19. Giersch C (2000) Mathematical modelling of metabolism. Curr Opin Plant Biol 3:249–253
20. Pettersson G, Ryde-Pettersson U (1988) A mathematical model of the Calvin photosynthesis cycle. Eur J Biochem 175:661–672
21. Pettersson G (1997) Control properties of the Calvin photosynthesis cycle at physiological carbon dioxide concentrations. Biochim Biophys Acta 1322:173–182
22. Pokhilko A, Bou-Torrent J, Pulido P et al (2015) Mathematical modelling of the diurnal regulation of the MEP pathway in Arabidopsis. New Phytol 206:1075–1085
23. Funahashi A, Morohashi M, Kitano H, Tanimura N (2003) CellDesigner: a process diagram editor for gene-regulatory and biochemical networks. Biosilico 1:159–162
24. Schomburg I, Hofmann O, Baensch C et al (2000) Enzyme data and metabolic information: BRENDA, a resource for research in biology, biochemistry, and medicine. Gene Funct Dis 1:109–118
25. Hoops S, Sahle S, Gauges R et al (2006) COPASI – a COmplex PAthway SImulator. Bioinformatics 22:3067–3074
26. Schmidt H (2007) SBaddon: high performance simulation for the Systems Biology Toolbox for MATLAB. Bioinformatics 23:646–647
27. Steuer R, Gross T, Selbig J, Blasius B (2006) Structural kinetic modeling of metabolic networks. Proc Natl Acad Sci U S A 103:11868–11873
28. Reznik E, Segre D (2010) On the stability of metabolic cycles. J Theor Biol 266:536–549
29. Henkel S, Nägele T, Hörmiller I et al (2011) A systems biology approach to analyse leaf carbohydrate metabolism in *Arabidopsis thaliana*. EURASIP J Bioinform Syst Biol 2011:2
30. Fürtauer L, Nägele T (2016) Approximating the stabilization of cellular metabolism by compartmentalization. Theory Biosci 135:73
31. Jiao W-B, Schneeberger K (2017) The impact of third generation genomic technologies on plant genome assembly. Curr Opin Plant Biol 36:64–70
32. Vivek-Ananth RP, Samal A (2016) Advances in the integration of transcriptional regulatory information into genome-scale metabolic models. Biosystems 147:1–10
33. Thiele I, Palsson BO (2010) A protocol for generating a high-quality genome-scale metabolic reconstruction. Nat Protoc 5:93–121
34. Nägele T, Mair A, Sun X et al (2014) Solving the differential biochemical Jacobian from metabolomics covariance data. PLoS One 9: e92299
35. Kügler P, Yang W (2014) Identification of alterations in the Jacobian of biochemical reaction networks from steady state covariance data at two conditions. J Math Biol 68:1757–1783
36. Doerfler H, Lyon D, Nägele T et al (2013) Granger causality in integrated GC-MS and LC-MS metabolomics data reveals the interface of primary and secondary metabolism. Metabolomics 9:564–574
37. Sun XL, Weckwerth W (2012) COVAIN: a toolbox for uni- and multivariate statistics, time-series and correlation network analysis and inverse estimation of the differential Jacobian from metabolomics covariance data. Metabolomics 8:S81–S93
38. Nukarinen E, Nägele T, Pedrotti L, Wurzinger B, Mair A, Landgraf R, Börnke F, Hanson J, Teige M, Baena-Gonzalez E, Dröge-Laser W, Weckwerth W (2016) Quantitative phosphoproteomics reveals the role of the AMPK plant ortholog SnRK1 as a metabolic master regulator under energy deprivation. Sci Rep 6(1):31697
39. Wang L, Nägele T, Doerfler H, Fragner L, Chaturvedi P, Nukarinen E, Bellaire A, Huber W, Weiszmann J, Engelmeier D, Ramsak Z, Gruden K, Weckwerth W (2016) System level analysis of cacao seed ripening reveals a sequential interplay of primary and secondary metabolism leading to polyphenol accumulation and preparation of stress resistance. Plant J 87(3):318–332
40. Schelter B (2006) Handbook of time series analysis recent theoretical developments and applications. Weinheim, Wiley-VCH

41. Derryberry DR (2014) Basic data analysis for time series with R. Wiley, Hoboken, NJ
42. Smilde AK, Westerhuis JA, Hoefsloot HCJ et al (2010) Dynamic metabolomic data analysis: a tutorial review. Metabolomics 6:3–17
43. Xia J, Sinelnikov IV, Wishart DS (2011) MetATT: a web-based metabolomics tool for analyzing time-series and two-factor datasets. Bioinformatics 27:2455–2456
44. Xia J, Wishart DS (2016) Using MetaboAnalyst 3.0 for comprehensive metabolomics data analysis. Curr Protoc Bioinformatics 55:14.10.1–14.10.91
45. Girbig D, Selbig J, Grimbs S (2012) A MATLAB toolbox for structural kinetic modeling. Bioinformatics 28:2546–2547
46. Schmidt H, Jirstrand M (2006) Systems Biology Toolbox for MATLAB: a computational platform for research in systems biology. Bioinformatics 22:514–515
47. Aurich MK, Fleming RMT, Thiele I (2016) MetaboTools: a comprehensive toolbox for analysis of genome-scale metabolic models. Front Physiol 7:327
48. Fitzpatrick MA, McGrath CM, Young SP (2014) Pathomx: an interactive workflow-based tool for the analysis of metabolomic data. BMC Bioinformatics 15:396
49. Granger CWJ (1969) Investigating causal relations by econometric models and cross-spectral methods. Econometrica 37:414–426
50. van den Berg RA, Hoefsloot HC, Westerhuis JA et al (2006) Centering, scaling, and transformations: improving the biological information content of metabolomics data. BMC Genomics 7:142
51. Mintz-Oron S, Meir S, Malitsky S et al (2012) Reconstruction of Arabidopsis metabolic network models accounting for subcellular compartmentalization and tissue-specificity. Proc Natl Acad Sci U S A 109:339–344
52. Bogart E, Myers CR (2016) Multiscale metabolic modeling of C4 plants: connecting nonlinear genome-scale models to leaf-scale metabolism in developing maize leaves. PLoS One 11:e0151722
53. Shaw R, Kundu S (2015) Flux balance analysis of genome-scale metabolic model of rice (Oryza sativa): aiming to increase biomass. J Biosci 40:819–828
54. Nägele T, Fürtauer L, Nagler M et al (2016) A strategy for functional interpretation of metabolomic time series data in context of metabolic network information. Front Mol Biosci 3:6

Index

A

B

C

D

Carla António (ed.), *Plant Metabolomics: Methods and Protocols*, Methods in Molecular Biology, vol. 1778, https://doi.org/10.1007/978-1-4939-7819-9, © Springer Science+Business Media, LLC, part of Springer Nature 2018

N

O

P

T

U

V

W

Z

Printed by Printforce, the Netherlands